GASIFICATION

GASIFICATION

Chris Higman and Maarten van der Burgt

ELSEVIER

AMSTERDAM BOSTON HEIDELBERG LONDON
NEW YORK OXFORD PARIS SAN DIEGO
SAN FRANCISCO SINGAPORE SYDNEY TOKYO

Gulf Professional Publishing is an imprint of Elsevier

Recognizing the importance of preserving what has been written, Elsevier prints its books on acid-free paper whenever possible.

Library of Congress Cataloging-in-Publication Data
Higman, Chris.
  Gasification / Chris Higman and Maarten van der Burgt.
    p. cm.
  Includes bibliographical references and index.
  ISBN-13: 978-0-7506-7707-3   ISBN-10: 0-7506-7707-4
  1. Coal gasification. I. Burgt, Maarten van der. II. Title.
  TP759.H54 2003
  665.7′72—dc21
    ISBN-13: 978-0-7506-7707-3
    ISBN-10: 0-7506-7707-4 (hc: alk. paper)
                                                                        2003049012

British Library Cataloguing-in-Publication Data
A catalogue record for this book is available from the British Library.

The publisher offers special discounts on bulk orders of this book. For information, please contact:

Manager of Special Sales
Elsevier
200 Wheeler Road
Burlington, MA 01803
Tel: 781-313-4700
Fax: 781-313-4882

For information on all Gulf Professional Publishing publications available, contact our World Wide Web home page at: http://www.gulfpp.com

10   9   8   7   6   5   4

Printed in the United States of America

Cover Photos: 250 MW Coal IGCC, Buggenum, Netherlands (Courtesy of Nuon Power); Oil gasifiers in ammonia plant, Brunsbüttel, Germany (Courtesy of Hydro Agri Brunsbüttel GmbH)

Contents

Preface

Gasification, at least of coal, is in one sense an old technology, having formed the heart of the town gas industry until the widespread introduction of natural gas. With the decline of the town gas industry, gasification became a specialized, niche technology with limited application. After substantial technical development, gasification is now enjoying a considerable renaissance. This is documented by the more than thirty projects that are in various stages of planning or completion at the present time. The reasons for this include the development of new applications such as gas-to-liquids (Fischer-Tropsch) projects, the prospect of increased efficiency and environmental performance including CO_2 capture through the use of integrated gasification combined-cycle (IGCC) in the power industry, as well as the search for an environmentally benign technology to process low-value or waste feedstocks, such as refinery residues, petroleum coke, or biomass or municipal waste.

The literature of gasification is extremely fragmented with almost all recent (post-1990) contributions being confined to conference papers or articles in the appropriate journals. In the coal literature it is mostly relegated to a single chapter, which is unable to do the subject proper justice.

The knowledge of gasification is mostly confined to commercial process licensors and the operators of existing plants. Therefore there is little opportunity for outsiders to acquire an independent overview before embarking on a project of their own.

In discussing these issues between ourselves, we concluded that there was a need for a book that collected and collated the vast amount of information available in the public domain and provided a "single point-of-entry" to the field of gasification without necessarily answering all the questions that might arise. In fact, we felt that the most important task is to communicate an understanding for the questions to put in a given situation. This book may supply some of those answers directly; others will require further follow-up. This approach is no doubt colored by our own professional experience, where the very flexibility of gasification technology, with its differing feedstocks, differing end products, differing economic situations, and continual development has inevitably led to project-specific solutions for certain issues. Individual solutions will, we believe, continue to prevail in gasification technology, rather than a global standard after Henry Ford's philosophy of "any color they want, so long as it's black." For gasification, standardization, which is certainly an indispensable requisite to its economic competitiveness, must, in our opinion, first be introduced as a structuralized approach to the issues to be faced. And in developing this book, we have aimed at providing a structure that we hope can help in this process.

We trust that in doing so we can be of assistance to a broad audience, including

- Staff of companies who might want to build a plant and need to acquire know-how quickly in a compact form but independent of process licensors.
- Engineers of potential project financiers or insurers wanting to have an understanding of the technical risks involved in such a project, or those working for government departments and agencies involved in the licensing and permitting of gasification projects.
- People in the power industry who otherwise have little access to data on the subject of gasification.
- Established workers in the field looking for a reference work with a broad theoretical and practical overview.
- University students needing a book that combines the elements of academic theory and industrial practice.

After a brief historical introduction to gasification and its relevance to the development of our modern technological society in Chapter 1, there follow two chapters of theory. In order to have a good understanding of the practicalities of gasification, it is necessary to have a sufficient theoretical background. Chemical engineers will have this anyway, but many project engineers who become involved in gasification projects may have an educational background in mechanical or some other branch of engineering, and for such readers a brief summary is sure to be of use. The main emphasis of Chapter 2 is on thermodynamics, since this is generally sufficient for understanding and calculating the results of synthesis gas generation processes. But the development of computational fluid dynamics is beginning to make kinetics accessible to calculation in a manner hardly thinkable 20 years ago so that we have included a basic treatment of kinetic aspects of gasification in Chapter 3.

Chapter 4 reviews the wide variety of feedstocks that can be gasified, ranging from coal, through oils and gas, to biomass and waste. It discusses their properties as it affects both the gasification process itself and the downstream synthesis gas treatment and end usage.

The heart of the book lies in Chapters 5, 6, and 7. Chapter 5 discusses actual processes. The emphasis is on processes in commercial use today, such as those of Shell, Texaco, Lurgi, Noell, and others, such as the circulating fluid bed processes of Foster Wheeler and Lurgi. It includes brief mentions of some of the important forerunner processes, such as Winkler and Koppers-Totzek. A number of promising new processes, such as the Japanese CCP and EAGLE gasifiers, are also handled.

Chapter 6 looks at a broad selection of practical issues, including the drying and pressurizing of coal, syngas cooling and particulate removal, equipment issues, process control, trace components in synthesis gas, choice of oxidant, and corrosion aspects.

Typical applications are reviewed in Chapter 7. This includes the production of chemicals ranging from ammonia and methanol, through hydrogen to carbon monoxide, and synthesis gas for the production of oxo-alcohols. The section on synfuels production covers gas-to-liquids (GTL) and substitute natural gas (SNG). The discussion on power applications includes state-of-the-art IGCCs as well as a look at the potential for increasing efficiency with advanced cycles.

No gasification plant stands alone. Most processes require a source of oxygen, and the product synthesis gas needs treating and conditioning before it can be used. The principle auxiliary technologies for these tasks and the principal issues surrounding their selection are discussed in Chapter 8.

Every project stands or falls on its economics. Gasification is no exception, and economic aspects are addressed in Chapter 9. This chapter also looks at the environmental impact of gasification, particularly its superior performance in power generation. Its innate ability to provide a means of CO_2 capture with only minor additional cost is an important aspect of this subject. This chapter also addresses those safety issues that can be considered specific to the technology.

By way of an epilogue, we have tried to look into the crystal ball to see what part gasification can play in our futures. We discuss the potential contribution that gasification of fossil fuels can make to the transition to a hydrogen economy. Even in an ideal "fully sustainable" world, gasification of biomass may help us in the provision of some of the petrochemical products we so take for granted today.

At a number of different points in the text we have deliberately questioned current practice or thinking. We hope that the one or other idea produced may stimulate others and help further the technology as a whole.

COMPANION WEBSITE

As an accompaniment to this book, we have built a website (www.gasification. higman.de), which includes a number of computer programs arising out of the work involved in preparing this book. They include a complete gasification calculation based on the content of Chapter 2 and also a literature databank with keyword search capability.

TERMINOLOGY

A preliminary word on terminology may be in order. Gasification has a place in many industries, each with its own specific linguistic tradition. Recognizing this, we have not tried to impose our own language on the reader, but have used whatever synonym appears appropriate to the context. Thus the words fuel, feed, and feedstock are used interchangeably without any attempt to distinguish between them. Similarly, oxidant, blast, or gasification agent are used with the same meaning in different places.

ACKNOWLEDGMENTS

We would like to thank all our friends in the industry who have helped and encouraged us in this project, in particular Neville Holt of EPRI, Dale Simbeck of SFA Pacific, and Rainer Reimert of Universität Karlsruhe. We would also like to thank Nuon Power and Hydro Agri Brunsbüttel for the use of the cover photographs. A complete list would be too long to include at this point, but most will find their names somewhere in the bibliography, and we ask them to accept that as a personal thank you. Chris would also like to thank Lurgi for the time and opportunity to research and write this book. We would both like to thank our extremely tolerant wives, Pip and Agatha, who have accompanied us through our careers and this book and who have meanwhile come to know quite a lot about the subject too.

Finally, we hope that this book will contribute to the development of a better understanding of gasification processes and their future development. If it is of use to those developing new gasification projects, then it will have achieved its aim.

Chris Higman
Maarten van der Burgt

Chapter 1

Introduction

The manufacture of combustible gases from solid fuels is an ancient art but by no means a forgotten one. In its widest sense the term *gasification* covers the conversion of any carbonaceous fuel to a gaseous product with a useable heating value. This definition excludes combustion, because the product flue gas has no residual heating value. It does include the technologies of pyrolysis, partial oxidation, and hydrogenation. Early technologies depended heavily on pyrolysis (i.e., the application of heat to the feedstock in the absence of oxygen), but this is of less importance in gas production today. The dominant technology is partial oxidation, which produces from the fuel a synthesis gas (otherwise known as syngas) consisting of hydrogen and carbon monoxide in varying ratios, whereby the oxidant may be pure oxygen, air, and/or steam. Partial oxidation can be applied to solid, liquid, and gaseous feedstocks, such as coals, residual oils, and natural gas, and despite the tautology involved in "gas gasification," the latter also finds an important place in this book. We do not, however, attempt to extend the meaning of gasification to include catalytic processes such as steam reforming or catalytic partial oxidation. These technologies form a specialist field in their own right. Although we recognize that pyrolysis does take place as a fast intermediate step in most modern processes, it is in the sense of partial oxidation that we will interpret the word *gasification*, and the two terms will be used interchangeably. Hydrogenation has only found an intermittent interest in the development of gasification technologies, and where we discuss it, we will always use the specific terms *hydro-gasification* or *hydrogenating gasification*.

1.1 HISTORICAL DEVELOPMENT OF GASIFICATION

The development of human history is closely related to fire and therefore also to fuels. This relationship between humankind, fire, and earth was already documented in the myth of Prometheus, who stole fire from the gods to give it to man. Prometheus was condemned for his revelation of divine secrets and bound to earth as a punishment. When we add to fire and earth the air that we need to make fire and the water to keep it under control, we have the four Greek elements that play such an important role in the technology of fuels and for that matter in gasification.

1

The first fuel used by humans was wood, and this fuel is still used today by millions of people to cook their meals and to heat their homes. But wood was and is also used for building and, in the form of charcoal, for industrial processes such as ore reduction. In densely populated areas of the world this led to a shortage of wood with sometimes dramatic results. It was such a shortage of wood that caused iron production in England to drop from 180,000 to 18,000 tons per year in the period of 1620 to 1720. The solution—which in hindsight is obvious—was coal.

Although the production of coal had already been known for a long time, it was only in the second half of the eighteenth century that coal production really took hold, not surprisingly starting in the home of the industrial revolution, England. The coke oven was developed initially for the metallurgical industry to provide coke as a substitute for charcoal. Only towards the end of the eighteenth century was gas produced from coal by pyrolysis on a somewhat larger scale. With the foundation in 1812 of the London Gas, Light, and Coke Company, gas production finally became a commercial process. Ever since, it has played a major role in industrial development.

The most important gaseous fuel used in the first century of industrial development was town gas. This was produced by two processes: pyrolysis, in which discontinuously operating ovens produce coke and a gas with a relatively high heating value ($20,000–23,000 \, kJ/m^3$), and the water gas process, in which coke is converted into a mixture of hydrogen and carbon monoxide by another discontinuous method (approx. $12,000 \, kJ/m^3$ or medium Btu gas).

The first application of industrial gas was illumination. This was followed by heating, then as a raw material for the chemical industry, and more recently for power generation. Initially, the town gas produced by gasification was expensive, so most people used it only for lighting and cooking. In these applications it had the clearest advantages over the alternatives: candles and coal. But around 1900 electric bulbs replaced gas as a source of light. Only later, with increasing prosperity in the twentieth century, did gas gain a significant place in the market for space heating. The use of coal, and town gas generated from coal, for space heating only came to an end—often after a short intermezzo where heating oil was used—with the advent of cheap natural gas. But one should note that town gas had paved the way to the success of the latter in domestic use, since people were already used to gas in their homes. Otherwise there might have been considerable concern about safety, such as the danger of explosions.

A drawback of town gas was that the heating value was relatively low, and it could not, therefore, be transported over large distances economically. In relation to this problem it is observed that the development of the steam engine and many industrial processes such as gasification would not have been possible without the parallel development of metal tubes and steam drums. This stresses the importance of suitable equipment for the development of both physical and chemical processes. Problems with producing gas-tight equipment were the main reason why the production processes, coke ovens, and water gas reactors as well as the transport and storage were carried out at low pressures of less than 2 bar. This resulted in relatively voluminous equipment, to which the gasholders that were required to cope with variations in demand still bear witness in many of the cities of the industrialized world.

Until the end of the 1920s the only gases that could be produced in a continuous process were blast furnace gas and producer gas. Producer gas was obtained by partial oxidation of coke with humidified air. However, both gases have a low heating value (3500–6000 kJ/m^3, or low Btu gas) and could therefore only be used in the immediate vicinity of their production.

The success of the production of gases by partial oxidation cannot only be attributed to the fact that gas is easier to handle than a solid fuel. There is also a more basic chemical reason that can best be illustrated by the following reactions:

$$C + \tfrac{1}{2}O_2 = CO \qquad\qquad -111\,\text{MJ/kmol} \qquad\qquad (1\text{-}1)$$
$$CO + \tfrac{1}{2}O_2 = CO_2 \qquad\qquad -283\,\text{MJ/kmol} \qquad\qquad (1\text{-}2)$$
$$C + O_2 = CO_2 \qquad\qquad -394\,\text{MJ/kmol} \qquad\qquad (1\text{-}3)$$

These reactions show that by "investing" 28% of the heating value of pure carbon in the conversion of the solid carbon into the gas CO, 72% of the heating value of the carbon is conserved in the gas. In practice, the fuel will contain not only carbon but also some hydrogen, and the percentage of the heat in the original fuel, which becomes available in the gas, is, in modern processes, generally between 75 and 88%. Were this value only 50% or lower, gasification would probably never have become such a commercially successful process.

Although gasification started as a source for lighting and heating, from 1900 onwards the water gas process, which produced a gas consisting of about equal amounts of hydrogen and carbon monoxide, also started to become important for the chemical industry. The endothermic water gas reaction can be written as:

$$C + H_2O \leftrightarrows CO + H_2 \qquad\qquad +131\,\text{MJ/kmol} \qquad\qquad (1\text{-}4)$$

By converting part or all of the carbon monoxide into hydrogen following the CO shift reaction,

$$CO + H_2O \leftrightarrows H_2 + CO_2 \qquad\qquad -41\,\text{MJ/kmol} \qquad\qquad (1\text{-}5)$$

it became possible to convert the water gas into hydrogen or synthesis gas (a mixture of H_2 and CO) for ammonia and methanol synthesis, respectively. Other applications of synthesis gas are for Fischer-Tropsch synthesis of hydrocarbons and for the synthesis of acetic acid anhydride.

It was only after Carl von Linde commercialized the cryogenic separation of air during the 1920s that fully continuous gasification processes using an oxygen blast became available for the production of synthesis gas and hydrogen. This was the time of the development of some of the important processes that were the forerunners of many of today's units: the Winkler fluid-bed process (1926), the Lurgi moving-bed pressurized gasification process (1931), and the Koppers-Totzek entrained-flow process (1940s).

With the establishment of these processes little further technological progress in the gasification of solid fuels took place over the following forty years. Nonetheless, capacity with these new technologies expanded steadily, playing their role partly in Germany's wartime synthetic fuels program and on a wider basis in the worldwide development of the ammonia industry.

This period, however, also saw the foundation of the South African Coal Oil and Gas Corporation, known today as Sasol. This plant uses coal gasification and Fischer-Tropsch synthesis as the basis of its synfuels complex and an extensive petrochemical industry. With the extensions made in the late 1970s, Sasol is the largest gasification center in the world.

With the advent of plentiful quantities of natural gas and naphtha in the 1950s, the importance of coal gasification declined. The need for synthesis gas, however, did not. On the contrary, the demand for ammonia as a nitrogenous fertilizer grew exponentially, a development that could only be satisfied by the wide-scale introduction of steam reforming of natural gas and naphtha. The scale of this development, both in total capacity as well as in plant size, can be judged by the figures in Table 1-1. Similar, if not quite so spectacular, developments took place in hydrogen and methanol production.

Steam reforming is not usually considered to come under the heading of gasification. The reforming reaction (allowing for the difference in fuel) is similar to the water gas reaction.

$$CH_4 + H_2O \leftrightarrows 3H_2 + CO \qquad +206\,MJ/kmol \qquad (1\text{-}6)$$

The heat for this endothermic reaction is obtained by the combustion of additional natural gas:

$$CH_4 + 2O_2 = CO_2 + 2H_2O \qquad -803\,MJ/kmol \qquad (1\text{-}7)$$

Unlike gasification processes, these two reactions take place in spaces physically separated by the reformer tube.

Table 1-1 **Development of Ammonia Production Capacity 1945–1969**		
Year	**World ammonia production (MMt/y)**	**Maximum converter size (t/d)**
1945	5.5	100
1960	14.5	250
1964	23.0	600
1969	54.0	1400
Source: Slack and James 1973		

An important part of the ammonia story was the development of the secondary reformer in which unconverted methane is processed into synthesis gas by partial oxidation over a reforming catalyst.

$$CH_4 + \tfrac{1}{2}O_2 = CO + 2H_2 \qquad\qquad -36\,MJ/kmol \qquad\qquad\qquad (1\text{-}8)$$

The use of air as an oxidant brought the necessary nitrogen into the system for the ammonia synthesis. A number of such plants were also built with pure oxygen as oxidant. These technologies have usually gone under the name of autothermal reforming or catalytic partial oxidation.

The 1950s was also the time in which both the Texaco and the Shell oil gasification processes were developed. Though far less widely used than steam reforming for ammonia production, these were also able to satisfy a demand where natural gas or naphtha were in short supply.

Then, in the early 1970s, the first oil crisis came and, together with a perceived potential shortage of natural gas, served to revive interest in coal gasification as an important process for the production of liquid and gaseous fuels. Considerable investment was made in the development of new technologies. Much of this effort went into coal hydrogenation both for direct liquefaction and also for so-called hydro-gasification. The latter aimed at hydrogenating coal directly to methane as a substitute natural gas (SNG). Although a number of processes reached the demonstration plant stage (Speich 1981), the thermodynamics of the process dictate a high-pressure operation, and this contributed to the lack of commercial success of hydro-gasification processes. In fact, the only SNG plant to be built in these years was based on classical oxygen-blown moving-bed gasification technology to provide synthesis gas for a subsequent methanation step (Dittus and Johnson 2001).

The general investment climate in fuels technology did lead to further development of the older processes. Lurgi developed a slagging version of its existing technology in a partnership with British Gas (BGL) (Brooks, Stroud, and Tart 1984). Koppers and Shell joined forces to produce a pressurized version of the Koppers-Totzek gasifier (for a time marketed separately as Prenflo and Shell coal gasification process, or SCGP, respectively) (van der Burgt 1978). Rheinbraun developed the high-temperature Winkler (HTW) fluid-bed process (Speich 1981), and Texaco extended its oil gasification process to accept a slurried coal feed (Schlinger 1984).

However, the 1980s then saw a renewed glut of oil that reduced the interest in coal gasification and liquefaction; as a result, most of these developments had to wait a further decade or so before getting past the demonstration plant stage.

1.2   GASIFICATION TODAY

The last ten years have seen the start of a renaissance of gasification technology, as can be seen from Figure 1-1. Electricity generation has emerged as a large new market for these developments, since gasification is seen as a means of enhancing the

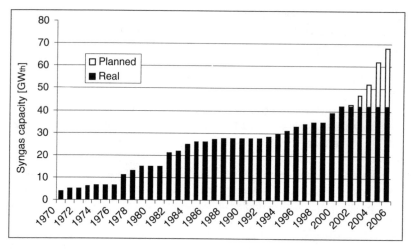

Figure 1-1. Cumulative Worldwide Gasification Capacity (*Source: Simbeck and Johnson 2001*)

environmental acceptability of coal as well as of increasing the overall efficiency of the conversion of the chemical energy in the coal into electricity. The idea of using synthesis gas as a fuel for gas turbines is not new. Gumz (1950) proposed this already at a time when anticipated gas turbine inlet temperatures were about 700°C. And it has largely been the development of gas turbine technology with inlet temperatures now of 1400°C that has brought this application into the realm of reality. Demonstration plants have been built in the United States (Cool Water, 100 MW, 1977; and Plaquemine, 165 MW, 1987) and in Europe (Lünen, 170 MW, 1972; Buggenum, 250 MW, 1992; and Puertollano, 335 MW, 1997).

A second development, which has appeared during the 1990s, is an upsurge in gasification of heavy oil residues in refineries. Oil refineries are under both an economic pressure to move their product slate towards lighter products, and a legislative pressure to reduce sulfur emissions both in the production process as well as in the products themselves. Much of the residue had been used as a heavy fuel oil, either in the refinery itself or in power stations and as marine bunker fuel. Residue gasification has now become one of the essential tools in addressing these issues. Although heavy residues have a low hydrogen content, they can be converted into hydrogen by gasification. The hydrogen is used to hydrocrack other heavy fractions in order to produce lighter products such as gasoline, kerosene, and automotive diesel. At the same time, sulfur is removed in the refinery, thus reducing the sulfur present in the final products (Higman 1993). In Italy, a country particularly dependent on oil for power generation, three refineries have introduced gasification technology as a means of desulfurizing heavy fuel oil and producing electric power. Hydrogen production is incorporated into the overall scheme. A similar project was realized in Shell's Pernis refinery in the Netherlands. Other European refineries have similar projects in the planning phase.

An additional driving force for the increase in partial oxidation is the development of "Gas-to-liquids" projects. For transport, liquid fuels have an undoubted advantage. They are easy to handle and have a high energy density. For the consumer, this translates into a car that can travel nearly 1000 km on 50 liters of fuel, a range performance as yet unmatched by any of the proposed alternatives. For the energy company the prospect of creating synthetic liquid fuels provides a means of bringing remote or "stranded" natural gas to the marketplace using existing infrastructure. Gasification has an important role to play in this scenario. The Shell Middle Distillate Synthesis (SMDS) plant in Bintulu, Malaysia, producing some 12,000 bbl/d of liquid hydrocarbons, is only the first of a number of projects currently in various stages of planning and engineering around the world (van der Burgt 1988).

REFERENCES

Brooks, C. T., Stroud, H. J. F., and Tart, K. R. "British Gas/Lurgi Slagging Gasifier." In *Handbook of Synfuels Technology*, ed. R. A. Meyers. New York: McGraw-Hill, 1984.

Dittus, M., and Johnson, D. "The Hidden Value of Lignite Coal." Paper presented at Gasification Technologies Conference, San Francisco, October 2001.

Gumz, W. *Gas Producers and Blast Furnaces*. New York: John Wiley & Sons, 1950.

Higman, C. A. A. "Partial Oxidation in the Refinery Hydrogen Management Scheme." Paper presented at AIChE Spring Meeting, Houston, March 1993.

Schlinger, W. G. "The Texaco Coal Gasification Process." In *Handbook of Synfuels Technology*, ed. R. A. Meyers. New York: McGraw-Hill, 1984.

Simbeck, D., and Johnson, H. "World Gasification Survey: Industry Trends and Developments." Paper presented at Gasification Technologies Conference, San Francisco, October 2001.

Slack, A. V., and James, G. R. *Ammonia, Part I* New York: Marcel Dekker, 1973.

Speich, P. "Braunkohle—auf dem Weg zur großtechnischen Veredelung." *VIK-Mitteilungen* 3/4 (1981).

van der Burgt, M. J. "Shell's Middle Distillate Synthesis Process." Paper presented at AIChE Meeting, New Orleans, 1988.

van der Burgt, M. J. "Technical and Economic Aspects of Shell-Koppers Coal Gasification Process." Paper presented at AIChE Meeting, Anaheim, CA 1978.

Chapter 2

The Thermodynamics of Gasification

In Chapter 1 we defined gasification as the production of gases with a useable heating value from carbonaceous fuels. The range of potential fuels, from coal and oil to biomass and wastes, would appear to make the task of presenting a theory valid for all these feeds relatively complex.

Nonetheless, the predominant phenomena of pyrolysis or devolatilization and gasification of the remaining char are similar for the full range of feedstocks. In developing gasification theory it is therefore allowable to concentrate on the "simple" case of gasification of pure carbon, as most authors do, and discuss the influence of specific feed characteristics separately. In this work we will be adopting this approach, which can also be used for the partial oxidation of gases such as natural gas.

In the discussion of the theoretical background to any chemical process, it is necessary to examine both the thermodynamics (i.e., the state to which the process will move under specific conditions of pressure and temperature, given sufficient time) and the kinetics (i.e., what route will it take and how fast will it get there).

The gasification process takes place at temperatures in the range of 800°C to 1800°C. The exact temperature depends on the characteristics of the feedstock, in particular the softening and melting temperatures of the ash as is explained in more detail in Chapter 5. However, over the whole temperature range described above, the reaction rates are sufficiently high that modeling on the basis of the thermodynamic equilibrium of the main gaseous components and carbon (which we will assume for the present to be graphite) gives results that are close enough to reality that they form the basis of most commercial reactor designs. This applies unconditionally for all entrained slagging gasifiers and may also be applied to most fluid-bed gasifiers and even to moving-bed gasifiers, provided the latter use coke as a feedstock.

One exception to the above assumption that one can model with thermodynamic equilibria alone is the moving-bed gasifier where coal is used as a feedstock and where the blast (oxygen and steam) moves counter-currently to the coal as in, for example, the Lurgi gasifier that is described in further detail in Section 5.1. In such gasifiers pyrolysis reactions are prevalent in the colder upper part of the reactor, and therefore a simple description of the process by assuming thermodynamic equilibrium is not allowable for that region of the reactor. The reactions in the hot bottom section in which coke reacts with the blast can, however, be described well by thermodynamic equilibrium. A second exception is the gasification of biomass at temperatures of about 850°C (Kersten 2002).

In this discussion of the theory of gasification the emphasis will be limited to gasifiers that operate at temperatures of 850°C and higher. Below 850°C is, on the one hand, the realm of pyrolysis reactions that are extremely complex to model, whereas, on the other hand, the partial oxidation reactions proceed at so slow a rate that they become of little practical value.

2.1   REACTIONS

During the process of gasification of solid carbon whether in the form of coal, coke, or char, the principle chemical reactions are those involving carbon, carbon monoxide, carbon dioxide, hydrogen, water (or steam), and methane. These are:
Combustion reactions,

$$C + \tfrac{1}{2} O_2 = CO \qquad\qquad -111\,\text{MJ/kmol} \qquad\qquad (2\text{-}1)$$
$$CO + \tfrac{1}{2} O_2 = CO_2 \qquad\qquad -283\,\text{MJ/kmol} \qquad\qquad (2\text{-}2)$$
$$H_2 + \tfrac{1}{2} O_2 = H_2O \qquad\qquad -242\,\text{MJ/kmol} \qquad\qquad (2\text{-}3)$$

the Boudouard reaction,

$$C + CO_2 \leftrightarrows 2\ CO \qquad\qquad +172\,\text{MJ/kmol} \qquad\qquad (2\text{-}4)$$

the water gas reaction,

$$C + H_2O \leftrightarrows CO + H_2 \qquad\qquad +131\,\text{MJ/kmol} \qquad\qquad (2\text{-}5)$$

and the methanation reaction,

$$C + 2\ H_2 \leftrightarrows CH_4 \qquad\qquad -75\,\text{MJ/kmol} \qquad\qquad (2\text{-}6)$$

As reactions with free oxygen are all essentially complete under gasification conditions, reactions 2-1, 2-2, and 2-3 do not need to be considered in determining an equilibrium syngas composition. The three heterogeneous (i.e., gas and solid phase) reactions 2-4, 2-5, and 2-6 are sufficient.

In general, we are concerned with situations where also the carbon conversion is essentially complete. Under these circumstances we can reduce equations 2-4, 2-5, and 2-6 to the following two homogeneous gas reactions:
CO shift reaction:

$$CO + H_2O \leftrightarrows CO_2 + H_2 \qquad\qquad -41\,\text{MJ/kmol} \qquad\qquad (2\text{-}7)$$

and the steam methane reforming reaction:

$$CH_4 + H_2O \leftrightarrows CO + 3\,H_2 \qquad\qquad +206\,\text{MJ/kmol} \qquad\qquad (2\text{-}8)$$

Note that by subtracting the moles and heat effects from reaction 2-4 from those in reaction 2-5 one obtains reaction 2-7, and by subtracting reaction 2-6 from 2-5 one obtains reaction 2-8. Thus reactions 2-7 and 2-8 are implicit in reactions 2-4, 2-5, and 2-6. But not the other way around! Three independent equations always contain more information than two.

Reactions 2-1, 2-4, 2-5, and 2-6 describe the four ways in which a carbonaceous or hydrocarbon fuel can be gasified. Reaction 2-4 plays a role in the production of pure CO when gasifying pure carbon with an oxygen/CO_2 mixture. Reaction 2-5 takes a predominant role in the water gas process. Reaction 2-6 is the basis of all hydrogenating gasification processes. But most gasification processes rely on a balance between reactions 2-1 (partial oxidation) and 2-5 (water gas reaction).

For real fuels (including coal, which also contains hydrogen) the overall reaction can be written as

$$C_n H_m + n/2\,O_2 = n\,CO + m/2\,H_2 \qquad\qquad (2\text{-}9)$$

where for gas, as pure methane, $m=4$ and $n=1$, hence $m/n=4$, and
for oil, $m/n \approx 2$, hence $m=2$ and $n=1$, and
for coal, $m/n \approx 1$, hence $m=1$ and $n=1$.

Gasification temperatures are in all cases so high that, thermodynamically as well as in practice, apart from methane, no hydrocarbons can be present in any appreciable quantity.

Thermodynamic Equilibrium

As indicated by the double arrows, equations 2-4 to 2-8 are all reversible reactions, that is to say, they may proceed both from right to left as well as from left to right. In general, the forward and the reverse reactions take place simultaneously and at different rates. For any given temperature these reaction rates are proportional to the quantity of reactants available to drive the reaction in the direction under consideration.

If we take the CO shift reaction (2-7) as an example, the forward reaction rate, r_f, is proportional to the molar concentrations of CO and H_2O per unit volume, or

$$r_f = k_f \cdot [CO] \cdot [H_2O]$$

where the constant of proportionality k_f is temperature dependant. Similarly, for the reverse reaction,

$$r_r = k_r \cdot [CO_2] \cdot [H_2]$$

Over a period of time these two reaction rates will tend to reach a common value and the gas composition will have reached a state of equilibrium. Under these circumstances

$$K_p = \frac{k_f}{k_r} = \frac{[CO_2] \cdot [H_2]}{[CO] \cdot [H_2O]}$$

where K_p is the temperature dependant equilibrium constant for the CO shift reaction. Assuming ideal gases this can also be expressed as

$$K_p = \frac{P_{CO_2} \cdot P_{H_2}}{P_{CO} \cdot P_{H_2O}} = \frac{v_{CO_2} \cdot v_{H_2}}{v_{CO} \cdot v_{H_2O}} \qquad (2\text{-}10)$$

where P_{CO} is the partial pressure and v_{CO} is the volume fraction $\frac{P_{CO}}{P}$ of CO in the gas, and so on.

Similarly the equilibrium constants for the other reactions can be expressed as

$$K_p = \frac{P^2_{CO}}{P_{CO_2}} = \frac{(v_{CO})^2}{v_{CO_2}} \cdot P \qquad (2\text{-}11)$$

for the Boudouard reaction (2-4),

$$K_p = \frac{P_{CO} \cdot P_{H_2}}{P_{H_2O}} = \frac{v_{CO} \cdot v_{H_2}}{v_{H_2O}} \cdot P \qquad (2\text{-}12)$$

for the water gas reaction (2-5), and

$$K_p = \frac{P_{CO} \cdot P^3_{H_2}}{P_{CH_4} \cdot P_{H_2O}} = \frac{v_{CO} \cdot v^3_{H_2}}{v_{CH_4} \cdot v_{H_2O}} \cdot P^2 \qquad (2\text{-}13)$$

for the reforming reaction (2-8), where P is the total absolute pressure of the gas.

The temperature dependency of these equilibrium constants can be derived from fundamental data but are usually expressed as a correlation of the type

$$\ln (K_{p,T}) = \ln (K_{p,T0}) + f(T)$$

where T is the absolute temperature in Kelvin. The derivation of the equilibrium constants is described in the file gasify.hlp on the companion website. The above (albeit nonlinear) equations 2-10 to 2-13 provide us with a means of determining the relative concentrations of the gas components in the syngas on the assumption that the reactions have reached equilibrium.

Note that in all the above discussions it has been assumed that the gases are ideal gases and no fugacities have been taken into account. Although many processes operate at pressures of 30–70 bar, this assumption is justified because of the very high temperatures in the processes, which lie very far from the critical temperature of each compound. Note that even where the calculations are used for the low temperature CO shift reaction, which operates at temperatures of 200–250°C, this approximation gives sufficiently accurate results for basic designs.

Other Compounds

Most fuels contain additional material beyond the carbon, hydrogen, and oxygen discussed above. Sulfur in the fuel is converted into H_2S and COS, and the nitrogen into elemental nitrogen, NH_3, and HCN. Generally speaking, the quantities of sulfur and nitrogen in the fuel are sufficiently small that they have a negligible effect on the main syngas components of hydrogen and carbon monoxide. Nonetheless it is necessary to consider the fate of sulfur and nitrogen because of the effect of the resulting compounds downstream of the gas production, for example, environmental emissions, catalyst poisons, and so on.

Details concerning the formation of sulfur and nitrogen compounds are included in Section 6.9. Since the amounts of sulfur and nitrogen converted into the various product molecules are not large, the distribution of sulfur and nitrogen compounds is usually estimated in advance (for example, $H_2S/COS = 9-9.5$, $NH_3 = 25\%$, and $HCN = 10\%$ of fuel nitrogen, respectively), and interaction with carbon, hydrogen, and oxygen is then limited to mass and heat balance calculations.

2.2   THERMODYNAMIC MODELING OF GASIFICATION

Both designers and operators need to have some knowledge about thermodynamic modeling, although in developing models for gasification, it should be noted that the requirements of a designer and an operator are different.

The designer has the task of calculating a limited number of design cases and using these to size the plant equipment. He will be interested in throughputs of the different feedstocks, gas compositions, heat effects, quench requirements, startup and shutdown requirements, optimal conditions for the design feedstocks, and process control requirements.

The operator has his equipment as it is, but will need to optimize operations for feedstocks, which are seldom identical with the formal design case. He will therefore be more interested in what he can expect when feeding a specific cocktail of feedstocks, how to interpret gas compositions, and, for example, the steam make in a syngas cooler. Once the unit runs stably he will become interested in optimizing the process.

A good model will therefore be so built that both requirements can be readily met without the user having to perform an undue number of iterative calculations to perform his task.

The purpose of gasification modeling is:

- The calculation of the gas composition.
- The calculation of the relative amounts of oxygen and/or steam and/or heat required per unit fuel intake.

- The optimization of the energy in the form of the heat of combustion of the product gas or, alternatively, of the synthesis gas production per unit fuel intake.
- To provide set points for process control.

Calculations comprising the gasification proper are based on thermodynamics, mass and energy balances and process conditions, such as temperature, pressure, and the addition or subtraction of indirect heat. In all these calculations it is essential that the elemental composition and the temperature of the feed streams are known. For coal, both the proximate analysis (fixed carbon, volatile matter, moisture, ash) and the ultimate analysis (elemental, apart from ash) must be known.

In gasification, use is made of a variety of reactions of which some are exothermic and some are endothermic, as was shown in Section 2.1. In virtually all cases the desired operating temperature is obtained by judiciously playing with the exothermic and endothermic reactions. The reaction of the fuel with oxygen is always complete and exothermic, whereas the reaction with steam or carbon dioxide is always endothermic and never complete because of thermodynamic limitations.

In gasifiers where both oxygen and steam are used to control the temperature, the role of steam is that of a moderator. Some other methods to moderate the temperature are to add nitrogen or carbon dioxide to the oxygen, or to remove heat indirectly from the gasification reactor.

In all cases the fuel to a gasifier will contain carbon. The blast or gasifying agent is the mixture of oxygen-containing gas and/or steam and/or carbon dioxide. Hence the blast always contains oxygen either as free oxygen or bound in the form of, for example, steam.

The companion website to this book includes a simple gasification model (gasify.exe) that illustrates these principles. The following discussion of specific aspects of gasifier modeling includes both general points, applicable to any model as well as an explanation of the particular approach that we have adopted for our own model.

2.2.1 Basic Data

A model for gasification is only as good as the basic thermodynamic data used. To ensure consistency of mass and heat balances it is of fundamental importance to limit the data set to a minimum and generate all other data from this.

In the calculations used in this book and in the website programs, use has been made of various sources, the most important being Barin (1989). For the thermodynamic calculations standard enthalpy and entropy data have been used in combination with a curve-fit for the temperature dependence of the specific heat for each type of molecule.

For the moisture-and-ash-free (maf) coal the default coefficients in the formula giving the temperature dependence of the specific heat have been based on graphite.

For the ash temperature dependence the default coefficients in the formula giving the temperature dependence of the specific heat have been based on an average ash in which the heat of fusion has been taken into account. In the case where the ash and/or the fluxing agent added to the ash to lower the melting point contain carbonates, the heat of decomposition may have to be taken into account. The easiest way to do this, though, is to calculate the heat required and put it into the calculations as an additional heat loss from the gasifier.

Where alternative data, in particular for coal and ash, are desired, it would be necessary to modify this appropriate data. In the case of gasify.exe, this would need to be done in the source code.

Feedstocks

Feedstocks for gasification may vary from natural gas to heavy oil residues and coal. Furthermore, waste streams and biomass may be used. In order to carry out proper calculations it is essential that the elemental composition and the standard heat of formation of the fuels are known.

The standard heat of formation of the fuel may be calculated from the heat of combustion and the elemental composition of the fuel without the ash. Ash in the fuel is generally considered as being inert in the calculations. By subtracting the heat of combustion of the various elements in the fuel (in practice, carbon, hydrogen, and sulfur) from the heat of combustion of the fuel, the standard heat of formation is obtained. Care should be taken to normalize the quantities of the fuel and the relevant elements, as well as to ensure that they are all based on either the lower heating value (LHV) or the higher heating value (HHV) of the fuel.

Where no measured combustion values of the fuel are available, the heat of combustion of the fuel may be calculated from the proximate and ultimate analyses. Gasify.exe includes some suitable empirical correlations. It is always recommended to check whether the measured data (when available) are in agreement with the calculated data. If these deviate by more than 2–3%, it is recommended to use the calculated data for coal and heavy oil fuels. For natural gas fuels it is recommended always to use the calculated combustion values. The most difficult fuels are biomass and waste feedstocks, where the calculated and measured values may easily deviate by over 5%. Careful scrutiny of the measured data is then recommended.

Coal

Composition and combustion data for coal are often very confusing, as the composition data may be based on an as-received (ar), moisture-free (mf), ash-free (af), or ash-and-moisture-free (maf) basis. Moreover, the heating value can be given as LHV or HHV, possibly on another basis than the proximate and ultimate analysis. For this reason, gasify.exe includes a module that readily converts all these data into each other. For the ash and maf coal it has been arbitrarily assumed that these components have a molecular mass of 100.

Moderator

The most common moderator used in gasification processes is steam. The steam must have a minimum temperature corresponding to that of saturated steam at the pressure prevailing in the gasifier, otherwise condensation in the lines to the gasifier will occur. In general, steam is used that is superheated to a temperature of 300–400°C. At pressures above 40 bar this superheat is mandatory, since otherwise the steam becomes wet on expansion.

The use of carbon dioxide as a moderator is unusual but not entirely unknown. Some plants use it to influence the H_2/CO ratio of the synthesis gas where carbon monoxide or a CO-rich gas is the product. A process has been published using an oxygen-blown gasifier for the production of pure carbon monoxide from coke, incorporating carbon dioxide as a moderator (Lath and Herbert 1986).

Where CO_2 is used as a transport gas for pulverized coal in entrained gasifiers, it will also act as moderator or part thereof.

2.2.2   Equations

Forgetting for the moment the presence of other compounds and elements such as sulfur, nitrogen, argon, ash, and so on, the following equations will apply in virtually all gasification processes.

1. Carbon balance.
2. Hydrogen balance.
3. Oxygen balance.
4. Dalton equation, stating that the sum of the mole fractions in the product gas equals unity.
5. Heat balance, stating that the sum of the heat of formation and the sensible heat of the product(s) equals that of the corresponding data in the feed stream(s), provided it is corrected for heat that is indirectly added to or subtracted from the process.
6. Reaction constants of the relevant reactions. In general, 3 for the heterogeneous case where carbon is present and 2 for the homogeneous case.

Note that in selecting the reactions for both heterogeneous and homogeneous reactions, it is essential that the relevant compounds in the set of reactions are present. In the heterogeneous case, the reactions 2-5 and 2-6 could just as well have been replaced by reactions 2-7 and 2-8; whereas in the homogeneous case, reaction 2-8 could be replaced by the CO_2 reforming reaction:

$$CH_4 + CO_2 \leftrightarrows 2CO + 2H_2 \qquad +247 \, MJ/kmol \, CH_4 \qquad (2\text{-}14)$$

When pure carbon is to be gasified with a blast of oxygen and carbon dioxide, then the hydrogen balance and reactions 2-5 and 2-6 fall away.

In practice, fuels such as coal and heavy oils will also contain sulfur, nitrogen, and ash, and the oxygen may contain argon. This implies that the material balances have to be extended with the following equations:

7. Sulfur balance.
8. Nitrogen balance.

9. Ash balance.
10. Argon balance.

Of course, these additional elements will also have to be considered in the heat balance (see point #5 above).

2.2.3 Variables

Again ignoring sulfur, nitrogen, and others initially, it is necessary to define eight variables for the heterogeneous case and seven for the homogeneous one to provide a mathematically soluble problem. Five variables that virtually always apply in gasification are the gas component fractions in the synthesis gas for:

CO_2, CO, H_2, CH_4, and H_2O.

The remaining three variables in case of heterogeneous gasification and two in the case of homogeneous gasification may be selected from the following list:

- Fuel used per kmole product gas.
- Blast (oxidant) used per kmole product gas.
- Moderator (mostly steam) used per kmole product gas.
- Heat loss from the gasifier reactor or heat required for the gasification.
- Gasification temperature (though any result would need to be checked in respect of the ash properties).

2.3 DEDUCTIONS FROM THE THERMODYNAMIC MODEL

2.3.1 Effect of Pressure

There is considerable advantage to gasifying under pressure, sufficiently so that practically all modern processes are operated at pressures of at least 10 bar and up to as high as 100 bar. The reasons for this are savings in compression energy and reduction of equipment size. To appreciate the realities of savings in compression energy, we can compare the energy required to provide 100,000 Nm³/h raw synthesis gas at 45 bar by either

1. gasifying at a relatively low pressure (5 bar) and compressing the synthesis gas, or alternatively,
2. compressing the feedstocks to 55 bar (allowing for pressure drop in the system) and gasifying at the higher pressure.

For the calculations we will use an oil feedstock, so as to include simply the energy for raising the fuel pressure in case 2. It is also assumed that oxygen is available from an air separation unit at atmospheric pressure in both cases. The energy is given as shaft energy on the machines. The enthalpy difference of the moderating steam in the two cases is neglected (see Table 2-1).

		5 Bar Gasification	50 Bar Gasification
Feed pumping energy	35,450 kg/h	0.03 MW	0.09 MW
Oxygen compression	21,120 Nm^3/h	2.85 MW	4.97 MW
Syngas compression	100,000 Nm^3/h	19.70 MW	0.00 MW
Total		22.58 MW	5.05 MW

Table 2-1
Comparison of Compression Energy for Low and High Pressure Gasification

The pressure in a gasifier is therefore generally selected in accordance with the requirements of the process or equipment upstream or downstream of the gasifier. For extremely high pressures, as required by ammonia synthesis (130–180 bar), for example, this argument ceases to be dominant, as gasification at pressures above 70–100 bar become impractical for equipment reasons. When the gas is to be used in a combined cycle (CC) power station where the gas turbine requires a pressure of say 20 bar, the gasifier pressure may operate at a somewhat higher pressure than this to allow for pressure losses between the gasifier and the gas turbine.

Strictly speaking, allowance should be made in preparing the above figures for the fact that the syngas ($CO+H_2$) yield and the heating value of the gas are somewhat different at the two gasification pressures. This slight difference does not, however, alter the conclusion that, other things being equal, compression of the reactants is energetically superior to compression of the gasifier product gas. Other considerations when selecting a gasification pressure are discussed in Section 6.1.

It is instructive to see how the gas composition changes with pressure, and this is shown in Figures 2-1 and 2-2 with the calculations all performed at 1000°C. The increase in methane and CO_2 content in the synthesis gas with increasing pressure can be seen clearly in Figure 2-1. In Figure 2-2 it is plain that the yield of synthesis gas (as H_2+CO) drops with pressure, whereas the heat content yield increases (reflecting the higher methane content). Similarly, the variation in oxygen demand goes in opposite directions depending on whether it is expressed per unit of syngas or per unit of heating value.

We will return to these effects later in the chapter, when looking at the differences between optimizing for IGCC and for synthesis gas applications.

If we repeat the above calculations at, say, 1500°C, we see in principle the same trends with increasing pressure. However, if we look at the actual numbers in Table 2-2, then we notice that at this temperature the actual changes of gas composition with pressure are almost negligible.

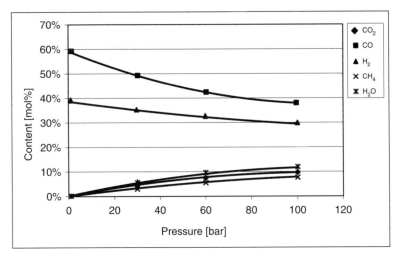

Figure 2-1. Variation of Syngas Compositions with Temperature at 1000°C

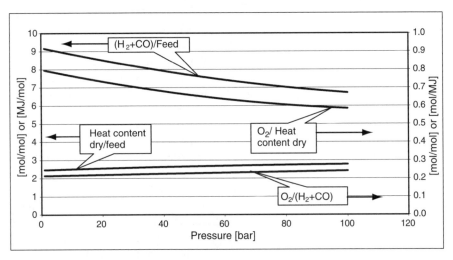

Figure 2-2. Variations of Yields with Temperature at 1000°C

2.3.2 Effect of Temperature

The temperature is generally selected on the basis of the ash properties (i.e., below the softening point of the ash for fluid-bed and dry ash moving-bed gasifiers and above the melting point for slagging gasifiers). For coals with very high ash melting points it is often advantageous to add flux to the coal feed in order to lower the ash melting point. As will be discussed later, gasifying at very high temperatures

Table 2-2
Variations of Syngas Compositions and Yields at 1500°C

	1 Bar	30 Bar	60 Bar	100 Bar
CO_2, mol%	0.00	0.01	0.21	0.34
CO, mol%	63.42	63.33	62.88	62.88
H_2, mol%	34.37	33.89	33.07	33.07
CH_4, mol%	0.01	0.27	0.85	0.85
H_2O, mol%	0.01	0.21	0.67	0.67
Others, mol%	2.19	2.20	2.20	2.19
Total, mol%	100.00	100.00	100.00	100.00
H_2+CO/feed, mol/mol	8.61	8.52	8.45	8.36
LHV dry/feed, MJ/mol	2.31	2.31	2.32	2.32
$O_2/(H_2+CO)$, mol/mol	0.27	0.27	0.27	0.27
O_2/LHV dry, mol/MJ	0.99	0.99	0.98	0.97

will increase the oxygen consumption of a gasification process and will reduce the overall process efficiency.

For process control purposes where ratios between fuel, oxygen, and/or steam are known, the temperature can be calculated. This is an important aspect, as temperatures in slagging gasifiers can only be measured with great difficulty and are generally not very trustworthy.

Since most modern gasification processes operate at pressures of 30 bar or higher, temperatures of above 1300°C are required in order to produce a synthesis gas with a low methane content. The fact that such a high temperature is required in any case for thermodynamic reasons is why there is little scope for the use of catalysts in gasifiers. The use of catalysts is restricted to clean gasification environments that are only encountered in the partial oxidation of natural gas and in steam methane or naphtha reforming.

This observation leads us on to perform the same exercise of investigating the variations of gas compositions and yield with temperature as shown in Figures 2-3 and 2-4.

In Figures 2-3 and 2-4 we can see that with increasing temperature the gas becomes increasingly CO rich. In Figure 2-4 the increased oxygen demand at high temperature is apparent. The H_2+CO yield goes through a mild maximum between 1200°C and 1300°C.

2.3.3 Fuel Footprint

For a good understanding of the gasification of any particular feedstock, whether coal, heavy oil, waste biomass, or gas, it is important to calculate the gasification characteristics of the fuel in order to be able to understand what is going on in the reactor and to optimize the reaction conditions.

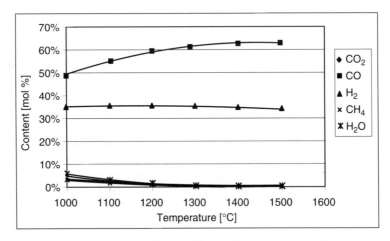

Figure 2-3. Variation of Syngas Compositions with Pressure at 30 bar

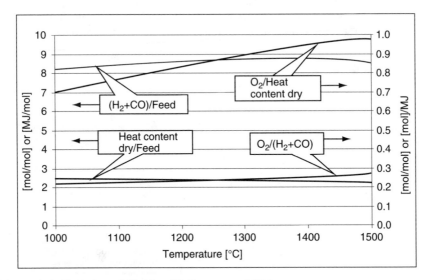

Figure 2-4. Variation of Yields with Pressure at 30 bar

Instead of discussing this topic in the general terms of fuel, oxygen-containing gas, and moderator, we will do so in terms of carbon, oxygen, and steam. But remember, it holds for any fuel!

A most instructive representation of the fuel gasification characteristic is the "footprint," which comprises a graph where for a specified heat loss from the gasification reactor and a specific set of reactants—coal, oxygen, and steam, each with a fixed temperature and composition—the steam consumption per kmole product gas is plotted along the ordinate, and the oxygen consumption per kmole of product gas is plotted along the abscissa. Such a graph made for various heat losses forms the fuel footprint. In Figure 2-5 an example is given in which for a fixed heat loss various isotherms are drawn together with the borderline below which no complete conversion of the carbon is thermodynamically possible.

In the lower left-hand corner of Figure 2-5 the straight line represents the minimum amount of gasifying agents (oxygen and steam) that are required to gasify the coal, assuming that no H_2O, CO_2, and CH_4 are present in the gas. Thermodynamically this is not possible; the real line assuming thermodynamic equilibrium, and the addition of just sufficient oxygen and steam so as to gasify all carbon, is represented by the curved line. The lower temperature part of this line is typical for fluid-bed gasifiers and, for example, the hot lower part of a dry ash moving-bed gasifier. The departure from the straight line implies that part of the steam remains unconverted for thermodynamic reasons. It also means that the CO_2 content becomes higher. It is observed that at temperatures of about 1500°C, which are typical for dry coal feed entrained-flow slagging gasifiers, the straight line and the thermodynamic equilibrium line almost touch each other.

To generate the isotherms, the steam-to-product gas ratio is fixed above the minimum required for complete gasification at the given temperature. It is seen that in order to carry the additional steam ballast to the same high temperatures prevailing in the gasifier, more oxygen is required. This is hence a less efficient operation.

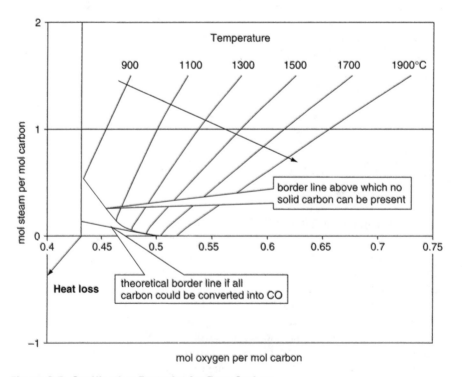

Figure 2-5. Gasification Footprint for Pure Carbon

The surplus of gasifying agent implies that no carbon can be present. Therefore the two equilibrium constants of the CO shift and the steam methane reforming equation will have to be used in the calculations instead of those of the three heterogeneous equations (Boudouard, water gas, and hydrogasification). Because the steam-to-product gas ratio is known, the number of equations and variables remain identical and the algorithm can be solved.

In practice, one will never want to operate at the thermodynamic equilibrium line, as carbon formation may occur at the slightest operational disturbance. Typically, for dry coal feed slagging gasifiers it is assumed that less than 1% of the carbon remains unconverted, hidden in ash particles, and 1–1.5 mol% of CO_2 is present in the gas. This gives some room for process control.

2.3.4 Surprises in Calculations

When calculations are carried out for a gasifier assuming heterogeneous reactions, it may come as a surprise that sometimes the quantity of moderator required is negative. The reason may be that the fuel and/or the blast contain so much ballast in the form of ash, water, steam, or nitrogen that the desired temperature can only be obtained by consuming more oxygen than is required to gasify all carbon in the fuel. What happens in the calculations is that the algorithm tries to reduce the amount of moderator, but only when the moderator becomes negative is a mathematically valid solution obtained. In other words, one has reached the area below the abscissa in Figure 2-5. The logical way out is to set the quantity of moderator at zero, but then there is one equation too many, as the ratio of moderator to product gas is set to zero. Again, the solution is found in using the two equilibrium constants of the CO shift and the steam-methane reforming reactions instead of the three for the heterogeneous gasification. Examples where this situation may occur are:

• The use of low-rank coal in a dry coal feed entrained-flow slagging gasifier.
• The use of a coal/water slurry in an entrained-flow slagging gasifier.
• The use of air as blast in entrained-flow gasifiers.

In all cases where no moderator is required, the process control becomes much simpler, just as in a furnace where only two streams have to be controlled instead of three.

2.4 OPTIMIZING PROCESS CONDITIONS

Dry coal feed entrained-flow slagging gasifiers operate at temperatures of typically 1500°C. At this temperature the oxygen consumption is high, but nevertheless there is still some moderator required. The challenge is to operate with close to the minimum amount of gasifying agent required, as this reduces the amount of expensive oxygen per unit product gas. But there is more: the less oxygen that is used, the more steam we need. This is not so bad, as oxygen is more costly than steam. However, there is a more important advantage that becomes clear when one looks at the two most relevant gasification reactions for this type of gasifier:

$$C + \tfrac{1}{2}\,O_2 = CO \qquad\qquad -111\,MJ/kmol \qquad\qquad (2\text{-}1)$$

and

$$C + H_2O \leftrightarrows CO + H_2 \qquad\qquad +131\,MJ/kmol \qquad\qquad (2\text{-}5)$$

The heat balance dictates that most of the gasification will be accomplished via reaction 2-1, but every carbon that can be gasified should be gasified with steam via reaction 2-5, as this reaction yields two molecules of synthesis gas per atom of carbon with cheap steam, but reaction 2-1 yields only one with expensive oxygen. More about optimization will be discussed in Section 6.8.

In order to optimize the process conditions in a dry coal feed entrained-flow slagging gasifier, first the equations are solved for the heterogeneous case, as this gives the minimum amount of oxygen per mole synthesis gas. Then in order to obtain a realistic operating window, the calculations are repeated for a somewhat higher CO_2 content in the gas for the homogeneous case, and that is then made the set point for the further operation.

Although most entrained-flow gasifiers operate with a surplus of gasifying agent (blast), there are and have been exceptions. One example is the first slagging stage of a two-stage gasifier where, also because of a lack of gasifying agent, carbon slips to a second nonslagging stage where it is further converted with steam. Such a gasifier is discussed in Chapter 5.

2.4.1 Process Indicators

When calculating the coal footprint it is very useful to draw also the iso-CO_2 and iso-CH_4 lines (Figure 2-6). The fact that the iso-CO_2 lines run more or less perpendicular to the isotherms and the iso-CH_4 lines more or less parallel, shows that the CH_4 content of the product gas is a better indicator for the temperature at which the gas leaves the gasifier than the CO_2 content. The latter can only be used in cases where the oxygen/steam ratio is fixed, or where no steam is required as moderator for the gasification.

Another good indicator of the gasification temperature is the heat flux through the reactor wall. Where a tube-wall or a jacket is used to protect the reactor wall, the steam make in the protection is a very valuable indicator for the reactor temperature (provided that this steam production is not integrated with other steam systems in such a way that it can not be measured properly).

For the gasifier performance, the measurement of the CO_2 and CH_4 content in the product gas, together with the heat flux through the reactor wall, are the best indicators. In practice, data will be used that are gathered from experience. The calculations will

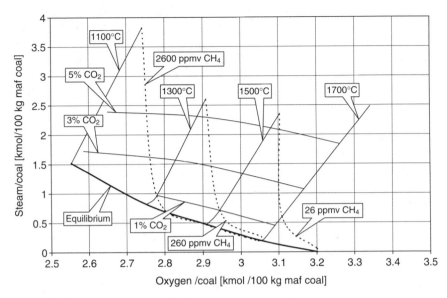

Figure 2-6. Iso-CO$_2$ and Iso-CH$_4$ Lines

give good leads during start-up and the initial nonoptimal operation of the unit, but for the fine-tuning the real data are more valuable, as will be explained in Section 6.8.

2.4.2 Optimum Operating Point

Efficiencies

There are a number of different criteria that are frequently quoted for gasification processes (Reimert 1989). The two most commonly encountered are cold gas efficiency (CGE) and carbon conversion. The definitions are:

$$\text{Cold gas efficiency } [\%] = \frac{\text{Heating value in product gas [MW]}}{\text{Heating value in feedstock [MW]}} \times 100$$

whereby it is important always to clarify whether the heating values are on a higher heating value (HHV) or lower heating value (LHV) basis.

Carbon conversion is defined as:

$$\text{Carbon conversion } [\%] = \left\{ 1 - \frac{\text{Carbon in gasification residue [kmol/h]}}{\text{Carbon in feedstock [kmol/h]}} \right\} \times 100$$

Despite the frequency with which these figures are quoted for different gasification systems, care is required with their interpretation, since both only provide a limited statement about the process efficiencies and gas quality. A process that produces a gas with a relatively high methane content and therefore has a high cold gas efficiency will be good in a power application, but it may not be the optimum choice for a synthesis gas application, in which case the $H_2 + CO$ yield will provide a better guide to process selection.

IGCC Applications

The reason why it can be advantageous to gasify coal in power plants is twofold: for efficiency reasons and environmental reasons.

The efficiency advantage is attributed to the fact that use can be made of the more advanced combined cycle where gas is fired in a gas turbine (Brayton or Joule cycle) and the hot gases leaving the turbine are used to raise steam for a conventional steam (Rankine) cycle. The alternative is firing coal in a conventional steam plant using only a Rankine cycle. The efficiency for gas firing in a state-of-the-art combined cycle is about 58%, a figure that has to be multiplied with the gasification efficiency of, for example, 80%, resulting in an overall efficiency of 46.4%, whereas the efficiency of coal firing in a conventional state-of-the-art steam plant is about 45%. A simple calculation shows that the CGE has to be above 78% in order to make gasification attractive in terms of overall process efficiency. This is all that we will say at this point as it is sufficient in the context of optimization. More will follow in Section 7.3.

The environmental advantage of gasification-based power stations has always been used as an important argument in their favor based mainly on the fact that in the past the sulfur compounds in the fuel gas could be removed with a higher efficiency than from flue gases from conventional coal or heavy oil fired power stations. Moreover, there was much optimism that high-temperature sulfur removal would be possible, which would enhance the efficiency of the IGCC. As will be explained in Section 7.3, there is a case to be made for reconsidering flue gas treating for IGCC, as many more compounds have to be removed apart from sulfur. Furthermore, the efficiency of flue gas desulphurization has improved considerably over the last twenty years. However, one has to accept that environmental arguments are a permanently moving target, since the recent discussion of CO_2 capture and sequestration may provide a new advantage for fuel-gas treating. This is, however, still dependent on politics.

For fuel gas applications the gasification temperature has to be as low as possible, as this will result in the highest CGE (and in the lowest oxygen consumption). In Figure 2-7 iso-CGE lines have been drawn that clearly illustrate this point, as the isotherms run essentially parallel to the iso-CGE lines. Although fluxing may help to enlarge the operating window for certain coals, as will be discussed in Section 5.3, the minimum temperature will always be determined by the reactivity of the coal and the ash-melting characteristics in slagging gasifiers.

In general, temperatures of below 1400°C for low-rank coals and below 1500°C for high-rank coals are impractical. As the data in Figure 2-7 show, the operation

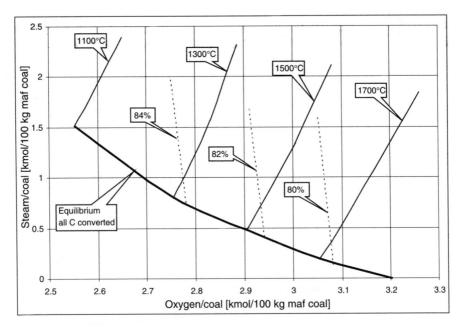

Figure 2-7. Cold Gas Efficiencies

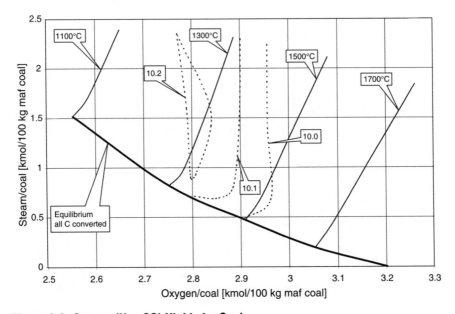

Figure 2-8. Syngas (H$_2$+CO) Yields for Coal

should be as close to the thermodynamic equilibrium line for the heterogeneous reaction as is practically possible.

Syngas and Hydrogen Applications

For synthesis-gas application the situation is somewhat different. In this case the moles of $H_2 + CO$ per unit feedstock have to be optimized. The effect of the operating conditions on this maximum is illustrated in Figure 2-8.

The optimum along the abscissa is caused by the effect that at lower temperatures more carbon is converted into CO_2 and CH_4, whereas the aim is to convert it as much as possible in CO. There is also a maximum along the ordinate. This is caused by the fact that by adding somewhat more steam less CH_4 will be formed and some additional heat will be generated by the CO shift reaction.

The latter effects are small, but they are there. The result is that for a fixed heat loss and reactant composition and temperature, there is one point that yields the maximum amount of synthesis gas. The effect is of more importance to heavy oil gasification than for coal gasification, as the ash limitations for coal do not allow an operation at such low temperatures where these effects become relevant.

REFERENCES

Barin, I. *Thermo-Chemical Data of Pure Substances*. Weinheim: VCH Verlagsgesellschaft, 1989.

Gumz, W. *Gas Producers and Blast Furnaces*. New York: John Wiley & Sons., 1950.

Kersten, S. R. A. *Biomass Gasification in Circulating Fluidized Beds*. Enschede: Twente University Press, 2002.

Lath, E., and Herbert, P. "Make CO from Coke, CO_2, and O_2." *Hydrocarbon Processing* 65(8) (August 1986):55–56.

Reimert, R. "Gas Production." In *Ullmann's Encyclopedia of Industrial Chemistry*, 5th ed., vol. A 12. Weinheim: VCH Verlagsgesellschaft, 1989, pp. 218–220.

van der Burgt, M. J. "Techno-Historical Aspects of Coal Gasification in Relation to IGCC Plants." Paper presented at 11th EPRI Conference on Gas-Fired Power Plants, San Francisco., 1992.

Chapter 3

The Kinetics of Gasification and Reactor Theory

The kinetics of gasification is as yet not as developed as is its thermodynamics. Homogeneous reactions occurring, for example, in the gas phase can often be described by a simple equation, but heterogeneous reactions are intrinsically more complicated. This is certainly the case with the gasification of solid particles such as coal, (pet) coke, or biomass because of their porous structure. The latter complication causes mass transfer phenomena to play an important role in the gasification of solids.

3.1 KINETICS

The kinetics of coal gasification has been and still is a subject of intensive investigation. Despite this, the results of such investigations have to date flowed into the design procedures for commercial gasification reactors to only a limited extent. In contrast to Chapter 2, therefore, the presentation of kinetic theory in this chapter is restricted to the basic ideas and an indication of how and where the appropriate application of kinetic theory could help in the design of future reactors.

A simplified reaction sequence for coal or biomass gasification can be described, as in Figure 3-1.

For coke gasification, where the volatiles have already been removed in a separate process step, only the char reactions apply. As is discussed in other parts of this book, counter-current moving-bed gasifiers such as the Lurgi gasifier are an exception to this model, since the oxygen reacts with the coke and the devolatilization takes place using the hot synthesis gas. The volatiles in such a process do not come into contact with free oxygen.

For oil gasification, where the feedstock consists almost entirely of pure volatiles, the pattern is also slightly different.

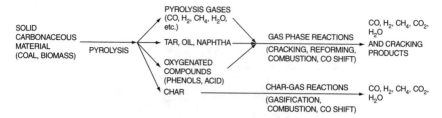

Figure 3-1. Reaction Sequence for Gasification of Coal or Biomass (*Source: Adapted from Reimert and Schaub 1989*)

3.1.1 Devolatilization

The first step, heating up of the coal particles, is in one sense the simplest part of the process. Nonetheless, the speed at which it takes place has an influence on the subsequent steps, so that it is of great importance in any accurate model.

Devolatilization takes place already at low temperatures (350–800°C) and in parallel with the heating up of the coal particles. The rate of heating of the coal particles influences the way in which the devolatilization takes place (Jüntgen and van Heek 1981, p. 65). The rate of devolatilization is dependant not only on the rate of heating, however, but also on the particle size and the rate of gasification by the water gas reaction, and hence on the reaction temperature and the partial pressure of steam.

The interplay between pyrolysis and gasification under different heating conditions is shown in Figure 3-2. If the heating up is slow then the pyrolysis reactions set in from about 350°C. The gasification reaction of both volatiles and char with steam is very slow at this temperature. The concentration of volatiles outside the particle increases rapidly, and gasification only sets in after devolatilization is complete. If, however, the rate of heating is high, then both pyrolysis and gasification take place simultaneously, so that a high concentration of volatiles is never allowed to build up. This is one reason why high-temperature entrained-flow reactors produce a clean gas in such a short time. Compare this with a counter-current moving-bed process where lump coal is used. The heating up rate is slow and a high volatiles concentration is built up and removed unreacted from the reactor by the syngas.

For finely pulverized coal particles at high temperature the residence time is very short (10–200 ms) (Smoot and Smith 1985, p. 55 ff). The extent of devolatilization is highly dependant on the final temperature and can vary considerably from that found by performing a proximate analysis in accordance with ASTM (American Society for Testing and Materials), DIN (Deutsches Institut für Normung), or other standard methods. The product distribution of the devolatilization process also varies significantly with changes in the pyrolysis temperature and the speed of heating up.

Although devolatilization processes during gasification and combustion are thought to be generally similar, the fact that many gasification processes operate at

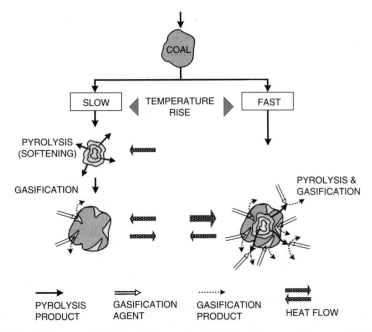

Figure 3-2. Influence of Heating Rate on Gasification Process (*Source: Jüntgen and van Heek 1981*)

higher pressures has to be taken into account. The weight loss due to devolatilization can be of the order of 10% less at typical gasifier pressures of 30 bar.

3.1.2   Volatiles Combustion

The devolatilization of coal produces a variety of species, including tars, hydrocarbon liquids, and gases, including methane, CO, CO_2, H_2, H_2O, HCN, and so forth. (Smoot and Smith 1985). This material reacts with the oxidant surrounding the coal particle. The extent to which the oxidant is completely or only partially depleted depends on the amount of volatiles produced.

In a combustion environment, where there is an overall excess of oxygen, the combustion of the volatiles is complete. In a gasification environment, this is not necessarily the case, especially where the coal has a high volatiles content. There is a recirculation of synthesis gas in many gasification reactors, not only in fluid-bed but also to some degree in entrained-flow reactors. To the extent that this occurs in the vicinity of the burner, the effects are very different from the combustion situation. Recirculated combustion flue gas consists mainly of carbon dioxide, water vapor and (in the air-blown situation) nitrogen. The carbon dioxide and water vapor have a moderating effect, thus reducing temperature. In the case of gasification the recycled gas contains significant quantities of carbon monoxide and hydrogen (up to 90% for an oxygen-blown gasifier) and will cause locally very high temperatures, should it come into contact with the oxidant.

There is not much kinetic data available on volatiles combustion. It is, however, clearly established that this process, being a reaction between gases, is much more rapid than the heterogeneous char gasification where mass transport limitations play a more important role.

3.1.3   Char Gasification

The slowest reactions in gasification, and therefore those that govern the overall conversion rate, are the heterogeneous reactions with carbon, namely the water gas, Boudouard, and hydrogenation reactions already discussed in Chapter 2. The rates of reaction for the water gas and Boudouard reactions with char are comparable and are several orders of magnitude faster than for the hydrogenation reaction (Smoot and Smith 1985, p. 79).

There are several different models describing the Boudouard and water gas reactions (Williams et al. 2000). A widely utilized model for the Boudouard reaction is attributable to Ergun (1956) and proposes a two-step process.

Step 1 $$C_{fas} + CO_2 \leftrightarrows C(O) + CO$$

Step 2 $$C(O) \rightarrow CO + C_{fas}$$

In the first step, CO_2 dissociates at a carbon free active site (C_{fas}), releasing carbon monoxide and forming an oxidized surface complex ($C(O)$). In the second step the carbon-oxygen complex produces a molecule of CO and a new free active site. The rate limiting step is the desorption of the carbon-oxygen surface complex.

The model for the water gas reaction is basically similar:

Step 1 $$C_{fas} + H_2O \leftrightarrows C(O) + H_2$$

Step 2 $$C(O) \rightarrow CO + C_{fas}$$

In this case the first step is the dissociation of a water molecule at a carbon-free active site (C_{fas}), releasing hydrogen and forming an oxidized surface complex ($C(O)$). In the second step, the carbon-oxygen complex produces a molecule of CO and a new free active site. In some models the rate-limiting step is the desorption of the carbon-oxygen surface complex as for the Boudouard reaction. Other models include the possibility of hydrogen inhibition by the inclusion of a third step:

Step 3a $$C_{fas} + H_2 \leftrightarrows C(H)_2$$

or

Step 3b $\qquad\qquad\qquad C_{fas} + \frac{1}{2}\,H_2 \leftrightarrows C(H)$

whereby some of the sites can become blocked by hydrogen.

Fundamental work continues in this field to develop a detailed understanding of the mechanisms of gasification reactions (Williams et al. 2000, p. 237).

Rate of Reaction. For the Boudouard reaction (2-4),

$$C + CO_2 \leftrightarrows 2CO$$

being an equilibrium reaction, the reaction rate of carbon conversion,

$$r_m = \frac{dC}{dt} \tag{3-1}$$

is assumed to be proportional to the concentration of CO_2 in the gas, so that

$$r_m = k_m \cdot c_{co_2} \tag{3-2}$$

where k_m is the mass-related reaction-rate constant, c_{CO_2} the concentration of CO_2 in the gas and the order of the reaction may be assumed to be 1.

The temperature dependency of the rate constant can be expressed in Arrhenius form as

$$k_m = A \cdot e^{\frac{-E}{RT}} \tag{3-3}$$

where A is a pre-exponential factor and E is the activation energy for the reaction. This can be expressed alternatively as

$$\ln(k_m) = -\frac{E}{R} \cdot \frac{1}{T} + \ln(A) \tag{3-4}$$

which provides a convenient form for comparing the reactivities of different chars.

Comparison of Different Types of Solid Feedstocks. The reactivity of different coals and chars depends on a number of factors, in particular

- The porosity of the coal, that is, its inner structure, surface, and active sites.
- The crystal structure of the fixed carbon.
- Catalytic effects of ash components in the coal.

Young (low-rank) coals such as browncoal have a large specific surface and thus a high reactivity. On the other hand, older coals, particularly anthracitic coals, have a poor reactivity. Reactivity is enhanced by alkalis, particularly potassium.

34   *Gasification*

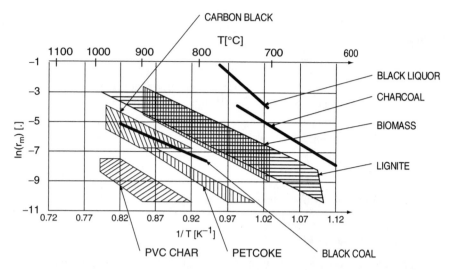

Figure 3-3. Reactivity of Various Materials as a Function of Temperature (*Source: Bürkle 1998*)

In a systematic review Bürkle (1998) has plotted the reactivity of different chars from various biomasses, coals, and other materials, as shown in Figure 3-3.

Effective Reactivities. In all cases it is necessary to distinguish between the physical and chemical steps involved and which effects control the measurable rate of reaction in different temperature zones.

Zone I,   the low-temperature zone, in which the chemical reaction is the rate-controlling step and the experimentally observed activation energy is the true activation energy.

Zone II,   a medium-temperature zone, in which the rate of chemical reaction is higher, but is limited by internal diffusion of the gaseous reactants through the pores of the individual particles. The observed activation energy is only about half the true value.

Zone III, a high-temperature zone in which external, bulk surface diffusion of the gaseous reactants is rate controlling and the apparent activation energy is very small.

This is illustrated in Figure 3-4.

Data for coal gasification is presented in Figure 3-5, from which it can be concluded that in any solid-fuel gasification process the progress of the reaction is determined by the mass transfer phenomena.

Another indication of the importance of mass transfer is given in Figure 3-6 where the time required for the gasification of a solid fuel is plotted as a function of the particle size.

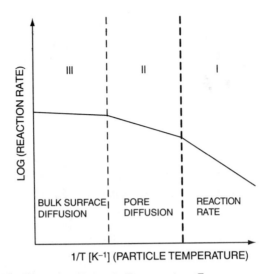

Figure 3-4. Effective Reaction Rates in Temperature Zones

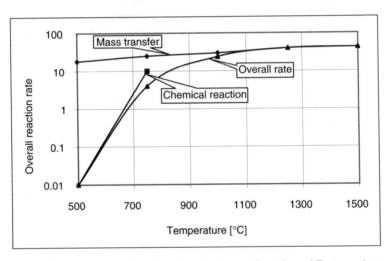

Figure 3-5. Overall Gasification Reaction Rate as a Function of Temperature (*Source: Hedden 1961* with permission from Elsevier)

3.2 REACTOR THEORY

Classical chemical engineering theory teaches us about a variety of idealized reactor types. The oldest type of chemical reactor is the batch reactor, of which the coke oven is currently the only example that is relevant to gasification (Thoenes 1994;

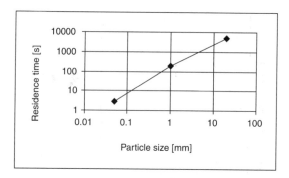

Figure 3-6. Time Required for the Gasification of a Solid Fuel as a Function of Particle Size

and Westerterp, van Swaaij, and Beenackers 1987). In the coke oven a batch of coal is indirectly heated via the side walls in a relatively flat vertical oven where devolatilization takes place. Such a batch process can also be made semicontinuous, as is the case in a coal stove for domestic use and a moving-bed gasifier (see Section 5.1), or fully continuous, as in large-scale grate-type furnaces as applied in power stations and industry. For the coal or other solid fuel the fully continuous version is a typical example of a so-called plug flow reactor.

Plug Flow Reactors. The idealized plug flow reactor (PFR) is characterized by the following properties:

- There is a continuous flow through the reactor.
- There are no radial gradients.
- There is no axial mixing, that is, there is no exchange of material or heat.
- In addition, a plug flow reactor will generally be operated in a steady state manner.

For a first-order kinetics and the generalized component A, equations 3-1 and 3-2 can be expressed as:

$$-\frac{dc_A}{dt} = kc_A \tag{3-5}$$

which after integration yields for the concentration after time t:

$$\frac{c_A}{c_{A,0}} = e^{-kt} \tag{3-6}$$

and for the conversion of component A:

$$\frac{c_{A,0} - c_A}{c_{A,0}} = 1 - e^{-kt} \tag{3-7}$$

These reactions also hold for a batch reactor.

Continuously Stirred Tank Reactors. The ideal continuously stirred tank reactor (CSTR) has the following characteristics:

- There is a continuous flow through the reactor.
- The contents of the reactor are ideally mixed; thus the conditions are the same at all points in the reactor. This implies that the effluent has the same composition as anywhere else in the reactor. In the CSTR the mixing time should be about two orders of magnitude lower than the average residence time to achieve this.

When we set the average residence time equal to τ we derive the relation from formula 3-5 for a first-order reaction at constant density:

$$c_{A,0} - c_A = kc_A\tau \tag{3-8}$$

and so:

$$\frac{c_A}{c_{A,0}} = \frac{1}{k\tau + 1} \tag{3-9}$$

For the conversion of component A we then get:

$$\frac{c_{A,0} - c_A}{c_{A,0}} = \frac{k\tau}{k\tau + 1} \tag{3-10}$$

Equations 3-9 and 3-10 show that increasing the residence time τ will increase the conversion. In some cases, this can lead to impractically large reactors. In order to avoid this the single CSTR can be replaced by two or more CSTRs in series, which each have the same volume and residence time, but that have a combined average residence time and volume that is equal to the single stage CSTR. For N stages we then obtain:

$$\frac{c_{A,N}}{c_{A,0}} = \left(1 + k \cdot \frac{\tau}{N}\right)^{-N} \tag{3-11}$$

As $N \to \infty$, this yields equation 3-6. An infinite number of CSTRs hence gives the same results as a PFR with the same overall residence time. In practice a number of about four CSTRs already gives results that approximate a PFR with the same residence time, provided the conversions required are not excessive. In Figure 3-7, some comparative results are given for various values of N.

The result that a large number of CSTRs gives the same result as a PFR can easily be understood when we consider that in a single stage CSTR there is a small chance that reactants leave the reactor without having the time to react. Adding a second CSTR diminishes this chance, although the overall residence time remains the same. Adding subsequent CSTRs virtually eliminate the chance of "short circuiting" of reactants, and for the conversions given in Figure 3-7 we see that the results of the PFR can hardly be distinguished from that of eight CSTRs in series having the same overall residence time.

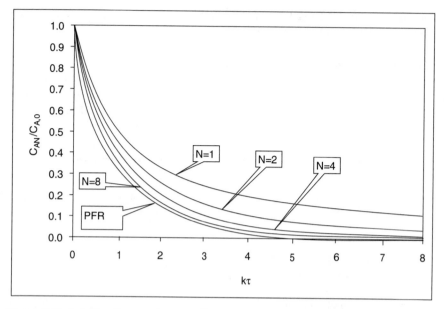

Figure 3-7. Relative Conversion as a Function of the Dimensionless Time $k\tau$

3.3 APPLICATIONS TO REACTOR DESIGN

If we include coke ovens as a typical batch reactor in our definition of gasification, then practically all of these idealized reactor forms have been applied at some time or other in the search for an optimum gasifier design. Of course, what is optimum in any particular case is heavily dependant on the application in hand (chemical or power), the nature of the feedstock, the size of the plant, and a number of other factors as well as the classical trade-off between investment and operating costs. It is therefore not surprising that representatives of most reactor forms still find commercial application today. The various basic gasification processes are discussed against this background and that of the basic theoretical models of PFR and CSTR in Chapter 5, where the implications for some of the commercial processes are also shown.

3.3.1 Modeling

The systems described in this chapter are extremely complex, which has to date placed limitations on the application of kinetic models to commercial gasification reactor design. Nonetheless "the availability of increasingly powerful computers with a deeper understanding of the physical and chemical processes of coal combustion

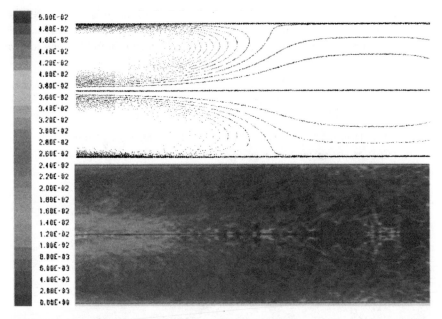

Figure 3-8. CFD Model of a Gasification Burner Outlet Zone (*Source: Hofmockel, Liebner, and Ulber 2000*)

now enables the development and use of greater sophistication in the models employed" (Williams et al. 2002). In particular, applications are emerging where their incorporation into computational fluid dynamics (CFD) is proving to be of practical material benefit.

The most important practical applications to date from the knowledge of actual coal kinetics are in the field of combustion, particularly in the field of NO_x prediction. However, although CFD already has an important place as a design aid in describing the combustion of coal in utility furnaces, increasing demands are being made to provide quantitative rather than qualitative results (Williams et al. 2002).

Progress is now being made in applying CFD to gasification. In their discussion of the application of CFD to burner developments for Lurgi's MPG oil gasification burner, Hofmockel, Liebner, and Ulber (2000) describe including kinetic models for bulk and pore diffusion. (This is further detailed in Ulber, 2003). An example from their results of CFD is given in Figure 3-8.

Bockelie et al (2002) have presented initial results for various entrained-flow coal gasifiers, illustrating, for example, performance variations when using different feedstocks. This work is part of an ongoing program to investigate generic improvements for the operation and design of such gasifiers.

The complexity of kinetic calculations ensures that any practical quantitative results are only likely to be accessible via CFD and will therefore remain the realm of the specialist. Current developments in this field are encouraging, even if the path ahead is still a long one. Optimization of burners and reactor design, as well as their influence on one another, is the area most likely to be a fruitful field of application in the near future.

REFERENCES

Bockelie, M. J., Denison, M. K., Chen, Z., Linjewile, T., Senior, C. L., Sarofim, A. F., Holt, N. A. "CFD Modeling for Entrained Flow Reactors." Paper presented at Gasification Technologies Conference, San Francisco, October 27–30, 2002.

Bürkle, S. "Reaktionskinetische Charakterisierung abfalltypischer Stoffe und deren Verbrennung in einem Drehrohrofen unter Sauerstoffanreicherung." Ph.D. diss., Engler-Bunte-Institut der Universität Karlsruhe (TH), 1998.

Ergun, S. B. "Kinetics of the Reaction of Carbon Dioxide with Carbon." *J. Phys. Chem.* 60 (1956):480–485.

Hedden, K., "Die Bedeutung der Reaktionsfähigkeit des Brennstoffs für koksbeheizte Schachtöfen." Chemical Engineering Science 14(1961):317–330.

Hofmockel, J., Liebner, W., and Ulber, D. "Lurgi's Multipurpose Gasification Application and Further Development." Paper presented at IChemE Conference, "Gasification for the Future," Noordwijk, The Netherlands, April, 2000.

Jüntgen, H., and van Heek, K. H. *Kohlevergasung*. Munich: Thiemig, 1981.

Reimert, R. and Schaub, G. "Gas Production." In *Ullman's Encyclopedia of Industrial Chemistry*, 5th ed., vol. A12. Weinheim: VCH Verlagsgesellschaft, 1989, p. 215.

Smoot, L. D., ed. *Fundamentals of Coal Combustion*. Amsterdam: Elsevier, 1993.

Smoot, L. D., and Smith, P. J. *Coal Combustion and Gasification*. New York: Plenum, 1985.

Thoenes, D. *Chemical Reactor Development*. Dordrecht: Kluwer Academic Publishers, 1994.

Ulber, D. "Modellierung der Flugstromdruckvergasung von Öl-Rückständen." Ph.D. diss., Rheinisch-Westfälische Technische Hochschule, Aachen: 2003.

Westerterp, K. R., van Swaaij, W. P. M., and Beenackers, A. A. C. M. *Chemical Reactor Design and Operation*. New York: John Wiley & Sons, 1987.

Williams, A., Backreedy, R., Habib, R., Jones, J. M., Pourkashanian, M. "Modeling Coal Combustion: The Current Position," *Fuel* 81 (March 2002):605–618.

Williams, A., Pourkashanian, M., Jones, J. M., Skorupska, N. *Combustion and Gasification of Coal*. New York: Taylor and Francis, 2000.

Chapter 4
Feedstocks and Feedstock Characteristics

4.1 COALS AND COKE

Approximately 2250 MMtoe/y of coal is consumed each year (BP 2002). Of this about 35 MMt/y is gasified to produce 150 million Nm^3/d of synthesis gas (Simbeck and Johnson 2001). Roughly half of this gas is generated in the Sasol synfuels plants in South Africa where the synthesis gas is used for the production of liquid hydrocarbons and other chemicals. Most of the remainder of this gas is used for ammonia production and—in China—for the production of town gas.

The total proven reserves of coal amount to 984×10^9 t worldwide. Coal consumption has been stable over the last ten years. Nonetheless it plays an important part in the thinking of many long-term energy strategies, despite its contentious role in the production of CO_2 as a greenhouse gas. The reason for this can be seen from the figures in Table 4-1. The ratio of reserves to current production (R/P ratio) is 216. In other words, at current consumption rates the world's reserves would last 216 years.

Table 4-1 World Coal Reserves by Region		
Region	**% of Total Reserves**	**R/P Ratio**
North America	26.2%	234
South and Central America	2.2%	381
Europe	12.7%	167
Former Soviet Union	23.4%	>500
Africa and Middle East	5.8%	246
Asia/Pacific	29.7%	147
World	100%	216
Source: BP 2002		

Compare this with 62 years for natural gas and 40 years for oil. Furthermore the reserves are more evenly distributed than oil (65% of reserves in the Middle East) or natural gas (72% in the Middle East and Russia).

The composition of coals is very complex, and the types of coal differ considerably. The detailed petrographic composition of the organic part of coals, often characterized by a so-called maceral analysis, has little influence on most gasification processes, and the interested reader is referred to the many treatises on this subject (e.g., Speight 1983; Smoot and Smith 1985; and Kural 1994). Important for gasification are the age of the coal, its caking properties, its water content, and its ash properties.

4.1.1   Formation of Coal

All coal has been formed from biomass. Over time this biomass has been turned into peat. When covered under a layer of overburden, the influence of time, pressure, and temperature convert this material into browncoal or lignite. Subsequently, the latter material will turn into sub-bituminous coal, then into bituminous coal, and finally into anthracite. Coal is often classified in terms of its rank, which increases from browncoal to anthracite. The classification of coal by rank for ash and moisture-free coal is given in Tables 4-2 and 4-3. Figure 4-1 provides an alternative presentation. Browncoal, lignite, and sub-bituminous coals are called low-rank coals, whereas higher-rank coals are often called hard coals. The terms *browncoal* and *lignite* are essentially synonymous, lignite being used more often in the United States and browncoal in Europe and Australia.

4.1.2   Coal Analysis

The methods generally used for specifying the analysis of coals have developed along pragmatic lines and are aimed at providing a useful guide to coal users rather than a purely chemical approach. The two types of analysis for any coal are the proximate analysis and the ultimate analysis.

Table 4-2 Classification of Coals			
Class	**Volatile Matter**	**Fixed Carbon**	**Heating Value**
	wt%	**wt%**	**MJ/kg**
Anthracite	<8	>92	36–37
Bituminous	8–22	78–92	32–36
Sub-bituminous	22–27	73–78	28–32
Browncoal (Lignite)	27–35	65–73	26–28

Table 4-3
Classification of Coals

Class		Fixed Carbon wt%	Heating Value BTU/lb	Agglomerating Character
I. Anthracitic	1. Meta-anthracitie	≥98		Nonagglomerating
	2 Anthracite	92–98		
	3. Semianthracite	86–92		
II. Bituminous	1. Low-volatile bituminous coal	78–86		Commonly agglomerating
	2. Medium-volatile bituminous coal	69–78		
	3. High-volatile A bituminous coal	<69	≥14,000	
	4. High-volatile B bituminous coal		13,000–14,000	Agglomerating
	5. High-volatile C bituminous coal		11,500–13,000	
III. Sub-bituminous	1. Sub-bituminous A coal		9,500–10,500	
	2. Sub-bituminous B coal		8,300–9,500	Nonagglomerating
	3. Sub-bituminous C coal		8,300–9,500	
IV. Lignitic	1. Lignite A		6,300–8,300	
	2. Lignite B		<6,300	

Source: ASTM D 388

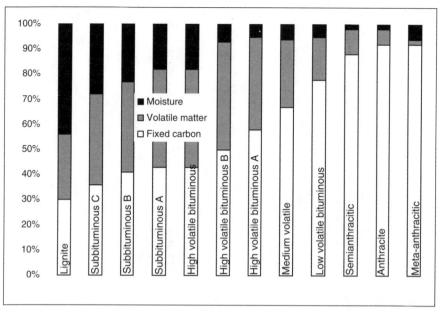

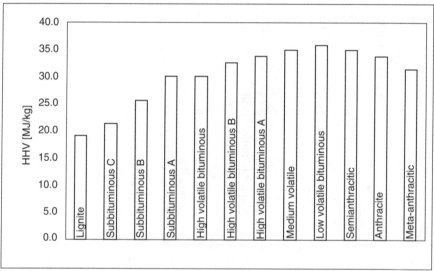

Figure 4-1. Proximate Analyses of Coals by Rank (*Source: Baughman, 1978*)

Proximate Analysis

The proximate analysis determines the moisture, volatile matter, fixed carbon, and ash in the coal (see Figure 4-1). The analysis is an essentially practical tool providing an initial indication of the coal's quality and type. The methods for performing these analyses have been standardized by all the major standards institutions (e.g., ASTM, ISO, DIN, BS, and others). These standards, though similar in nature, are different from one another in, for example, the temperature specified for determining the volatiles content, so it is important when providing data to specify the method used.

Moisture is determined by drying the coal under standard conditions for 1 h at 104–110°C. The method determines the sum of all moisture; that is, both the surface moisture caused by rain and so on, and the inherent moisture. The inherent moisture is the water that is very loosely bound in the coal. It can vary from a few percent in anthracite to 60–70% in browncoal. (Equilibrium moisture is defined by ASTM D1412, BS1016 part 21; bed/inherent moisture by ASTM D3173, DIN 51 748).

Volatile matter is determined by heating the coal in a covered crucible for a defined time at a defined temperature (e.g., 7 min at 950°C to ASTM). The loss in mass, minus the mass of the moisture, represents the mass of the gaseous constituents formed by the pyrolysis under the conditions mentioned (ASTM D3175, BS 1016 part 104, DIN 51 720).

Ash is the inorganic residue that remains after combustion of the coal. It consists mainly of silica, alumina, ferric oxide, lime, and of smaller amounts of magnesia, titanium-oxide, and alkali and sulfur compounds (ASTM D3174, BS 1016 part 104, DIN 51 719).

Fixed carbon is determined by subtracting from 100 the mass percentages of moisture, volatile matter, and ash (ASTM D3172). It should be remarked that fixed carbon is an artificial concept and does not mean that this material was present in the coal as pure C in the beginning.

Although the proximate analysis already tells the expert a lot about the coal, for gasification it is mandatory to have also the ultimate analysis at our disposal, which tells us more about the elemental composition of the hydrocarbonaceous part of the coal.

Ultimate Analysis

For the ultimate analysis the percentages of carbon, hydrogen, oxygen, sulfur, and nitrogen are determined. In the past, oxygen was sometimes reported as "by difference." If at all possible this should not be accepted, as it makes it impossible to have any control over the quality of the analysis. Proper balances are the basis for a good process design and a good operation of plants, but a good balance is equally dependant on a good elemental analysis. (Carbon and hydrogen ASTM D3178, BS1016 part 106: Nitrogen ASTM D3179, BS1016: Sulfur ASTM D3177, BS1016 part 106)

Table 4-4 lists proximate and ultimate analyses of various types of coal.

The relevance of *sulfur* in the coal for gasification is the same as for oil-derived heavy residual feedstocks, which generally contain more sulfur than most coals, and which are described in Section 4.2. Sulfur contents in coal range from 0.5–6 wt%. It

Table 4-4
Analyses of Various Coals

Coal			Proximate Analysis, % Mass ar				Ultimate Analysis, % Mass on maf Coal					LHV
Country	Region	Class	Fixed carbon	Vol. Mater	Water	Ash	C	H	O	N	S	maf coal MJ/kg
Germany	Rhein	Browncoal	17.3	20	60	2.7	67.5	5.0	26.5	0.5	0.5	26.2
U.S.A.	North Dakota	Lignite	27.8	24.9	36.9	10.4	71.0	4.3	23.2	1.1	0.4	26.7
U.S.A.	Montana	Sub-bituminous	43.6	34.7	10.5	11.2	76.4	5.6	14.9	1.7	1.4	31.8
U.S.A.	Illinois	Bituminous	39.3	37.0	13.0	10.7	78.4	5.4	9.9	1.4	4.9	33.7
Poland	Typical	Bituminous	54.9	35.6	5.3	4.2	82.8	5.1	10.1	1.4	0.6	36.1
South Africa	Typical	Bituminous	51.3	32.7	2.2	13.8	83.8	4.8	8.4	2.0	1.0	34.0
China	Datung	Bituminous	50.9	28.1	11.9	9.1	84.4	4.4	9.5	0.9	0.8	33.4
India	Typical	Bituminous	30	23	7	40	75.5	6.4	15.2	1.5	1.4	32.1
Australia	Typical	Bituminous	44.9	41.1	4.5	9.5	81.3	5.8	10.0	2.0	0.9	33.8
Germany	Ruhr	Anthracite	81.8	7.7	4.5	6.0	91.8	3.6	2.5	1.4	0.7	36.2

may be present in coal in three forms: pyritic sulfur, inorganic sulfates, and as sulfur in organic compounds. These three forms of sulfur can be determined using ASTM D 2492. In coals with a high sulfur content, most of the sulfur is generally present in the form of pyrite. Note that the quantity of pyritic sulfur is an indicator for the potential abrasiveness of the coal.

The *nitrogen* content in coals ranges from 0.5–2.5 wt%. Only part of the nitrogen in the coal is converted into ammonia and HCN upon gasification, whereas the remainder is converted into elemental nitrogen. The presence of the coal-derived nitrogen in the product gas is one reason why it is not always essential to gasify coal with very pure oxygen (>99 mol%), even when the gas is used for the production of syngas or hydrogen. The percentage of the nitrogen in the coal that is converted into elemental nitrogen upon gasification will depend on the type of nitrogen compounds in the coal.

4.1.3 Other Minerals in Coal

Beyond the elements described above, which are provided with every ultimate analysis of coal, it will be found that a substantial part of the periodic table can be shown to be present in coals. These other elements can be divided into macro components, the presence of which is usually given in wt%, and the micro or trace elements that are only present at ppm levels. Values from a sample coal within the typical range are given later in Table 4-7.

The *chlorine* content in coal is mostly well below 1 wt%. However, in some coals it may be as high as 2.5 wt%. In combination with a low nitrogen content in the coal, this will result in a high caustic consumption in the wash section of a gasifier.

Chlorides have three possible detrimental effects in the plant:

1. Chlorides have a melting point in the range 350–800°C; they deposit in the syngas cooler and foul the exchanger surface. The first indication of this is an increase in the syngas cooler outlet temperature.
2. In the reactor chlorides can react with the hydrogen present to form HCl, which will decrease the pH of the wash water or condensate.
3. Chlorides may also form NH_4Cl with high nitrogen feeds. With such feedstocks the chloride deposits as NH_4Cl in the economizers at temperatures below about 280°C. Further, as an aqueous solution this leads to severe chloride stress corrosion in stainless steels that are used, for example, in burners and instrument lines.

Coals also contain *phosphorus*, but this has less significance for gasification than, for instance, for the steel industry.

Basis of Reporting Analyses

It is important for the users of the analytical data of coal to be able to convert various types of analysis into each other. Apart from the difference in units used that can readily be transferred into each other, there are problems associated with the basis for the analysis. The proximate analysis and the ultimate analysis may be given based on the so-called as-received (ar) coal; that is, the coal including moisture and ash. When doing so it is also important to record whether the sample was taken direct from storage or air dried before analysis, since the difference between total and inherent moisture cannot be determined. But it is also possible to get these analyses on a moisture-and-ash-free (maf) basis, a moisture-free (mf) basis, or an ash-free (af) basis. Furthermore, the heating value may be the higher heating value (HHV) or the lower heating value (LHV). In relation to the latter it should be kept in mind that the HHV of steam and water are positive and zero, respectively, and that the LHV of steam and water are zero and negative, respectively. A program is included on the companion website to convert given data for proximate and ultimate analysis as well as for the heating value on any basis (ar, af, mf, maf, and HHV or LHV) into any other basis.

4.1.4   Other Properties

Heating Value

The heating value is obtained by combustion of the sample in a calorimeter. If not available, the heating value can be calculated with, for example, the Dulong formula (Perry and Chilton 1973, p. 9–4) from the ultimate analyses:

$$HHV \text{ in MJ/kg} = 33.86*C + 144.4*(H - O/8) + 9.428*S$$

where C, H, O, and S are the mass fractions of the elements obtained from the ultimate analysis. There are other formulae for calculating the heating value from the ultimate and/or proximate analyses (e.g., Channiwala and Parikh 2002):

$$HHV \text{ in MJ/kg} = 34.91*C + 117.83*H - 10.34*O - 1.51*N + 10.05*S - 2.11*Ash$$

It is always useful to calculate the heating value from these analyses, as it is a good crosscheck on measured values. If the deviation is more than a few percent, all analyses must be checked.

Caking and Swelling Properties

Another important property of a coal is the swelling index. The swelling index is determined by heating a defined sample of coal for a specified time and temperature, and comparing the size and shape taken by the sample with a defined scale. There are a number of different scales defined in, for example, ASTM D 720-91, BS 1016 (Gray-King method), or ISO 335 (Roga method). The swelling index is an indicator for the caking properties of a coal and its expansion on heating. Softening/caking does not occur at a precise temperature but over a temperature range. It is an important variable for moving-bed and fluid-bed gasifiers. For the gasifiers of entrained-flow systems, the coal softening point has no relevance. However, the softening point may limit the amount of preheating of the pulverized coal feedstock used in dry coal feed gasifiers.

Hardness

Physical properties are not very relevant for the operation of a gasifier as such. The hardness of the coal is, for example, mainly important for the milling and grinding upstream of the gasifier. The hardness of a coal is usually dependant on the nature and quantity of its ash content, although some coals, such anthracites, are also hard. A high ash content or a very high hardness of the ash in the coal can make a feedstock unattractive for gasification because of the high cost of milling and grinding. Ashes with high silica and/or alumina contents have a high hardness. The hardness is generally characterized by the Hardgrove grindability index (ASTM D 409).

Density

The density is primarily of importance for the transport of the coal. In this connection, it is important to discriminate between the particle density and the bulk density of the coal. The bulk density is always lower, as is shown in Table 4-5.

Table 4-5
Comparison between Particle and Bulk Density for Various Coals

Fuel	Density (kg/m^3)	
	Particle (true)	**Bulk (apparent)**
Anthracite	1450–1700	800–930
Bituminous coal	1250–1450	670–910
Lignite	1100–1250	550–630

4.1.5 Ash Properties

Melting Properties

For all gasifiers the ash-softening and ash-melting or fusion temperatures are important variables. For fluid-bed gasifiers these properties govern the upper operating temperature at which agglomeration of the ash is initiated. For entrained-flow gasifiers it is essential to ensure that the ash flows continuously and that the slag tap does not freeze up. The method for determining these temperatures is specified in ASTM D1857, "Fusibility of Coal and Coke Ash," or similar specifications, such as ISO 540. In these methods the temperatures measured relate to the behavior of an ash sample under specified conditions and are reported as IDT (initial deformation temperature), ST (softening temperature), HT (hemispherical temperature), and FT (fluid temperature). For gasifier applications the ash-melting characteristics should be determined under reducing conditions, as these data may differ considerably from data for oxidizing conditions (generally, but not universally lower).

An additional property required for slagging gasifiers is the slag viscosity-temperature relationship. It is generally accepted that for reliable, continuous slag-tapping a viscosity of less than 25 Pa.s (250 Poise) is required. The temperature required to achieve this viscosity (T_{250}) is therefore sometimes used in the literature (Stultz and Kitto 1992). Some slags are characterized by a typical exponential relationship between viscosity and temperature over a long temperature range. For others this relationship is foreshortened at a critical temperature (T_{cv}) at which the viscosity increases very rapidly with decreasing temperature. For a slagging gasifier to operate at a reasonable temperature, it is necessary for the slag to have a $T_{cv} < 1400°C$.

The relationship between ash-melting characteristics and composition is a complicated one and is dependant largely on the quaternary SiO_2-Al_2O_3-CaO-FeO (Patterson, Hurst, and Quintanar 2002). In general, slags that are high in silica and/or alumina will have high ash-melting points, but this is reduced by the presence of both iron and calcium—hence the use of limestone as a flux. However, the SiO_2/Al_2O_3 ratio is also important, and where the calcium content is already high (as in some German browncoals), there can be some advantage to lowering the ash melting point by adding SiO_2. Properties of some typical ashes are given in Table 4-6.

In dry ash moving-bed gasifiers and in fluid-bed gasifiers, coals with a high ash melting point are preferred, whereas in slagging gasifiers, coals with a low ash melting point are preferred.

The caking properties of a coal and the melting characteristics of its ash are the reason that there are forbidden temperature ranges that have to be taken into account, both in design and during operation. In entrained-flow gasifiers only the ash properties are important.

The ash that is produced in gasifiers always has a lower density than the minerals from which they originate, due to loss of water, decomposition of carbonates, and other factors, and the presence of some carbon. The bulk density of the ash in particular may be low due to the formation of hollow ash particles (cenospheres). This means that special attention has to be given to the transport of such ashes.

Slag is very different from ash as it has been molten and is in fact a fusion-cast material similar to glass. Ideally, slag becomes available as an inert, fine, gritty material with sharp edges due to the sudden temperature drop upon contact with a water bath. Because lumps of solid slag will form during process upsets, a slag breaker is sometimes installed between the water bath and the slag depressurizing system.

Chemical Composition of Ash

In Table 4-6 some of the major components of various ashes are given. Apart from these there are many trace components present that do not contribute much to the

Table 4-6
Ash Fusion Temperatures and Analysis

Coal	Ash-Melting Point		Analysis of Ash as Oxides (wt%)									
	reducing °C	oxidizing °C	SiO_2	Al_2O_3	TiO_2	Fe_2O_3	CaO	MgO	Na_2O	K_2O	SO_3	P_2O_5
Pocahontas No. 3 West Virginia Bituminous	IDT>1600	IDT>1600	60.0	30.0	1.6	4.0	0.6	0.6	0.5	1.5	1.1	0.1
Ohio No. 9 Bituminous	1440	1470	47.3	23.0	1.0	22.8	1.3	0.9	0.3	2.0	1.2	0.2
Illinois No. 6 Bituminous	1270	1430	47.5	17.9	0.8	20.1	5.8	1.0	0.4	1.8	4.6	0.1
Pittsburgh, WV Bituminous	1300	1390	37.6	20.1	0.8	29.3	4.3	1.3	0.8	1.6	4.0	0.2
Utah Bituminous	1280	1320	61.1	21.6	1.1	4.6	4.6	1.0	1.0	1.2	2.9	0.4
Antelope Wyoming Sub-bituminous	1270	1260	28.6	11.7	0.9	6.9	27.4	4.5	2.7	0.5	14.2	2.3
Texas Lignite	1230	1250	41.8	13.6	1.5	6.6	17.6	2.5	0.6	0.1	14.6	0.1

Source: Stultz and Kitto 1992

melting characteristics of the ash but can have a major effect on the environmental problems associated with coal use: for example, mercury, arsenic, zinc, lead, cadmium, chromium, chlorine, and fluorine. Most of the elements present in coal as given in Table 4-7 can appear at least in part in the ash.

Table 4-7 Detailed Chemical Analysis of Minerals in Coal	
As, ppmw	2.1
B, ppmw	35
Ba, ppmw	130
Be, ppmw	1.2
Br, ppmw	1.5
Cd, ppmw	0.07
Ce, ppmw	24
Co, ppmw	3.5
Cr, ppmw	7.0
Cs, ppmw	0.30
Cu, ppmw	9.2
Eu, ppmw	0.32
F, ppmw	227
Ge, ppmw	0.50
Hf, ppmw	1.8
Hg, ppmw	0.13
La, ppmw	17
Mn, ppmw	84
Mo, ppmw	1.6
Ni, ppmw	10
Pb, ppmw	14
Rb, ppmw	2.1
Sb, ppmw	0.57
Sc, ppmw	2.9
Se, ppmw	3.1
Sm, ppmw	1.4
Sr, ppmw	316
Th, ppmw	8.4
Ti, ppmw	0.43
U, ppmw	2.1
V, ppmw	17
W, ppmw	0.73

4.1.6   Coke

Coke is a material consisting essentially of the fixed carbon and the ash in the coal. It was in the past a common fuel in water gas plants, but as it is more expensive than coal, anthracite is now often the preferred fuel. It is virtually never used in gasification plants. Coke plays a very important role in blast furnaces, which may be considered to be very large gasifiers (Gumz 1950). One of the main reasons to use coke in blast furnaces is that it is much stronger than coal.

4.1.7   Petroleum Coke

Petroleum coke, more often named petcoke, is increasingly considered as an attractive feedstock for gasification, in particular as it becomes more and more difficult to fire this high-sulfur material as a supplemental fuel in coal-fired power stations. Apart from the feed systems that are similar to those for pulverized coal, the behavior of petcoke in gasifiers is very similar to that of heavy oil fractions. When gasified in certain entrained-flow slagging coal gasifiers, it is essential to add ash, because otherwise the build-up of a proper slag layer on the membrane wall will not be successful (Mahagaokar and Hauser 1994). In the Elcogas plant in Puertollano, Spain petcoke is processed together with (high-ash) coal (Sendin 1994).

4.2   LIQUID AND GASEOUS FEEDSTOCKS

In 2002 some 154 million Nm^3/d of synthesis gas was produced by partial oxidation of liquid or gaseous feeds (Simbeck and Johnson 2001; SFA Pacific 2001). If fed to state-of-the-art IGCC units, this would have generated some $11,500\,MW_e$. By far the largest portion of this synthesis gas (about 80%) is generated from refinery residues, typically visbreaker vacuum bottoms or asphalt. The most important product from these plants is ammonia. Methanol is also important, but refinery hydrogen and power applications via IGCC are rapidly increasing (Figure 4-2). Most plants with gaseous feed are small units for the production of CO-rich synthesis gases, particularly for the production of oxo-alcohols. The largest single gas-fed plant is the Shell unit at Bintulu, Malaysia, which serves as the front end for a synfuels plant using Fischer-Tropsch technology. The number of projects recently announced indicates that this type of application is likely to gain increasing importance in the near future. However in the long term one must recognize that neither oil nor natural gas availability is as great as that of coal. The reserves-to-production ratios are 40 and 62 years, respectively, compared with 216 years for coal.

4.2.1   Refinery Residues

Over 95% of liquid material gasified consists of refinery residues. Being a high temperature, noncatalytic process, partial oxidation is by and large flexible with

Table 4-8
Liquid and Gaseous Gasification Capacities

	Number of Reactors	Syngas Production (MM Nm³/d)	Approximate Feed Rate
Liquid feed	142	124	42,000 t/d
Gaseous feed	41	31	10.9 MM Nm³/d
Total	183	155	

Source: SFA Pacific 2001

Table 4-9
Applications for Gasification of Liquid and Gaseous Feedstocks

	Synthesis Gas from		Total Syngas (MM Nm³/d)	Approximate Product Rate
	Liquid Feeds (MM Nm³/d)	Gaseous Feeds (MM Nm³/d)		
Ammonia	50.5	0.8	53.7	6.93 MMt/y
Methanol	15.3	6.3	21.7	3.01 MMt/y
Hydrogen	7.1	2.0	9.0	12.50 MMNm³/d
Synfuels	–	7.6	7.6	
Other chemicals	9.4	12.4	21.9	
Power	34.3	–	34.3	2554 MW$_e$
Other	4.7	1.6	6.2	
Total	122.1	30.6	154.3	

Note: Totals do not tally exactly due to some double counting on multiproduct facilities.
Source: SFA Pacific 2001

regard to feedstock quality and does not make great demands on the specification of a liquid feed. However, every feedstock needs to be evaluated to ensure that design aspects are correctly selected for its particular requirements. Generally, any feedstock would need to fulfill the following requirements:

• The fuel per se must be single phase at the reactor burner inlet. Small amounts of gas in a predominantly liquid feed may ignite close to the burner, causing damage at that point. Droplets of liquid in a predominantly gaseous phase feed can lead to erosion and premature wear of equipment. This does not pose a limitation for most feed-

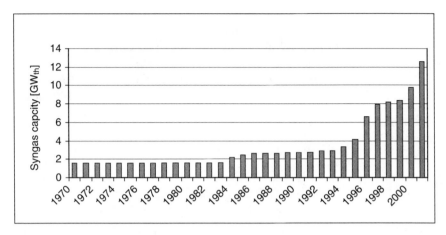

Figure 4-2. Growth of Hydrogen and Power Applications for Gasification of Liquid and Gaseous Feedstocks (*Source: SFA Pacific 2001*)

stocks, but must be considered in the design of the preheat train. It is important in this connection, however, to ensure that abrasive solids are kept to a minimum.
• The feedstock must be maintained within certain viscosity limits at critical points in the plant, such as at the burner. Again, this is generally a matter that can be addressed in the design of the feed transport and preheat systems and will not place any limitation on choice of feed.
• Too high a level of certain components in the ash, such as a high sodium content, limits the use of a syngas cooler, since the sodium salts deposit on the exchanger surface. Details are discussed under Sodium below.

The term "refinery residue" covers a wide range of material, some of which is solid at ambient temperatures, and others that are liquid under these conditions. Common to all is their origin in being the product of the distillation and possible subsequent treatment of crude oil. The most common residues are obtained by thermally cracking (visbreaking) vacuum residue or by subjecting it to solvent de-asphalting.

The specifications shown in Table 4-10 show typical gasifier feeds having a high sulfur content, high metal contents, and high viscosities (Posthuma, Vlaswinkel, and Zuideveld 1997). The data for tar sands bitumen, which are discussed in Section 4.2.2, are also included.

C/H Ratio. The C/H ratio of heavy refinery residues can vary between about 7 kg/kg (vacuum residue) and 10 kg/kg (asphalts), depending on the crude source and refining history. There are no specific limitations on the C/H ratio for gasification, but of course it does have an effect on syngas quality.

If all other conditions are maintained equal, the feeds with a higher C/H ratio will produce a synthesis gas with a lower H_2/CO ratio. Whether this is an advantage or a disadvantage will depend on the application.

Table 4-10
Typical Refinery Residue Specifications

Feedstock Type	Visbreaker Residue	Butane Asphalt	Tar Sands Bitumen
Elemental analysis/Units			
C, wt%	85.27	84.37	83.60
H, wt%	10.08	9.67	1.90
S, wt%	4.00	5.01	6.08
N, wt%	0.30	0.52	1.52
O, wt%	0.20	0.35	1.90
Ash, wt%	0.15	0.08	5.00
Total, wt%	100.00	100.00	100.00
C/H, kg/kg	8.3	8.7	44
Vanadium, mg/kg	270–700	300	250
Nickel, mg/kg	120	75	100
Iron, mg/kg	20	20	
Sodium, mg/kg	30	30	
Calcium, mg/kg	20	20	
Viscosity (100°C), cSt	10,000	60,000	110
Density (15°C), kg/m^3	1,100	1,070	1,030
LHV, MJ/kg	39.04	38.24	34.11

Sulfur Content. Typically, the sulfur content of a residue can vary between 1% and 7%. In exceptional cases (e.g., in parts of China) material with as little sulfur as 0.15% is gasified.

A sulfur content over this range has little effect on the design or operation of a gasifier, but it is an important issue for the design of the subsequent gas treating. A low sulfur content gives rise to a low H_2S/CO_2 ratio in the sour gas from the acid gas removal unit, and this can be decisive for the design of the sulfur recovery unit. Selective physical washes such as Rectisol can be designed to concentrate the H_2S in the sour gas, and with a low sulfur feed this would be necessary. Equally, low sulfur feeds can have surprising effects on a raw gas shift catalyst. This catalyst is required to operate in the sulfided state, and if insufficient sulfur is available to maintain this, then the catalyst can lose activity (BASF). Again, it is perfectly possible to design around this by including a sulfur recycle. It is therefore good practice to consider the low sulfur case carefully when specifying the basis of design for a new plant.

Corrosion effects in a residue gasifier, which can be connected with the presence of sulfur, are in most cases independent of the actual content of sulfur, and one needs to look at what other circumstances are contributing to the corrosion. In the event of, for example, high temperature sulfur corrosion, the solution to the problem will generally lie in avoiding the high temperatures rather than trying to lower the sulfur content of the feed.

Nitrogen Content. The refining process concentrates the nitrogen present in the crude oil into the residue. Nonetheless, there is seldom more than 0.6 wt% nitrogen in a gasifier feedstock. Much of the nitrogen entering the reactor as part of the feedstock is bound in organic complexes, and under the gasifier conditions it reacts with the hydrogen to form ammonia and hydrogen cyanide. The more nitrogen there is contained in the feed, the more ammonia and cyanide will be formed. For details see Section 6.9.2.

Ash Content. The ash content of typical feedstocks are summarized in Table 4-10. The evaluation of the feed must be done on the basis of both individual ash components as well as the total ash content. The most critical and most common components are discussed in the following sections.

Satisfactory experience with ash contents up to 2000 mg/kg has been achieved. Individual plants have run with even higher ash contents (Soyez 1988).

Vanadium. Experience of up to 700 mg/kg is available, and in one case of up to 3500 mg/kg at the reactor inlet. With the exception of residues from some Central and South American crudes, feedstocks with over about 350 mg/kg are unusual. Many residues from Far East crudes have an order of magnitude less than this.

Vanadium as a feed component has two undesirable properties:

- In an oxidizing atmosphere vanadium is present as V_2O_5, which has a melting point of 690°C (Bauer et al. 1989). At temperatures higher than this, V_2O_5 diffuses into refractory linings, whether of a reactor or a boiler fired with a high vanadium fuel such as carbon oil (see next bullet), and destroys the refractory binder. In a reducing atmosphere (i.e., in normal operation for a gasification reactor) vanadium is present as V_2O_3, which has a melting point of 1977°C and is therefore not critical. For plant operation the lesson is that special care should be taken during heat up and after shutdown, when the reactor could be subject to an oxidizing atmosphere at temperatures above 700°C (Collodi 2001).
- When combusted in a conventional boiler, a fuel with a high vanadium content will cause slagging and fouling on economizer heat-transfer surfaces. Where soot from the gasifier is admixed to oil (carbon oil) for external firing in an auxiliary boiler, this sort of fouling can be a particular problem. With 700 mg/kg vanadium in the boiler feed, cleaning of the surfaces may be required as often as once or more per year.

- Problems with burners and syngas coolers have also been reported when operating with over 6000 mg/kg vanadium in the reactor feed (Soyez 1988). This high level of vanadium in the feed is, however, extremely unusual.

Nickel. There is no generally recognized upper limit for nickel in gasifier feedstocks. Nickel does, however, have an important influence on the gas treatment. In the presence of carbon monoxide and under pressure, nickel and nickel sulfide both form nickel carbonyl, a gaseous compound that leaves the gasification as a component of the synthesis gas. This topic is handled in more detail in Section 6.9.8.

It is also worth noting that upon air contact nickel sulfide can react to form $NiSO_4$ in the water. The $NiSO_4$ goes into solution in the water and can then be recycled to the scrubber, where it can increase the amount of carbonyls formed.

Sodium. Sodium can be present in the gasification feedstock either as sodium chloride or as sodium hydroxide. The sodium leaves the reactor as sodium chloride, or as sodium carbonate where the origin is sodium hydroxide, which have melting points of 800°C and 850°C, respectively (Perry and Chilton 1973). In plants where the synthesis gas is cooled by steam raising in a heat exchanger, these salts deposit on the heat-exchange surface. The most serious effect is fouling of a large section of the surface, causing an increase in exchanger outlet temperature. The depositing can, however, also accumulate locally to the extent that there is a significant increase in pressure drop. It is therefore desirable to ensure that the sodium content does not exceed 30 mg/kg, or at the maximum 50 mg/kg. It is our experience that already as much as 80 mg/kg in a feed that had been contaminated with seawater caused the outlet temperature of a syngas cooler to rise at more than 1°C per day.

Fortunately, the sodium chloride fouling is a reversible phenomenon, and operation with a sodium free feed will cause the outlet temperature to reduce again even if not quite to the value prevailing before the sodium ingress. Full recovery requires a steam out.

Sodium compounds—and for that matter other alkali metals—have the additional unpleasant property that they diffuse into the refractory lining of the reactor, where they effect a change in the crystal structure of the alumina from α-alumina to β-alumina, which leads in turn to a gradual disintegration and loss of life of the refractory.

For most refinery applications the sodium limitation is not a restriction, since the desalting process in the refinery effectively maintains the sodium level within allowable limits. In cases where a high sodium content is expected on a regular and steady basis, then quench cooling is a preferable choice to a syngas cooler.

Calcium. Typically, there may be some 6–20 mg/kg calcium in a refinery residue. Calcium can react with the CO_2 in the synthesis gas to form carbonates. This does not normally present any problems. Where significantly more calcium is present, these carbonates can precipitate out of the quench or wash water depositing in the level indicators, possibly with disastrous results if not recognized in time. A certain degree of depositing on the surface of syngas coolers is also possible (Soyez 1988).

Iron. The iron content of the feed can be as high as 50 mg/kg, but is generally lower. The behavior of iron is similar to that of nickel. Carbonyl formation in the synthesis gas takes place at a lower temperature than for nickel. This is discussed in more detail in Section 6.9.8.

Silica. Silica can find its way into refinery residues from a number of sources, as sand, catalyst fines from a fluid catalytic cracker (FCC), or from abraded refractory. Typically, there may be between 20 and 50 mg/kg in the residue, which can be tolerated. Larger quantities can cause two types of problem.

On the one hand, silica is an abrasive material that largely passes into the water system, where it settles out to some extent. In recycle systems, it is partly mixed with the feedstock, increasing the quantity being fed to the reactor. Abrasion has been reported on the feedstock charge pumps as a result (Soyez 1988).

Silica entering the system as FCC catalyst fines can also deposit close to the burner area of the reactor, with the risk of disturbing the flame pattern in the reactor.

Additionally, there is the problem that under the reducing conditions in the reactor, silica is reduced to volatile SiO according to the reaction

$$SiO_2 + H_2 = SiO \uparrow + H_2O$$

The SiO condenses at about 800°C while cooling in the syngas cooler and deposits on the exchanger surface. This is one reason why all oil gasifiers use a low-silica high-alumina refractory lining (Crowley 1967).

Chloride. Chloride in the feedstock is mostly present as NaCl. Smaller quantities may also be present as K-, Fe-, Cr-, Ca-, and Mg- compounds.

Although large quantities of chlorides will damage the plant, the limitation of the main source, NaCl to 30 ppmw Na, is usually sufficient to limit the overall chloride intake.

The potential problems of fouling or corrosion associated with chlorides are described in Section 4.1.

Naphthenic Acids. Naphthenic acids can be present in residues to a greater (e.g., those derived from Russian crudes) or lesser extent. While this is not of importance for the gasification process itself, it can be an important evaluation criterion, since it would need to be considered in the selection of metallurgy for the feedstock transport system.

Viscosity. Many refinery residues have an extremely high viscosity and are (subjectively) solid, having the consistency of street bitumen at ambient temperatures. Since the viscosity decreases considerably with increased temperature, it is often desirable to transport and store this material in a heated condition.

The temperature dependency of viscosity takes the form of the Walther equation (Große et al. 1962, L4):

$$\log(\log(v + c)) = m*(\log(T_0) - \log(T)) + \log(\log(v_0 + c))$$

where v (cSt) is the kinematic viscosity at absolute temperature T (K), m is a constant for any given oil characterizing the temperature dependency of the viscosity, and c is a constant. This equation can be plotted as a straight line in a log(log(v)) versus log(T) diagram such as Figure 4-3. Ideally, m can determined from two reference temperatures for which the viscosity is known, since it is the gradient of the straight line joining the two points on the diagram. From this it is possible to use the value of m to determine the viscosity at any other desired temperature.

For cases where the viscosity at only one temperature is known, a correlation for m is required. Singh, Miadonye, and Puttagunta (1993) have developed such a correlation. A short program to calculate the viscosities of the feed on the basis of both one and two reference temperatures is included on the companion website.

The viscosity is an important parameter in the design of a gasification system, since the effectiveness of atomization at the burner is dependent on viscosity limits. Values of 20 cSt–300 cSt can be found in the literature (Supp 1990; Weigner et al. 2002.). The exact values depend on the individual burners, so licensors must be consulted for any specific project. In general, the desired temperatures required to achieve the prescribed temperature range can be achieved by steam heating, which is in most cases preferable to a fired heater, since the metal temperatures are lower and there is less tendency for the feed material to crack in the preheater.

If during operation the feed preheaters should fail, leading to an increase in viscosity, the tendency will be that the atomization deteriorates and an increase in the soot make can result.

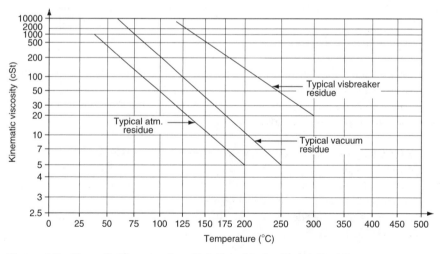

Figure 4-3. Viscosity-Temperature Relationship for Heavy Residues

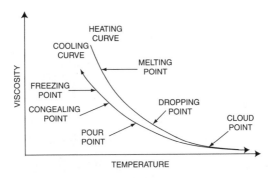

Figure 4-4. Pour Point (*Source: Baader 1942*)

Pour Point. The pour point is the second important property for the feedstock transport system. It provides an indication of the lowest temperature to ensure pumpability and avoid solidification of the feed in the line.

Since the transition from solid to liquid is gradual, there are a number of different defined points in the transition, which are shown in the diagram in Figure 4-4. The pour point is defined as the lowest temperature at which the oil will pour or flow under defined standard conditions (ASTM D 97).

When evaluating the transport properties of a feed, it is insufficient to look at the pour point in isolation. Two sample feeds illustrate this:

	Feed A	**Feed B**
Pour point	60°C	70°C
Viscosity at 100°C	50 cSt	2500–25,000 cSt

In the case of feed A, with only a slight increase of temperature above the pour point, the feed flowed sufficiently easily that no problems ever occurred on loss of steam tracing. Bringing the steam tracing back on line was sufficient to unblock the pipe. This would have been a considerable problem in the case of feed B, so a flushing system was included as part of the original design. The reason is that feed A is a feedstock with a high percentage of paraffins that once they are molten have a low viscosity. Feedstock B has a so-called viscosity pour point where heat must just be applied until the required low viscosity is reached, and no use can be made of a state transition.

Density. Densities of typical gasifier feedstocks lie between 970 and 1250 kg/m^3. There are no limitations imposed by gasifier performance or design.

Figure 4-5 shows the relationship between temperature and density for different oils. This correlation is also included in the companion website.

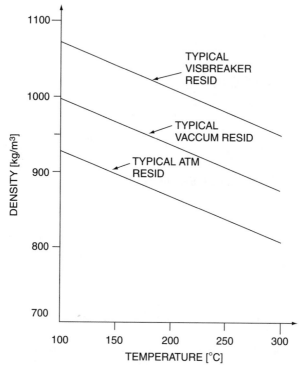

Figure 4-5. Density-Temperature Chart

Flash and Ignition Temperatures. The flash point is the temperature at which sufficient hydrocarbons have evaporated that an explosive mixture is formed that can be ignited by an external ignition source.

There are two methods commonly used for determination of the flash point: the Cleveland open cup method (ASTM D 92) and the Pensky-Martin closed cup method (ASTM D 93). Using the open cup method a certain amount of the evaporating hydrocarbons is lost to the surroundings, which leads to a relatively high value for the measured flash point. The closed cup method uses a closed vessel with a narrow neck so that all the hydrocarbons that evaporate remain part of the potentially ignitable mixture. The difference between the two methods is about 30°C, that is, the closed cup method will provide a value about 25–30°C lower than the open cup method.

In contrast to the flash point for which an external ignition source is used, the ignition point (ASTM D 874) is that temperature at which the hydrocarbon begins to burn without any external ignition source.

The flash point places an upper limit to the preheat temperature of a gasifier feedstock. If material is preheated to its flash point, then there is a danger that it will ignite immediately on exiting the burner and damage it. A damaged burner can in turn change the design flame pattern, causing local stoichiometric combustion with associated high temperatures and the potential for reactor containment failure. A suitable safety precaution is to maintain the preheat temperature 50–100°C lower than the flash point.

Generally, there is no need to preheat to anything like the flash point of normal residues, since the viscosity is already sufficiently low for good atomization at substantially lower temperatures. Problems can arise, however, with blended feedstocks. If, for instance, FCC light cycle oil is added to a heavy asphalt to reduce its viscosity, then the light material might have a flash point at a temperature required to achieve an acceptable viscosity of the blend at the burner.

Conradson Carbon. The Conradson carbon (ASTM D 189) is determined by placing a feed sample into a container and heating it to a given temperature so that it cracks. The Conradson carbon value is given by the amount of cracked residue expressed as a percentage of the original sample.

The Conradson carbon value is not used expressly in gasifier installation design. It does, however, provide an indication of the propensity for coke formation by the residue during preheat. It can also, in connection with the C/H ratio, provide a guide to the moderating steam requirement. Additionally, it can be of use when considering the consistency of other data received for a particular feed.

Typical values for the Conradson carbon are:

Propane asphalt	35%
Vacuum residue	20%
Atmospheric residue	10%

The Ramsbottom method (ASTM D 254) provides an alternative determination of the carbon residue. A conversion chart between the two methods can be found in Speight (1998, p. 335).

4.2.2 Other Liquid Feedstocks

Orimulsion. Orimulsion is the trade name for an emulsified bitumen-water mixture produced from the bitumen fields in the Orinoco belt of Venezuela. Orimulsion consists of about 70% bitumen and 30% water and contains about 1% surfactants. The emulsifying technology converts the bitumen into a transportable fuel with a pour point of 3°C and a viscosity of about 200 cP at 30°C. It has a lower heating value of 27.8 MJ/kg (Marrufo, Sarmiento, and Alcala 2001).

Technically, Orimulsion is a petroleum product; it contains sulfur, vanadium, and nickel in substantial quantities, and these behave just as in a conventional residue. The sodium content is low (12 mg/kg), assuming that there is no contamination in transport. The differences of Orimulsion when compared to a refinery residue are in the water content and the surfactant used. The water content causes a considerable loss of efficiency since energy is used in its heating and evaporation. The resulting syngas has a CO_2 content of about 10% compared with around 3% in unquenched syngas generated from a conventional residue. Early formulations of Orimulsion contained considerable quantities of magnesium, which could have presented problems similar to those caused by calcium. The newest formulation, Orimulsion 400, has a magnesium content of 6 mg/kg, which is sufficiently low to avoid these.

When designing for or operating with Orimulsion, it is necessary to take specific precautions (e.g., reduced preheat temperatures) to avoid the emulsion breaking. A handbook of suitable handling guidelines can be obtained from the suppliers.

Orimulsion has been tested as a gasification feedstock in Texaco's Montebello, California, pilot facility in 1989 with apparent success. The producer of Orimulsion, Bitumenes Orinoco S.A. (Bitor), claims to be able to supply the material at a price to allow competitive production in a gasification plant but there is no recorded commercial application at this time (2002).

Tar Sands Residues. Tar sands are deposits of heavy hydrocarbons located in a sandstone matrix that are not amenable to conventional pumping technology. The largest and most well-known deposits are in northern Alberta, Canada. Other deposits exist (in approximate order of size) in Venezuela, the United States (Utah, Texas, California, Kentucky), Russia (Olenek), Madagascar, and Albania, as well as in other locations in Canada (Melville island). A typical analysis is included in Table 4-10.

Tar sands represent a major hydrocarbon resource with an estimated 450 billion barrels of recoverable reserves (Speight 1998, p. 117), but the difficulties and cost of extraction have limited commercial exploitation. At present, there are only two commercially operating plants, both in the Athabasca River basin of northern Alberta. A number of pilot operations for the development of improved extraction techniques exist, also mostly in this area, and a number of commercial projects are currently under development (Parkinson 2002).

The processing of tar sands can be described in three principle steps:

1. Extraction, for which there are two fundamentally different approaches. One is to mine the bitumen-laden sandstone and transport it to a central extraction plant, where the sandstone and bitumen are separated by a hot-water extraction process (HWEP). The commercial operations of Syncrude and Suncor are both based on this method.

   Alternative methods have been developed for in situ extraction and separation, such as steam-assisted gravity drainage (SAGD). There are pilot plants in operation that demonstrate the possibilities of this approach, and at least one current project is based on it.
2. Primary conversion, for which conventional or modified coking, cracking, or solvent de-asphalting processes are applied.
3. Secondary conversion, which is essentially a hydrotreating step.

Table 4-11
Bitumen Solids Yields and Metals Analysis for BS Free Tar Sand Asphalt

Material and Recovery Process	BS wt%	V ppmw	Ni ppmw	Al ppmw	Ca ppmw
Athabasca mine (HWEP) before BS removal	0.9	820	210	640	1400
Athabasca mine (HWEP) after BS removal		880	280	200	110
Athabasca mine (Toluene extracted)	0.5	650	240	150	100
Athabasca (SAGD)	n.d.	1270	444	14	80
Nigeria	2.1	104	148	n.a.	n.a.
Utah	1.4	21	170	n.a.	n.a.

Source: Zhao et al. 2001 with permission from Elsevier

There is potential to use the residue from the primary conversion as gasifier feedstock to provide hydrogen for the secondary conversion. Existing operations do not do this; they generate their hydrogen by steam reforming of natural gas. But where or when natural gas availability is critical, gasification could become a serious option. A first project of this sort has recently been announced (Arnold et al. 2002).

Work has been performed on the characterization of residues from tar sands (Zhao et al. 2001).

From the point of view of gasification, these investigations have highlighted a number of important and interesting aspects (Table 4-11). The extremely high vanadium and nickel contents are a feature of the Canadian material. These values exceed current long-term experience for fresh feed in gasifier operation. In particular it would be important to avoid a recycle configuration for the carbon management system so as to avoid metals build-up in the circuit. The second interesting feature is the quantity of bitumen solids (BS) observed in the mined material, which is absent in that recovered by SAGD. These solids are typically ultra-fine aluminosilicate particles originating from clay inclusions in the sandstone structures that are brought into the processing plant by the inherently nonselective mining processes. Experience with gasification of conventional residues containing catalyst fines from an FCC unit has shown a tendency for such material to deposit in both the gasifier and the syngas cooler. The bitumen produced by the SAGD process is practically free of ultrafine solids, which makes it far more suited as a gasifier feedstock.

Liquid Organic Residues. Some gasifiers process organic residues from petrochemical processing, such as the manufacture of oxo-alcohols, and have done so successfully for many years. The only important consideration is that such residues may contain catalyst fines. Depending on the catalyst and/or carrier, this may have an abrasive effect on critical equipment or cause fouling or plugging as described above for refinery residues.

Coal Tar. The MPG process (see Section 5.4.3) was originally developed for coal-based tars generated in a plant using Lurgi fixed-bed gasifiers to gasify lignite. It has been in successful operation in such service since 1969 (Hirschfelder, Buttker, and Steiner 1997; Liebner 1998).

Other oil-processing gasifiers have taken in coal tar in order to reduce feed-stock costs. Such attempts at mixing coal tar and petroleum-derived residues have not generally been successful. The principle difficulty is the incompatibility of the different types of ash, which tend to form eutectica. The result is plugging of either the throat area in a quench reactor or of the tube bundle in a syngas cooler.

Spent Lubricating Oil. Spent lubricating oil is included here as a potential gasifier fuel more to warn against it than to encourage its use. Used lubricating oil can contain typically 1500 mg/kg each of lead and zinc. The lead content can be as much as 10,000 mg/kg. Lead and zinc sulfides solidify at temperatures of 700–800°C and will block syngas coolers and the throats of quench reactors. Soyez (1988) reports that "some 100 ppm was sufficient to plug the waste heat boilers completely within only seven days." Other similar cases are also known. The only sound advice concerning gasification of spent lubrication oil is: don't.

4.2.3   Natural Gas

Compared with the variety of aspects needing evaluation when dealing with liquid feeds, natural gas is relatively simple, and the principle issues to be considered are more of an economic rather than technical nature. For the production of hydrogen-rich synthesis gas it is generally more economic to employ steam reforming rather than partial oxidation. Partial oxidation of natural gas is only likely to demonstrate favorable economics for hydrogen production where no purpose-built oxygen plant is required, or where the hydrogen is a by-product of carbon monoxide production. The advantage of partial oxidation—namely, that as a noncatalytic process, no large amounts of steam are required to prevent carbon laydown on the catalyst—only comes into its own when a CO-rich syngas is required. Further details are discussed in Section 7.1.4.

Since most applications for partial oxidation of natural gas aim at a CO-rich synthesis gas, quenching hot gas with water—advantageous if a CO shift is desired—is economically unattractive, and most such plants employ a syngas cooler.

There is no specific requirement on the hydrocarbon content of the natural gas. Clearly, heavier gases with high ethane or propane content will produce a synthesis gas richer in CO than pure methane. When looking at the still heavier components of natural gas, however, it is really only necessary that they be gaseous at the burner.

Nitrogen is a component in natural gas, which passes through the reactor largely as an inert. The amount of nitrogen (or argon) allowable in the feedstock is governed purely by the synthesis gas specification. This is different from the CO_2 case. CO_2 is a partner in the partial oxidation reactions and will increase the CO yield from the gas. This will be favorable in many instances, but must be reviewed on a case-by-case basis.

In contrast to catalytic processes such as steam reforming or autothermal reforming, partial oxidation is tolerant of sulfur. In fact there are good reasons to accept sulfur into the partial oxidation reactor. Firstly the synthesis gas has a high partial pressure of carbon monoxide so that in the absence of large quantities of steam there is considerable potential for metal dusting corrosion (Posthuma, Vlaswinkel, and Zuideveld 1997), more so than with the equivalent steam reformer (see Section 6.11 for details). The most effective form of protection against metal dusting is sulfur in the gas (Gommans and Huurdeman 1994).

The second advantage of leaving the sulfur in the gas is to prevent a spontaneous methanation reaction in the synthesis gas.

4.2.4 Other Gaseous Feedstocks

Refinery Gas. Refinery gas has been used as a feedstock for partial oxidation. Although partial oxidation cannot usually compete with steam reforming for hydrogen production from natural gas, the situation can be different with a refinery gas feed. Refinery waste gas streams can contain considerable quantities of olefins, which would need to be hydrogenated upstream of a steam reformer. The partial oxidation route is not sensitive to the presence of unsaturates in the feed, and this flexibility can provide opportunities (Ramprasad et al. 1999).

Attention needs to be paid to the issues of metal dusting and methanation as with natural gas, but if there is no sulfur in the feed, then other solutions are possible, such as using a quench reactor if hydrogen is to be the end product.

FT Off-Gas. Fischer-Tropsch off-gas is essentially similar to many refinery off-gas streams in that it can contain significant quantities of unsaturated hydrocarbons. The same considerations apply. Where the main syngas generation for the FT synthesis is partial oxidation of natural gas, one only needs to be careful with the recycle of inerts (Higman 1990). This is, however, a limitation imposed by the synthesis process and not by the partial oxidation itself.

Coke Oven Gas. There are two recorded plants using coke oven gas as a feedstock. Coke oven gas is available only at low pressure. This makes the economics unfavorable in all but the most exceptional cases.

4.3 BIOMASS

The term *biomass* covers a broad range of materials that offer themselves as fuels or raw materials and that have in common that they are all derived from recently living organisms. This definition clearly excludes traditional fossil fuels, since although they also derive from plant (coal) or animal (oil and gas) life, it has taken millions of years to convert them to their current form. For the purpose of this book, we have chosen to include all agricultural and forestry wastes as well as purpose-grown material as biomass, thus clearly including animal refuse such as poultry litter. There is still a potential overlap between what is classified as waste and what as biomass. We have considered human sewage sludge as well as wastes from industrial processes as waste. Black liquor, an intermediate material in the paper industry with important fuel use, is treated here as biomass.

Although biomass is not a major industrial fuel, it supplies 15–20% of the total fuel use in the world. It is used mostly in nonindustrialized economies for domestic heating and cooking. In industrialized countries the use of biomass as a fuel is largely restricted to the use of by-products from forestry and the paper and sugar industries. Nonetheless, its use is being encouraged as part of a strategy for CO_2 abatement.

4.3.1   Properties of Biomass

The properties of biomass are as diverse as the sources from which they come. Typical data for some vegetable biomasses is included in Table 4-12.

Vegetable Biomass

Typical proximate and ultimate analyses are given in Table 4-13. Quaak, Knoef, and Stassen (1999) give a range for the bulk density of 150–200 kg/m^3 for straw shavings and 600–900 kg/m^3 for solid wood.

Ash Properties. The major difference between biomass and coal ashes is that for the majority of most biomasses the ash consists mainly of salts. Most of the biomass ash consists of potassium, calcium, phosphorus, and, further, sodium, magnesium, iron, silicon, and trace elements. Some examples are given in Table 4-14.

As a result biomass ashes have low ash-melting points of, for example, 800°C for some straws. Because these ashes are extremely aggressive towards refractory materials—K_2CO_3 can be used to dissolve minerals before further analysis—biomass does not lend itself for slagging gasification unless it is mixed with large quantities of coal.

Animal Biomass

Although biomass is often interpreted as vegetable biomass, animal-derived biomass remains an energy source that should not be totally ignored, even if its production is largely determined by developments in the agricultural and food sector. In order to provide an idea of the volumes involved, it should be noted that the annual production of waste by poultry is 8 kg/head. Pigs produce about 300 kg/y, beef cattle 900–1200 kg/y, and dairy cattle 1200–2000 kg/y (Smil 2001).

Table 4-12
Properties of Various Biomasses

Biomass	HHV MJ/kg	Moisture wt%	Ash wt%	Sulfur wt% dry	Chlorine wt% dry
Charcoal	25–32	1–10	0.5–6		
Wood	10–20	10–60	0.25–1.7	0.01	0.01
Coconut shell	18–19	8–10	1–4		
Straw	14–16	10	4–5	0.07	0.49
Ground nut shells	17	2–3	10		
Coffee husks	16	10	0.6		
Cotton residues (stalks)	16	10–20	0.1		
Cocoa husks	13–16	7–9	7–14		
Palm oil residues (shells)	15	15			
Rice husk	13–14	9–15	15–20		
Soya straw	15–16	8–9	5–6		
Cotton residue (gin trash)	14	9	12		
Maize (stalk)	13–15	10–20	2 (3–7)	0.05	1.48
Palm oil residues (fibers)	11	40			
Sawdust	11.3	35	2		
Bagasse	8–10	40–60	1–4		
Palm oil residues (fruit stems)	5	63	5		

Source: Derived from Arbon 2002; Quaak, Knoef, and Stassen 1999

In the SCGP coal gasifier in the Nuon plant in Buggenum in The Netherlands, trials have been carried out with the gasification of mixtures of coal with up to 12% poultry litter and/or municipal sewage sludge. There are plans to gasify up to 30% non-coal feedstocks in 2004 (Kanaar and Wolters 2002). It should be mentioned that the percentages of alternative feedstocks are generally given as a mass percentage. As an energy percentage, these figures are at least 50% lower.

Abattoir waste has also been co-gasified in the Puertollano plant up to an amount of 4.5% of total feed. The meat and bone meal that is classified as a risk material was divided into specific risk material (SRM), consisting of spinal marrow, bones, intestines, spleen, or brain, and high risk material (HRM), which is other parts of the animals, and blood material. When this material was fed as 4.5% of the total feed, the high chlorine content made itself noticeable in an increased chloride level in the

Table 4-13
Analysis of Typical Biomass

Proximate Analysis

Volatile matter, wt% maf	>70
Ash, wt% ar	1.5
Moisture, wt% ar	20
Fixed carbon, wt% ar	<15

Ultimate analysis

C, wt%	54.7
H, wt%	6.0
O, wt%	38.9
N, wt%	0.3
S, v	0.1

Table 4-14
Ash Components in Various Biomasses as Wt% of Total Ash

	Straw	Miscanthus	Wood
CaO	6.5	7.5	37.3
MgO	3.0	2.5	8.5
Na$_2$O	1.3	0.2	3.0
K$_2$O	23.7	12.8	8.6

Source: Klensch 2001

syngas scrubber (Schellberg and Peña 2002). Some typical properties of animal biomass are given in Table 4-15.

4.3.2 Black Liquor

Black liquor is an integral part of the papermaking process and represents about half the mass of the wood entering the pulping plant. It is the principal source of energy for the pulping process. In addition to the lignin and hemicellulose from the pulp wood, it also contains the cooking chemicals, which contain significant quantities of sodium and sulfur, the recovery of which for recycling to the pulp digesters is essential for the economy of the process.

Table 4-15
Properties of Animal Biomass

	HHV	Moisture	Volatiles	Ash
	MJ/kg	wt%	wt%	wt%
Meat and Bone Meal (SRM)	20.1	2.54	65.27	26.13
Meat and Bone Meal (HRM)	18.9	5.03	63.7	25.54
Meat and Bone Meal (Blood)	23.3	14.96	93.82	1.45
Meat and Bone Meal (MBM)	20.0–28.0			
Poultry Litter	13.0–14.0	63	25	20

Source: Schellberg and Peña 2002; Arbon 2002; and Bajohr 2002

Total black liquor production worldwide is an energy resource of about 600 TWh/y or 51 million toe/y. This represents about 0.5% of the total world energy consumption. In a country such as Sweden, with a strong pulping sector, black liquor can be as much as 6% of the national gross energy consumption. An average-sized mill produces 1000 air-dried tons of pulp per day, generating some 1800 tDS/d (tons dry solids per day) of black liquor. On an HHV basis this is about 300 MW$_{th}$.

In a typical pulping process (see Figure 4-6) wood chips are cooked with sodium hydroxide and NaHS in a pulp digester where lignin and other organic material, which contributes almost half the mass of the wood, is dissolved into the white liquor

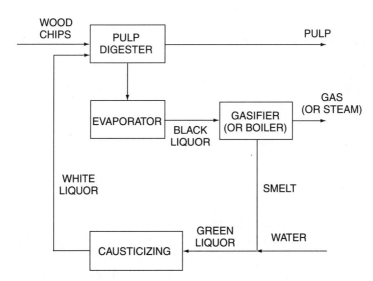

Figure 4-6. Gasification of Black Liquor

Table 4-16 Composition of Black Liquor (Dry)	
Element	**wt%**
C	36.40
H	3.50
O	34.30
N	0.14
S	4.80
Cl	0.24
Na	18.60
K	2.02
Total	100.00
Source: Marklund 2001	

containing the caustic. The solids content of the weak black liquor leaving the digester is about 15 wt%. This is increased to about 70 wt% in evaporators to leave a combustible pumpable material that in traditional processes is completely combusted in a special (black liquor) recovery or Tomlinson boiler. The inorganic material, mostly NaS and NaCO$_3$, leaves the boiler as a smelt that is recovered by causticizing for reuse in the digester.

The use of black liquor gasification is a relatively new development. It is located in the same position in the liquor circuit as the boiler (see Figure 4-6). In some applications black liquor gasifiers have been built in parallel with existing boilers as part of a de-bottlenecking project.

Typically, the strong (evaporated) black liquor has a viscosity of 70–120 cP and a density of 1400 kg/m^3. The heating value is about 13.5–14 MJ/kg (HHV).

A typical elemental analysis of the dry material in black liquor is shown in Table 4-16.

4.3.3 Biomass Production

Cropping

One of the most important limitations of biomass as a widespread fuel is the space required to grow it. Bridgwater (2002) estimates that "a sustainable crop of 10 dry t/ha/y of woody biomass can be produced in Northern Europe rising to perhaps 15 or maybe 20 t/ha/y in Southern Europe." Assuming that 1000 dry t/y will generate 150–300 kW depending on conversion efficiency then an area of the order of magnitude of 100 km^2 or more would be required to support a 30–40 MW power station.

It is unlikely that all transportation fuels and organic chemicals can in the future be produced from biomass. This would call for an immense acreage of biomass plantations of a few million square kilometers. This seems unrealistic for various reasons:

- The sheer size of such projects.
- It is far from certain that large-scale plantations of more or less conventional crops such as Eucalyptus, miscanthus, euphorbia, and other species are so green as the advocates of these plantations claim it to be. The large-scale biomass production in Minas Gerais in Brazil for making charcoal to be used in blast furnaces is a well-known environmental problem. Fertilizers, herbicides, and pesticides will be required; ash may have to be recycled in order to recycle alkali, phosphorus, and so on. Irrigation may be a problem, and so may soil erosion. Moreover, monocultures are very sensitive to pests. Last but not least, such developments are often difficult to reconcile with the wish for more biodiversity.
- Biomass is easy to transport when it comes to gathering firewood for nearby domestic use. However, it is expensive to transport in large quantities over any appreciable distance. The energy density of biomass is one-tenth that of liquid hydrocarbons, and further, it is a solid that cannot be pumped. In order to make wood and other biomass more amenable to transport, it can be chipped, but then the energy density will decrease even further. As a result, the energy required to get bulk biomass to usually distant markets is appreciable and constitutes a size-able percentage of the fuel market it wants to replace.

Biomass Waste

On the other hand, there are circumstances where large quantities of waste biomass arise as a result of some other economic activity such as forestry, papermaking, and sugar. Under these circumstances most of the costs associated with the collection of the biomass and its transport to a central location is borne by the main product. Thus in these areas somewhat larger biomass gasifiers (500–1000 t/d corresponding 80–160 MW_{th}) can and have been built. However, none of these are of a capacity to be compared with world-scale plants based on fossil fuel feedstocks of 200–1000 MW_{th}.

4.3.4 Development Potential

Bio-Oil

The major drawback of biomass is that the energy density is an order of magnitude lower than that of crude oil (Table 4-17). When it is further considered that biomass for fuel is a difficult to handle solid (grain is an exception in that it almost flows like water), this implies that fuel biomass can never be shipped economically over long intercontinental distances.

Table 4-17
Energy Densities of Various Fuels

Fuel	Particle Density	Bulk Density	Energy Density
	kg/m^3	kg/m^3	GJ/m^3 bulk product
Crude oil		855	35.8
Coal	1350	700	21
Natural gas (80 bar)		57	2.9
Biomass	450	230	3.7
Bio-oil		1200	20
Gasoline		760	35
Methanol		784	19

Therefore, when contemplating really large biomass production, the first conversion that has to take place is to convert the biomass into a transportable product. The best way to accomplish this is by flash pyrolysis, which is a fast pyrolysis at a high temperature of 450–475°C. In this way the solid biomass is converted into a transportable liquid with an energy density of about 20 GJ/m³ (see Table 4-17), which is similar to that of methanol. The pyrolysis plants can be located anywhere, because they can be economically built on a reasonably small scale. A stand-alone gasifier is only likely to be economic for world-scale plants, and hence they always depend on biomass from a variety of sources. The energy conversion of flash pyrolysis is now 75% but could well increase to 80% in the future. The bio-oil product is a transportable material that has as its only disadvantage that it is corrosive due to the presence of organic acids. A big advantage is that it can be gasified with oxygen in the same commercial gasifiers that can process heavy oil fractions. An added advantage of bio-oil is that it is easily homogenized. Holt and van der Burgt (1997) proposed such a concept of decentralized flash pyrolysis plants feeding a single large gasification installation. Research on details of actual implementation is continuing (Henrich, Dinjus, and Meier 2002).

Table 4-18 lists some of the most important properties of bio-oils. Additional data, in particular of some important organic compounds in bio-oil, is contained in Henrich, Dijus, and Meier (2002). In terms of its suitability for gasification in a standard oil gasifier, the most important aspects are the pH (material selection of the feed train) and the alkaline ash content. This latter may favor the use of a radiation screen instead of simple refractory lining.

Marine Cropping

A possible solution to large-scale biomass production in an environmentally acceptable manner could be plantations of salt-water organisms in desert areas,

Table 4-18
Typical Properties of Bio-Oil

Property	Typical Values
Moisture content	20–30%
pH	2–3
Density	$1200\,kg/m^3$
C	56 wt%
H	7 wt%
O	37 wt%
N	0.1 wt%
Ash	0–0.2 wt%
Viscosity (40°C, 20% H_2O)	40–100 cP
Particulates	< 0.3–1 wt%
Heating value	16–18 MJ/kg

Source: Meier 2002

which are located close to the sea. The advantage of this approach is that problems associated with fertilizers, irrigation, and so on are virtually absent. Moreover, aquatic crops are probably easier and safer to harvest.

Such farms should of course be located where there is a lot of sunshine. These are also locations where solar power is available and hence hydrogen can also be generated by electrolysis of water. This in turn enables the conversion of the biomass into true hydrocarbons for transport fuels or organic chemicals to be carried out close to the biomass production, thus grossly diminishing the transport problems. In addition, the option of first converting the biomass into bio-oil by flash pyrolysis could have advantages.

For countries in the Middle East and North Africa, which are now heavily dependent on crude oil exports for their income, this would suggest that they could continue to supply the world with hydrocarbon fuels. But other countries, such as Australia, also offer opportunities for such schemes.

Before considering biomass as a fuel it must be stressed that its primary function in a renewables-based world is that it is the only source of concentrated carbon. This aspect is often overlooked. It is important to remember, however, because when biomass is our only make-up source for concentrated carbon, it must be made possible to convert biomass into a material that can be an intermediate for all organic chemicals and transport fuels. The solution is then to have a process available that can convert biomass into synthesis gas efficiently. This can provide an alternative to ethylene as the main source of organic bulk chemicals as well as of Fischer-Tropsch products. This opens another an opportunity for gasification (van der Burgt 1997).

4.4 WASTES

"Waste" as a gasification feedstock covers a wide range of materials, both solid and liquid. Some materials, which we include in this section, could as easily have been included under biomass. Other chemical liquid wastes described here could also have been included in Section 4.2. The selection of materials described here is therefore to some extent arbitrary, but it has mainly been guided by self-classification of specific projects and general tradition within the gasification industry.

One of the most difficult aspects of waste as a feedstock, whether for gasification or incineration, is its heterogeneous nature. In this connection the interconnected properties of heating value and moisture content play an important part for both solid and liquid feeds. For solids, particle size is another key parameter. The presence of a number of components in the waste such as sulfur, chlorides, or metals can also vary considerably. The variations can be seasonal or even daily. Even with well-defined industrial wastes, where several waste streams are fed together to the gasifier, differing proportions of the individual streams can have an influence on gasifier performance.

For these reasons feedstock preparation plays an important role in the design of any waste gasification project. Equally important, this wide variation tends to make each project unique.

The long-term supply of the feedstock is another aspect to be considered. For this and other reasons it is often important that the refuse is processed with a support feedstock such as coal or residual oil. Ideally, the economics of the plant should be robust even in the absence of the refuse feedstock.

In the European Union there are about 500 million tons per year of waste that is suitable for thermal treatment. Of this some 6% is classified as hazardous waste, which includes clinical wastes, some wastes derived from agriculture, and chemical industry wastes (Schwager and Whiting 2002).

4.4.1 Solid Waste

One of the difficulties with refuse is the variability of its chemical composition. For a study comparing various refuse incineration and gasification processes, the State Environmental Office of Nordrhein-Westfalen developed a "standard refuse" composition, which is given in Table 4-19. The lower heating value was specified as 10 MJ/kg.

Löffler (1998) reports the composition of other types of refuse as specified for a circulating fluid-bed gasifier in Rüdersdorf.

Care should be exercised, however, when evaluating such data, which is very locality specific. In particular, local regulations on separation and recycling of household waste including plastics can have a dramatic effect on the heating value.

Table 4-19
"Standard Refuse" in Nordrhein-Westfalen, Germany

Composition	Mass%	Ash Composition	kg/1000 kg
C	27.16	SiO_2	110
H	3.45	Al_2O_3	34
O	18.39	CaO	31
N	0.30	Fe	30
S	0.20	Na_2O	15.2
Cl	0.50	Fe_2O_3	15
Moisture (H_2O)	25.00	MgO	4.5
Ash	25.00	Al	4
Total	100.00	K_2O	3
		Zn	1.5
		Pb	1.0
		Cu	0.5
		Cr	0.2
		Ni	0.075
		Cd	0.01
		Hg	0.005
		As	0.005

Source: Berghoff 1998

Feedstock Sizing. Typically, solid wastes in the "as-received" state have a highly irregular and asymmetrical lump size. Both for reasons of transport into and in the reactor, size reduction to a fine homogeneous material is required. Drying is in most cases also a requirement to ensure a smooth-flowing pneumatic transport.

There are a number of different approaches to pre-preparation. Most of these involve mechanical shredding and metals removal using magnetic and electric devices. Much refuse cannot be ground to a sufficient degree of fineness to permit slurry or dense phase fluid transport. Thus mechanical transport or dilute fluidized transport are generally the only possibilities. The latter type of transport cannot be economically performed under pressure, for then relatively large amounts of transport gas would be required, which would render the process uneconomical.

Görz (1998) has proposed an initial pyrolysis of the refuse simply to ensure a material that can then be ground to a size suitable for an entrained-flow reactor.

Moving bed processes such as BGL have a different requirement from fluid-bed and entrained-flow reactors. The feedstock is transported mechanically and introduced into the reactor via lock hoppers (which allows pressure operation), but the reactor bed can tolerate only a limited amount of fines without excessive pressure drop.

Table 4-20
Various Feeds for a CFB Refuse Gasifier

	Lignite Char	Coal	Old Wood	Refuse	Rubber
C, wt%	9.05	78.00	43.40	40.00	64.40
H, wt%	0.55	4.61	4.77	5.69	10.62
O, wt%	1.56	4.12	30.88	29.25	5.63
N, wt%	0.13	1.14	0.23	0.79	0.76
S, wt%	0.54	0.45	0.10	0.22	1.50
Cl , wt%	0.00	0.08	0.02	0.75	0.79
Ash, wt%	78.17	9.60	1.50	17.30	14.10
Moisture, wt%	10.00	2.00	19.10	6.00	2.20
Total , wt%	100.00	100.00	100.00	100.00	100.00
LHV, MJ/kg	3.3	31.4	15.4	15.9	34

Source: Löffler 1998

Thus a lump size of 20–70 mm is desirable. Material preparation therefore includes a pelletizing step in addition to initial shredding, metals removal and drying (Hirschfelder, Buttker, and Steiner 1997).

Heating Value. As indicated in Tables 4-19 and 4-20, the heating value of waste material can vary considerably. The drying required to facilitate pneumatic transport will in many cases be sufficient to ensure satisfactory operation of the gasification process. One should recognize, however, that wastes with excessive inert material cannot be gasified without a support fuel. The lower limit is 7–8 MJ/kg (Görz 1998).

4.4.2 Liquid Wastes

Organic Chemical Waste. Organic wastes from chemical production vary as widely as the processes from which they originate. One published example is the feedstock to a waste gasification plant at Seal Sands (United Kingdom), which processes streams from caprolactam and acrylonitrile production. These waste streams include mixtures of nitriles and amines as well as cyanide and ammonia compounds. The differences in the characteristics of these streams, the flow rates of which can vary independently from one another, can be observed in Table 4-21 (Schingnitz et al. 2000).

Other Liquid Wastes. Liquid wastes such as spent transformer oil can best be gasified in oil gasifiers together with heavy residual oils. The advantage of this approach is

Table 4-21
Waste Streams from Caprolactam and Acrylonitrile Production

	Stream			
	1	2	3	4
C, wt%	62.5	36.5	18.9	25.3
H, wt%	7.4	8.5	9.4	6.6
N, wt%	23.8	19.8	7.1	11.0
S, mg/kg	90	<50	690	46,500
Cl, mg/kg	<50	<50	<50	125
Ash, wt%	<0.05	<0.05	<0.05	0.68
Moisture, wt %	6.1	27.4	62.6	35.3
Heating value, MJ/kg	30.7	21.2	8.0	8.2

Source: Schingnitz et al. 2000

that use can be made of efficient large installations and that the concentrations of, for example, chlorine remain moderate and acceptable in terms of corrosion.

Many attempts have been and are being made to dissolve polymers such as plastics and rubber in oil and gasify them as a liquid rather than as a solid (e.g., Leuna-Werke 1995; Curran and Tyree 1998). The attraction of gasifying plastics as a liquid with oxygen is that a synthesis gas quality gas is produced, which can either be used as a fuel gas in a combined cycle or as a synthesis gas for methanol production, for example. The main problems are technically to make the liquid sufficiently homogeneous that it can be gasified using residual oil burners and commercially to ensure a long-term supply. Recycled plastic is generally too contaminated to allow a stand-alone project possible. Offcut waste from manufacturing processes, however, does offer both uncontaminated material and a better prospect for security of supply.

REFERENCES

Arbon, I. M. "Worldwide Use of Biomass in Power Generation and Combined Heat and Power Schemes." *Proceedings of the Institution of Mechanical Engineers* 216(A1) (2002):41–57.

Arnold, J., Bronicki, Y., Rettger, P., Hennekes, R., de Graaf, J. D., Hooper, M. "Integrated Oil Development Using Orcrude Upgrading and Shell Gasification Process." Paper presented at Gasification Technologies Conference, San Francisco, 2002.

ASTM, "Standard Classification of Coals by Rank," ASTM Standard D 388–99, American Society for Testing and Materials, Philadelphia, 1999.

Baader, A. "Über 'Anomalien' im Kälteverhalten der Öle." Oel und Kohle 38 (1942):432–444.

Bajohr, S. Private communication. September, 2002.

BASF. Technical leaflet, BASF Catalyst K 8-11. Ludwigshafen: BASF AG, (undated).

Bauer, G., Güther, V., Hess, H., Otto, A., Roidl, O., Roller, H., and Sattelberger, S. *In Ullmann's Encyclopedia of Industrial Chemistry*, 5th ed. vol. A 27. Weinheim: VCH Verlagsgesellschaft, 1989.

Baughman, G. L. *Synthetic Fuels Data Handbook*. Denver: Cameron Engineers, 1978.

Beeg, K., Schneider, W., and Sparing, M. "Einfache Bestimmung umweltrelevanter Metall-carbonyle in technischen Gasen mittels Flammen-AAS." *Chemische Technik* 45(3) (June 1993):158–161.

Berghoff, R. "Thermodynamische Bewertung der Abfallvergasungsprozesse." In *Vergasungsverfahren für die Entsorgung von Abfällen*, eds. M. Born and R. Berghoff. Düsseldorf: Springer-VDI, 1998.

BP *BP Statistical Review of World Energy*. London: BP PLC, June 2002.

Bridgwater, A. V. "Thermal Conversion of Biomass and Waste: The Status." Paper presented at IChemE Conference, "Gasification: The Clean Choice for Carbon Management." Noordwijk, 2002.

Channiwala, S. A., and Parikh, P. P. "A Unified Correlation for Estimating HHV of Solid, Liquid, and Gaseous Fuels." *Fuel* 81(8) (May 2002):1051–1063.

Collodi, G. "Commercial Operation of ISAB Energy and Sarlux IGCC." Paper presented at Gasification Technologies Conference, San Francisco, 2001.

Crowley, M. S. "Hydrogen-Silica Reactions in Refractories." *Ceramic Bulletin* 46(7) (1967):679–682.

Curran, P., and Tyree, R. F. "Feedstock Versatility for Texaco Gasification." Paper presented at IChemE Conference, "Gasification: The Gateway to a Cleaner Future," Dresden, September 1998.

Gommans, R. J., and Huurdeman, T. L. "DSM's Experience with Metal Dusting in Waste Heat Boilers." Paper presented at AIChE Ammonia Safety Symposium, Vancouver, 1994.

Görz, J. "Anforderungen an Einsatzstoffe und Prozessparameter bei Nutzung der Flug-stromvergasung für die Vergasung von Abfallstoffen." In *Vergasungsverfahren für die Entsorgung von Abfällen*, eds. M. Born and R. Berghoff. Düsseldorf: Springer-VDI, 1998.

Große, L., Henning, F., Kirschbaum, E., Reitz, O., and Schumacher, R. *Arbeitsmappe für Mineralölingenieure* page L 4. Düsseldorf: VDI-Verlag, 1962

Gumz, W. *Gas Producers and Blast Furnaces*. New York: John Wiley & Sons, 1950.

Hannemann, F., Kanaar, M., Karg, J., and Schiffers, U. "Buggenum Experience and Improved Concepts for Syngas Application." Paper presented at Gasification Technologies Conference, San Francisco, 2002.

Henrich, E., Dinjus, E., and Meier, D. "Flugstromvergasung von flüssigen Pyrolyseprodukten bei hohem Druck: Ein neues Konzept zur Biomassevergasung." Paper presented at DGMK Conference, "Energetische Nutzung von Biomassen," Velen, Germany, 2002.

Higman, C. A. A. "Synthesis Gas Processes for Synfuels Production." Paper presented at Eurogas '90, Trondheim, June 1990.

Hirschfelder, H., Buttker, B., and Steiner, G. "Concept and Realization of the Schwarze Pumpe 'Waste to Energy and Chemicals Centre.'" Paper presented at IChemE Conference "Gasification Technology in Practice," Milan, 1997.

Holt, N., and van der Burgt, M. "Biomass Conversion: Prospects and Context." Paper presented at 16th EPRI Gasification Technology Conference, San Francisco, October 1997.

Kanaar, M., and Wolters, C. "Nuon Power Buggenum IGCC." Paper presented at Gasification Technologies Conference, San Francisco, October 2002.

Klensch, S. "Verhalten von Schwermetallen bei der thermischen Umwandlung von Sägespänen im Flugstrom." Ph.D. diss., Engler-Bunte-Institut der Universität Karlsruhe (TH), 2001.

Kural, O. *Coal*. Istanbul: Istanbul Technical University, 1994.

Leuna-Werke GmbH "Verfahren zur Herstellung von Synthesegas aus Erdölrückständen und Altkunststoff oder Altkunststoffgemischen." German Patent DE 43 29 436 C1, March 1995.

Liebner, W. "MPG–Lurgi/SVZ Multi-purpose Gasification: A New and Proven Technology." Paper presented at Tenth Refinery Technology Meet, Mumbai, February 1998.

Löffler, J. C. (1998) "Zirkulierende Wirbelschicht zur Erzeugung von Schwachgas." In *Vergasungsverfahren für die Entsorgung von Abfällen*, eds. M. Born and R. Berghoff. Düsseldorf: Springer-VDI, 1998.

Mahagaokar, U., and Hauser, N. "Gasification of Petroleum Coke in the Shell Coal Gasification Process." Paper presented at IMP Gasification Colloquium, Mexico City, February 1994.

Marklund, M. "Black Liquor Recovery: How Does It Work?" Available at www.etcpitea.se/blg/document/PBLG_or_RB.pdf, (December 2001).

Marruffo, F., Sarmiento, W., and Alcala, A. "Gasification of Orimulsion: Synergism of the Two Technologies." Paper presented at Gasification Technologies 2001 Conference, San Francisco, 2001.

Meier, D. "Flash-Pyrolyse zur Verflüssigung von Biomasse—Stand der Technik." Paper presented at DGMK Conference, "Energetische Nutzung von Biomassen," Velen, Germany, 2002.

Meyers, R. A. *Handbook of Synfuels Technology*. New York: McGraw-Hill, 1984.

Parkinson, G. "Combing Oil from Tar Sands." *Chemical Engineering* 109(5) (May 2002):27–31.

Patterson, J. H., Hurst, H. J., and Quintanar, A. "Slag Compositional Limits for Coal Use in Slagging Gasifiers." Paper presented at 19th International Pittsburgh Coal Conference, Pittsburgh, 2002.

Perry, R. H., and Chilton, C. H. *Chemical Engineer's Handbook*. 5th ed. Tokyo: McGraw-Hill-Kogakusha, 1973.

Posthuma, S. A., Vlaswinkel, P. L., and Zuideveld, P. L. "Shell Gasifiers in Operation." Paper presented at IChemE Conference "Gasification Technology in Practice," Milan, 1997.

Quaak, P., Knoef, H., and Stassen, H. *Energy from Biomass: A Review of Combustion and Gasification Technologies.* Washington, DC: World Bank, 1999.

Ramprasad, R., Vakil, T., Falsetti, J., and Islam, M., "Competitiveness of Gasification at the Bulwer Island, Australia Refinery." Paper presented at Gasification Technologies Conference, San Francisco, 1999.

Ricketts, B., Hotchkiss, R., Livingston, B., and Hall, M. "Technology Status Review of Waste/Biomass Co-Gasification with Coal." Paper presented at IChemE Conference, "Gasification: The Clean Choice for Carbon Management," Noordwijk, April 2002.

Schellberg, W., and Garcia Peña, F. "Commercial Operation of the Puertollano IGCC Plant." Paper presented at IChemE Conference, "Gasification: The Clean Choice for Carbon Management," Noordwijk, 2002.

Schingnitz, M., Gaudig, U., McVey, I., and Wood, K. "Gasifier to Convert Nitrogen Waste Organics at Seal Sands, UK." Paper presented at IChemE Conference "Gasification for the Future," Noordwijk, April 2000.

Schwager, J., and Whiting, K. J. "European Waste Gasification: Technical and Public Policy Trends and Developments." Paper presented at Gasification Technologies Conference, San Francisco, October 2002.

Sendin, U. "Proyecto de Gasificación Integrada en Ciclo Combinado de Puertollano." Paper presented at IMP Gasification Colloquium, Mexico City, February 1994.

SFA Pacific Gasification Database, SFA Pacific Inc., Mountain View, CA, 2001.

Simbeck, D., and Johnson, H. "World Gasification Survey: Industry Trends and Developments." Paper presented at Gasification Technologies Conference, San Francisco, 2001.

Singh, B., Miadonye, A., and Puttagunta, V. R. "Heavy Oil Viscosity Range from One Test." *Hydrocarbon Processing* 75(8) (August 1993):157–162.

Smil, V. *Enriching the Earth.* Cambridge, Mass.: MIT Press, 2001.

Smoot, L. D., and Smith, P. J. *Coal Combustion and Gasification.* New York: Plenum, 1985.

Soyez, W. "Slag-Related Risks in Partial Oxidation Plants." Paper presented at AIChE Ammonia Safety Meeting, Denver, 1988.

Spare, P. H. "Biomass: What about Health and Safety?" *Energy World* (June 1997).

Speight, J. G. *The Chemistry and Technology of Petroleum.* 3rd ed. New York: Marcel Dekker, 1998.

Speight, J. G. *Fuel Science and Technology Handbook.* New York: Marcel Dekker, 1990.

Speight, J. G. *The Chemistry and Technology of Coal.* New York: Marcel Dekker, 1983

Speight, J. G. *The Desulfurization of Heavy Oils and Residua.* New York: Marcel Dekker, 1981.

Stultz, S. C., and Kitto, J. B. *Steam: Its Generation and Use, 40th ed.* Barberton, Ohio: Babcock & Wilcox, 1992.

Supp, E. *How to Produce Methanol from Coal*. Berlin: Springer, 1990.

van der Burgt, M. J. "The Role of Biomass as an Energy Carrier for the Future." *Energy World* 246 (Feb. 1997):16–17.

Wiegner, P., Martens, F., Uhlenberg, J., and Wolff, J. "Increased Flexibility of Shell Gasification Plant." Paper presented at IChemE Conference, "Gasification: The Clean Choice for Carbon Management," Noordwijk, April 2002.

Zhao, S., Kotlyar, L. S., Sparks, B. D., Woods, J. R., Gao, J., and Chung, K. H. "Solids Contents, Properties, and Molecular Structures of Asphaltenes from Different Oilsands." *Fuel* 80 (2001):1907–1914.

Chapter 5

Gasification Processes

In the practical realization of gasification processes a broad range of reactor types has been and continues to be used. For most purposes these reactor types can be grouped into one of three categories: moving-bed gasifiers, fluid-bed gasifiers, and entrained-flow gasifiers. The gasifiers in each of these three categories share certain characteristics that differentiate them from gasifiers in other categories. Some of these characteristics are summarized in Table 5-1.

Moving-bed gasifiers (sometimes called fixed-bed gasifiers) are characterized by a bed in which the coal moves slowly downward under gravity as it is gasified, generally by a counter-current blast. In such a counter-current arrangement, the hot synthesis gas from the gasification zone is used to preheat and pyrolyse the downward flowing coal. With this process the oxygen consumption is very low but pyrolysis products are present in the product synthesis gas. The outlet temperature of the synthesis gas is generally low, even if high slagging temperatures are reached in the heart of the bed. Moving-bed processes operate on lump coal. An excessive amount of fines, particularly if the coal has strong caking properties, can block the passage of the upflowing syngas.

Fluid-bed gasifiers offer extremely good mixing between feed and oxidant, which promotes both heat and mass transfer. This ensures an even distribution of material in the bed, and hence a certain amount of only partially reacted fuel is inevitably removed with the ash. This places a limitation on the carbon conversion of fluid-bed processes. The operation of fluid-bed gasifiers is generally restricted to temperatures below the softening point of the ash, since ash slagging will disturb the fluidization of the bed. Some attempts have been made to operate into the ash-softening zone to promote a limited and controlled agglomeration of ash with the aim of increasing carbon conversion. Sizing of the particles in the feed is critical; material that is too fine will tend to become entrained in the syngas and leave the bed overhead. This is usually partially captured in a cyclone and returned to the bed. The lower temperature operation of fluid-bed processes means that they are more suited for gasifying reactive feedstocks, such as low-rank coals and biomass.

Entrained-flow gasifiers operate with feed and blast in co-current flow. The residence time in these processes is short (a few seconds). The feed is ground to a size of $100 \mu m$ or less to promote mass transfer and allow transport in the gas. Given the short residence time, high temperatures are required to ensure a good conversion,

Table 5-1
Characteristics of Different Categories of Gasification Process

Category	Moving-Bed		Fluid-Bed		Entrained-Flow
Ash conditions	Dry ash	Slagging	Dry ash	Agglomerating	Slagging
Typical processes	Lurgi	BGL	Winkler, HTW, CFB	KRW, U-Gas	Shell, Texaco, E-Gas, GSP, KT
Feed characteristics					
Size	6–50 mm	6–50 mm	6–10 mm	6–10 mm	<100 μm
Acceptability of fines	limited	better than dry ash	good	better	unlimited
Acceptability of caking coal	yes (with stirrer)	yes	possibly	yes	yes
Preferred coal rank	any	high	low	any	any
Operating characteristics					
Outlet gas temperature	low (425–650°C)	low (425–650°C)	moderate (900–1050°C)	moderate (900–1050°C)	high (1250–1600°C)
Oxidant demand	low	low	moderate	moderate	high
Steam demand	high	low	moderate	moderate	low
Other characteristics	hydrocarbons in gas	hydrocarbons in gas	lower carbon conversion	lower carbon conversion	pure gas, high carbon conversion

Source: Adapted from Simbeck et al. 1993

and therefore all entrained-flow gasifiers operate in the slagging range. The high temperature operation creates a high oxygen demand for this type of process. Entrained-flow gasifiers do not have any specific technical limitations on the type of coal used, although coals with a high moisture or ash content will drive the oxygen consumption to levels where alternative processes may have an economic advantage.

There are one or two processes that do not fit into any of these three main categories. This includes in situ gasification of coal in the underground seam as well as molten bath processes. These are discussed in Section 5.8.

One important point to note throughout all the above is the significance of the slagging behavior of the ash. At temperatures above the ash softening point, the ash becomes sticky and will agglomerate, causing blockage of beds or fouling of heat-exchange equipment. Once above the slagging temperature, at which point the ash has a fully liquid behavior with a low viscosity, it is possible again to remove it from the system reliably. Thus, for all processes, there is a feedstock-specific "no-go" temperature range between the softening and slagging temperatures of the ash.

5.1 MOVING-BED PROCESSES

Historically, moving-bed processes are the oldest processes, and two in particular, the producer gas process and the water gas process, played an important role in the early production of synthesis gas from coal and coke. The preferred feedstocks were in general coke or anthracite, as otherwise the gas needs extensive cleaning to remove tars from the gas. Both processes operate at atmospheric pressure.

Producer Gas

In the producer gas process, humidified air is blown upward through a deep bed of coal or coke. The coal is fed from the top and moves slowly downwards as it is consumed. Ash is drawn off at the bottom of the reactor from a rotating grate. The air reacts with the coal, thereby producing a gas with a lower heating value of 6500 kJ/m^3. The presence of about 50% nitrogen in the product gas is the main cause for this low value. The advantage of the process was that it was continuous. The main drawback is that because of its low heating value, producer gas cannot be transported over long distances. When using low-rank feedstocks such as wood, the heating value of the gas can be as low as 3500 kJ/m^3. Gas producers continue to be built in small numbers today where gas is required locally for firing industrial furnaces, particularly in locations where coal is the only available energy form as in parts of China (Wellman undated).

Water Gas

The water gas process is a discontinuous process in which steam reacts with red-hot coke to form hydrogen and carbon monoxide. First, the coal or coke bed is heated by blowing air upward through the bed to temperatures of about 1300°C. Then the air flow is stopped and steam is passed through the coal or coke bed, first upward and then downward, thereby producing synthesis gas. The reason for the two different directions of steam flow was to make optimum use of the heat in the bed, which is required do drive the endothermic water gas reaction. When the temperature has dropped to about 900°C, the steam "run" is stopped and the cycle is repeated. After purification the syngas can be used for the synthesis of ammonia or methanol. Before the advent of large-scale air separation plants, the water gas process was the only means for making high-grade synthesis gas for chemical purposes. In order to obtain a continuous gas flow, at least three water gas reactors are required.

5.1.1 The Lurgi Dry Ash Process

The patent for the Lurgi dry ash process, or "coal pressure gasification" as it was originally known ("Vergasung von Braunkohle unter Druck mit Sauerstoff-Wasser-dampf-Gemischen," or "Gasification of browncoal under pressure with oxygen-steam mixtures"), was granted in 1927. In 1931 Lurgi started to develop, a pressurized version of existing atmospheric producer gas technology, using oxygen as blast, initially for gasification of lignite (Lurgi 1970; Lurgi 1997). The development was made in close cooperation with the Technical University in Berlin under the direction of Professor Rudolf Drawe, one of the fathers of modern gasification. The first commercial application was built in 1936 (Rudolf 1984). In 1944 two large-scale plants were in operation in Böhlen (Saxony) and Brüx (now Most, Czech Republic) producing town gas. The position of the Lurgi dry ash process as the only pressure gasification system for many years, together with continued development, for example, in scale-up, automation, and operational optimization, have all contributed to the commercial success of this technology.

Standardized reactor sizes are shown in Table 5-2.

These figures for throughput and gas production are nominal figures, which are substantially influenced by the quality of coal processed. For instance, Sasol reports

Table 5-2
Sizes and Capacities of Lurgi Dry Ash Gasifiers

Type	Nominal Diameter (m)	Coal Throughput (t/h)	Gas Production (Dry) (1000 Nm^3/day)
MK III	3	20	1000
MK IV	4	40	1750
MK V	5	60	2750

Source: Rudolf 1984

achieving 30% excess throughput compared with the original rated design (Erasmus and van Nierop 2002).

Most commercial gasifiers operate at pressures of 25–30 bar. A demonstration plant has been operated at pressures of 100 bar (see Section 5.1.3).

Process Description

The heart of the Lurgi process is in the reactor, in which the blast and syngas flow upwards in counter-current to the coal feedstock (Figure 5-1).

Coal is loaded from an overhead bunker into a lock hopper that is isolated from the reactor during loading, then closed, pressurized with syngas, and opened to the reactor. The reactor is thus fed on a cyclic basis.

The reactor vessel itself is a double-walled pressure vessel in which the annular space between the two walls is filled with boiling water. This provides intensive cooling of the wall of the reaction space while simultaneously generating steam from the heat lost through the reactor wall. The steam is generated at a pressure similar to the gasification pressure, thus allowing a thin inner wall that enhances the cooling effect.

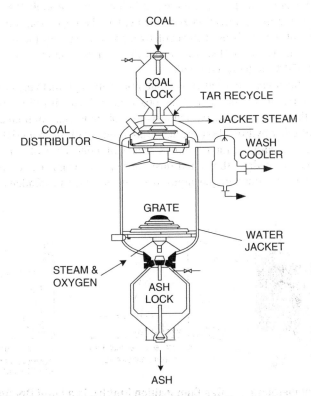

Figure 5-1. Lurgi Dry Ash Gasifier (*Source: Lurgi 1970*)

The coal from the lock hopper is distributed over the area of the reactor by a mechanical distribution device, and then moves slowly down through the bed undergoing the processes of drying, devolatilization, gasification, and combustion. The ash from the combustion of ungasified char is removed from the reactor chamber via a rotating grate and is discharged into an ash lock hopper. In the grate zone the ash is precooled by the incoming blast (oxygen and steam) to about 300–400°C.

The blast enters the reactor at the bottom and is distributed across the bed by the grate. Flowing upward it is preheated by the ash before reaching the combustion zone in which oxygen reacts with the char to CO_2. At this point in the reactor the temperatures reach their highest level (Figure 5-2). The CO_2 and steam then react with the coal in the gasification zone to form carbon monoxide, hydrogen, and methane. The gas composition at the outlet of the gasification zone is governed by the three heterogeneous gasification reactions: water gas, Boudouard, and methanation (for details see Section 2.1).

While describing the process in the form of these four zones, it should be stressed that the transition from one zone to the next is gradual. This is especially noticeable in the transition from the combustion zone to the gasification zone. The endothermic gasification reactions already begin before all the oxygen has been consumed in combustion. Thus the actual peak temperature is lower than that calculated by assuming a pure zonal model (see Figure 5-2) (Gumz 1950; Rudolf 1984).

This gas leaving the gasification zone then enters the upper zones of the reactor where the heat of the gas is used to devolatize, preheat, and dry the incoming coal. In this process the gas is cooled from about 800°C at the outlet of the gasification zone to about 550°C at the reactor outlet.

A result of the counter-current flow is the relatively high methane content of the outlet gas. On the other hand, part of the products of devolatilization are contained unreacted in the synthesis gas, particularly tars, phenols and ammonia, but also a wide spread of other hydrocarbon species. Bulk removal of this material takes place immediately at the outlet of the reactor by means of a quench cooler in which most of the high-boiling hydrocarbons and dust carried over from the reactor are condensed and/or washed out with gas liquor from the downstream condensation stage.

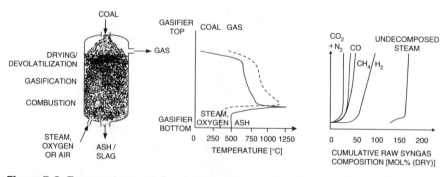

Figure 5-2. Temperature and Gas Composition Profiles in a Lurgi Dry Ash Gasifier (*Source: Adapted from Rudolf 1984; Supp 1990*)

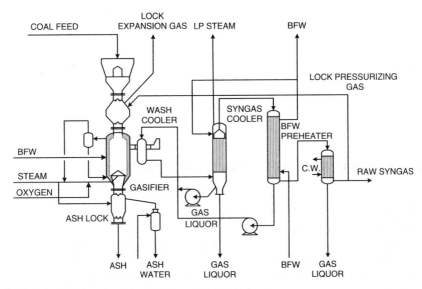

Figure 5-3. Process Flowsheet of Lurgi Dry Ash Gasification (*Source: Supp 1990*)

Gas Liquor Treatment

The gas liquor from the quench cooler typically contains the elements outlined in Table 5-3 (Supp 1990).

There are a number of alternate strategies for handling this material. Lurgi has developed two specific processes for treating it, in particular for allowing the recovery of pure ammonia.

Table 5-3 Gas Liquor Composition		
Suspended Matter	1000	mg/l
Sulfur	600	mg/l
Chloride	50	mg/l
NH_3 and NH_4 ions	10,000	mg/l
Cyanide	50	mg/l
Phenols	1000–5000	mg/l
COD	10,000	mg O_2/l
Source: Supp 1990		

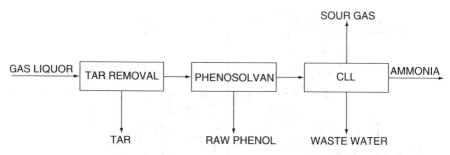

Figure 5-4. Gas Liquor Treatment for Lurgi Dry Ash Gasification

The process sequence (Figure 5-4) includes mechanical tar separation, followed by the Phenosolvan extraction process in which a raw phenol fraction is recovered. Sour gases and ammonia are then selectively and separately stripped from the dephenolated condensate in the CLL (Chemie Linz-Lurgi) unit. The sour gas stream is free of ammonia and so suitable for processing in a standard Claus sulfur recovery unit (see Section 8.4). The stripped condensate contains less than 50 ppm NH_3 and less than 1 ppm H_2S and can be discharged to a biotreater. The ammonia product from the CLL unit contains less than 4% water, less than 0.1 wt% CO_2, and less than 1 ppm H_2S. Further details on these processes can be found in Supp (1990) and Stönner (1989).

Equipment Issues

Both the Lurgi gasifier and the BGL gasifier described in Section 5.1.2 have moving parts that are absent in other gasifiers. For caking coals and sometimes for distribution reasons, a stirrer is installed in the top of the reactor. The distribution function is required to ensure an even depth of the coal bed over the whole cross-section of the reactor. With caking coals the distributor avoids pasting-up of the coal bed due to agglomeration of the coal particles after their temperature becomes higher than the softening point. The dry-bottom gasifier also has a rotating grate at the bottom for the ash outlet.

Moving-bed gasifiers require a relatively high amount of maintenance. However, in virtually all cases many gasifiers operate in parallel, and by proper scheduling the total unit will have a high availability.

Process Performance

An important feature of the Lurgi dry-bottom gasifier is its low oxygen consumption and high steam demand. The exact data depend on the feedstock. Table 5-4 provides data on the basis of three different coals that cover a broad range of coalification or rank.

Table 5-4
Typical Performance Data of Lurgi Dry Ash Gasifier

Type of Coal	Lignite	Bituminous	Anthracite
Components (maf)			
C, wt%	69.50	77.30	92.10
H, wt%	4.87	5.90	2.60
S, wt%	0.43	4.30	3.90
N, wt%	0.75	1.40	0.30
O, wt%	24.45	11.10	1.10
Raw gas composition (dry)			
$CO_2 + H_2S$, mol%	30.4	32.4	30.8
CO, mol%	19.7	15.2	22.1
H_2, mol%	37.2	42.3	40.7
CH_4, mol%	11.8	8.6	5.6
C_nH_m, mol%	0.4	0.8	0.4
N_2, mol%	0.5	0.7	0.4
Feed components			
per 1000 Nm³ CO+H₂			
Coal maf, kg	950	750	680
Steam, kg	1180	1930	1340
Oxygen, Nm³	170	280	300

Source: Supp 1990

Moving-bed gasifiers need graded coal, whereas the total mine output will always contain a large percentage of fines (the more modern the mine, the more fines). Although it is in principle possible to make briquettes with these fines, by binding them with heavy tar that is co-produced in the gasifier, in practice part of the fines cannot be processed.

Heavily caking coals cannot be processed in moving-bed gasifiers. Mildly caking coals require the assistance of the stirrer in order to avoid pasting-up of the bed. Tars and other oxygenated compounds are also produced as by-products. These products form about 25% of the total hydrocarbon feed input in terms of energy. Correcting the oxygen consumption and the CGE (cold gas efficiency) for this fact increases the former and reduces the latter by about one-third. When you add to this that the sensible heat in the gases cannot be used very effectively due to the presence of the tars, the overall efficiencies of the moving-bed processes are not much higher than that of fluid-bed or entrained-flow gasifiers.

Applications

The most notable commercial installation forms part of the Sasol synthetic fuels operation at two sites in South Africa, where in all thirteen MK III, eighty-three MK IV, and one MK V reactors produce a total of 55 million Nm^3/day syngas, the largest gasification complex in the world. The synthesis gas generated produces 170,000 bbl/day of Fischer-Tropsch liquid fuels as well as forming the basis for a substantial chemical industry (Erasmus and van Nierop 2002). The reason that Lurgi dry-ash gasifiers are used in the SASOL complex in South Africa is that at the time the complex was built it was the only pressurized gasifier available and was very suitable for the high ash-melting point coals to be processed.

The Lurgi type moving-bed dry ash gasifier is in widespread use around the world in, apart from South Africa, the United States, Germany, the Czech Republic, and China. The plants in Germany and the Czech Republic use some of the gas to fuel gas turbine combined cycle power plants. In the United States Lurgi dry ash gasifiers are used for the production of substitute natural gas (SNG). In China, the Lurgi process is used for the production of town gas, ammonia, and hydrogen. In the SNG, IGCC, and town gas applications, the high methane content (10–15%) of the product gas is an advantage, since methane is the desired product for SNG, and for IGCC and in town gas applications methane increases the heating value of the gas.

5.1.2   British Gas/Lurgi (BGL) Slagging Gasifier

The BGL slagging gasifier is an extension of the Lurgi pressure gasification technology developed by British Gas and Lurgi with the ash discharge designed for slagging conditions. Initial work took place in the 1950s and 1960s but ceased with the discoveries of natural gas in the North Sea. Work resumed in 1974 after the "oil crisis." An existing Lurgi gasifier in Westfield, Scotland was modified for slagging operation and operated for several years, proving itself with a wide range of coals and other solid feedstocks, such as petroleum coke. The motivation for the development of a slagging version of the existing Lurgi gasifier included a desire to:

- Increase CO and H_2 yields (at the expense of CO_2 and CH_4).
- Increase specific reactor throughput.
- Have a reactor suitable for coals with a low ash-melting point.
- Have a reactor suitable for accepting fines.
- Reduce the steam consumption and consequent gas condensate production.

The decline in interest for coal gasification generally in the 1980s prevented commercialization of the technology. In the mid-1990s the first commercial project was realized at Schwarze Pumpe in Germany, where a mixture of lignite and municipal solid waste (MSW) is gasified within a large complex and the syngas is used for methanol and power production (Hirschfelder, Buttker, and Steiner 1997). A number of other projects also for MSW gasification are currently under consideration.

Process Description

The upper portion of the BGL gasifier is generally similar to that of the Lurgi dry ash gasifier, although for some applications it may be refractory lined and the distributor and stirrer omitted. The bottom is completely redesigned, however, as can be seen in Figure 5-5.

In contrast to the Lurgi dry ash gasifier there is no grate. The grate of the Lurgi dry ash gasifier serves two purposes, distribution of the oxygen-steam mixture and ash removal. In the BGL gasifier the former function is performed by a system of

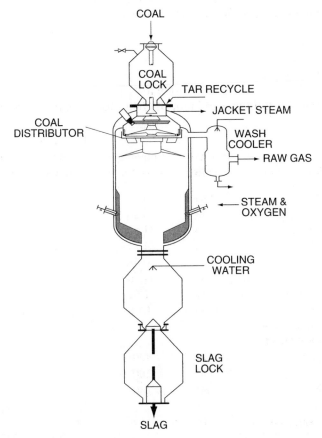

Figure 5-5. BGL Gasifier (*With permission: Lurgi*)

tuyères (water-cooled tubes) located just above the level of the molten-ash bath. The tuyères are also capable of introducing other fuels into the reactor. This includes the pyrolysis products from the raw gas, as well as a degree of coal fines that cannot be introduced at the top of the reactor for fear of blockage.

The lower portion of the reactor incorporates a molten slag bath. The molten ash is drained through a slag tap into the slag quench chamber, where it is quenched with water and solidified. The solid ash is discharged though a slag lock.

Process Performance

A comparison between the performance of the Lurgi dry ash and the BGL gasifier is given in Table 5-5.

The much lower steam and somewhat lower oxygen consumption of the slagging gasifier result in a much higher syngas production per unit of coal intake and a much lower yield of pyrolysis products compared with the dry bottom unit. Further, the CO_2 content of the gas is lower and the methane content in the gas is halved.

Table 5-5
Comparative Performance of Lurgi Dry Ash, BGL, and Ruhr 100 Gasifiers

Raw Gas Composition (Dry)	Dry Ash Gasifier	Slagging Gasifier	Ruhr 100 90 Bar Carb. Gas	Ruhr 100 90 Bar Clear Gas
CO_2, mol%	30.89	3.46	29.52	35.47
CO, mol%	15.18	54.96	18.15	17.20
H_2, mol%	42.15	31.54	35.11	39.13
CH_4, mol%	8.64	4.54	15.78	7.93
C_nH_m, mol%	0.79	0.48	1.02	0.04
N_2, mol%	0.68	3.35	0.35	0.19
H_2S+COS, mol%	1.31	1.31		
NH_3, mol%	0.36	0.36		
Feed components per 1000 Nm³ CO+H₂				
Coal maf, kg	750	520		
Steam, kg	1930	200		
Oxygen, Nm³	280	230		
kg pyrolysis products/1000 kg coal	81	19		

Source: Supp 1990; Lohmann and Langhoff 1982

Waste Gasification

As mentioned above, the BGL gasifier has found a commercial application in waste gasification at the SVZ Schwarze Pumpe plant near Dresden in Germany. A 3.6 m diameter BGL gasifier is fed with a mixed fuel of various wastes and coal. During test operation the unit has gasified a mixture of 25% hard coal, 45% refuse derived fuel (RDF) pellets, 10% plastic waste, 10% wood, and 10% tar sludge pellets.

During the performance test the following production and consumption figures were achieved (Greil et al. 2002):

Mixed Fuel (ar)	30	t/h
Specific oxygen consumption	<0.2	Nm^3/Nm^3 dry syngas
Gasification agent ratio steam/oxygen	<1.0	kg/Nm^3 dry syngas
Raw syngas (dry)	31,500	Nm^3/h

Gasification pressure is 25 bar. The ash content of the feed is discharged as a glassy frit. Analytical tests have demonstrated the nonleachability of the slag, and that disposal requirements, according to TA Siedlungsabfall, class 1 (German waste regulations), have been met.

5.1.3   Ruhr 100

Another development of the Lurgi dry-bottom gasifier was the high-pressure Ruhr 100 gasifier which was designed for operation at 100 bar. Apart from its high-pressure capability, the reactor contained a number of other new features. It had two coal lock hoppers, which reduces the lock gas losses by half by operating alternately. This was implemented to counter the rise in such losses, which accompanies the rise in operating pressure. It had the additional effect of simplifying the drive arrangement for the stirrer, which could now be mounted centrally. Another new feature was the addition of a second gas tapping at an intermediate level between the gasification and devolatilization zones. Not all the gas volume is required to achieve devolatilization of the feed, so the excess is drawn off from the bed as "clear gas." The measured compositions of the carbonization gas containing the volatiles and the clear gas are given in Table 5-5.

A 240 t/d pilot plant was started up in 1979. By September 1981 many of the most important goals of the tests had already been confirmed. Increasing the pressure from 25 bar to 95 bar, approximately doubled the throughput of the reactor. The pressure increase also increased the methane production from about 9 mol% in the product gas to 16 mol% (Lohmann and Langhoff 1982).

5.2   FLUID-BED GASIFIERS

The history and development of coal gasification and fluid-bed technology have been intimately linked since the development of the Winkler process in the early 1920s. Winkler's process operated in a fluidization regime, where a clear distinction exists between the dense phase or bed and the freeboard where the solid particles disengage from the gas. This regime is the classic or stationary fluid bed. With increasing gas velocity a point is reached where all the solid particles are carried with the gas and full pneumatic transport is achieved. At intermediate gas velocities the differential velocity between gas and solids reaches a maximum, and this regime of high, so-called "slip velocity" is known as the circulating fluid bed. Over the years gasification processes have been developed using all three regimes, each process exploiting the particular characteristics of a regime to the application targeted by the process development. These differences are portrayed in Figure 5-6.

In fluid-bed gasification processes the blast has two functions: that of a reactant and that of the fluidizing medium for the bed. Such solutions, where one variable has to accomplish more than one function, will tend to complicate or place limitations on the operation of the gasifier, as in for example, turndown ratio. These problems are especially severe during start-up and shutdown. In most modern gasification processes oxygen/steam mixtures are used as blast. However, when the gas is to be used for power generation, gasification with air may be applied—and in the case of biomass gasification, it often is.

Some simplified reactor sketches for bubbling fluid beds, circulating fluid beds, and transport reactors are given in Figures 5-9, 5-10, and 5-11. In Figure 5-7, the

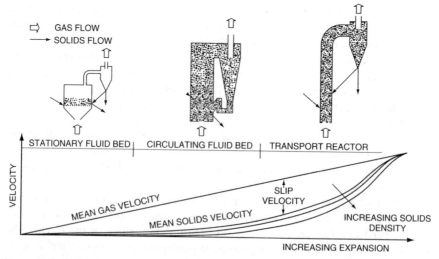

Figure 5-6. Fluid Bed Regimes (*Source: Greil and Hirschfelder 1998*)

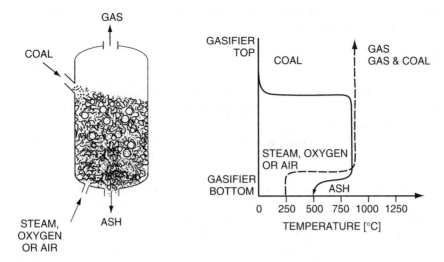

Figure 5-7. Bubbling Fluid-Bed Gasifier (*Source: Simbeck et al. 1993*)

temperature profiles for both the coal and gas are given. Temperature profiles for transport type gasifiers are very complex.

5.2.1   Common Issues

Operating Temperature

Any fluid bed depends on having the solid particles of a size that can be lifted by the upward flowing gas. A large portion (over 95%) of the solids content of the bed of a gasifier is ash, which remains in the bed while the carbon leaves the reactor as syngas. If the ash content of the fuel, be it coal, biomass, or other, should start to soften sufficiently that the individual particles begin to agglomerate, these newly formed larger particles will fall to the bottom of the bed, and their removal poses a considerable problem. For this reason fluid-bed gasifiers all operate at temperatures below the softening point of the ash, which is typically in the range of 950°C–1100°C for coal and 800–950°C for biomass.

On the other hand, the lower the temperature of the gasifier operation the more tar will be produced in the product synthesis gas. This is partly due to the fact that when the coal particles are heated slowly, more volatiles are produced, as discussed in Chapter 3. It is also due to the fact that at lower temperatures there is less thermal cracking of the tars produced.

The best known fluid-bed gasifiers that have no tar problems are regenerators of catalytic cracking units which operate under reducing—that is, gasification—conditions and can be found in some refineries. In these units the carbon residue that remains after gases and volatiles have been extracted from the feed, is gasified in a bubbling fluid-bed.

These operate at a few bar overpressure and temperatures of below 800°C. Maximum throughputs are 500 t/d carbon.

Feed Quality

Historically fluid-bed gasifiers have tended to be operated on low-rank coals such as lignite, peat, or biomass, though not exclusively. This is logical, since low-rank coals have a higher reactivity, which compensates to some degree for the lower temperature. To the extent that the ash properties allow it, operation at a higher temperature improves the ability to process high-rank but less reactive bituminous coals.

Fluid-bed gasifiers need ground coal and therefore have few limitations when it comes to coal size. Particles with a size above 10 mm should be avoided, just as coals with too high a percentage of very small particles. However, because coal particles may differ in both size and shape, it is not an easy material to fluidize. Bubbling beds especially have to be monitored constantly, on the one hand for de-fluidization due to deposition of large coal particles in the bed, and on the other hand for lifting out too many fine particles out of the bed. The latter problem is aggravated as coal particles reduce in size through gasification and are entrained in the hot raw gas as it leaves the reactor. Most of these char and ash particles are recovered in a cyclone and recycled to the reactor, but with too many fines these will choke the system.

Carbon Conversion

The intensive mixing in a fluid bed has its advantages in promoting excellent mass and heat transfer. In fact, a fluid bed approaches the ideal of a continuously stirred tank reactor. This does, however, have its disadvantages. There is wide range of residence times for the individual particles, which are distributed evenly over the whole volume of the bed. Thus removal of fully reacted particles, which consist only of ash, will inevitably be associated with removal of unreacted carbon. The best of existing fluid-bed processes only have a carbon conversion of 97%. This is in contrast to moving-bed and entrained-flow processes, where carbon conversions of 99% can be obtained. Only in pressurized biomass gasification have fluid-bed processes efficiencies of 99% been reported (Kersten 2002).

Many attempts (e.g., Synthane or Hy-Gas) have been and are being made to introduce some staging effect by which the last carbon in the ash to be discharged is converted with (part of) the incoming blast, but this remains difficult as fluid beds lend themselves poorly to staging because of the "no-go" temperature zones mentioned earlier in this chapter.

Ash

The ash of fluid-bed gasifiers comes available in a highly leachable form. This problem is exacerbated when limestone is added in order to bind the sulfur present in the gas and to avoid the need for wet sulfur removal processes. The limestone is never completely converted into gypsum, and hence the ash will always contain unconverted lime. Similar problems are associated with fluid-bed combustion processes. Adding limestone to the fluid bed is an example of doing two things at once. It looks elegant, but you are never free to optimize both desulfurization and gasification or combustion.

Ash particles are removed from below the bed and/or from the cyclones in the top of the bed. In the former case they are sometimes used for preheating the blast. Fluid-bed gasifiers may differ in the manner in which ash is discharged (dry or agglomerated) and in design aspects that improve the carbon conversion. Some "carbon stripping" of the char containing ash to be discharged with incoming oxygen is possible.

Equipment Issues

Although fluid-bed gasifiers operate under nonslagging conditions and no membrane walls are required, the temperatures can run up to 1100°C, and hence insulating brick walls are preferred. These consist of an outer layer of insulating bricks to protect the outer steel shell of the reactor from high temperatures, and an inner layer of more compact bricks that can better withstand the high temperatures and the erosive conditions of the gasifier. Especially in the case of circulating fluid-beds with their complex asymmetrical forms, attention has to be paid to the proper design of the brickwork in relation to thermal expansion.

Reactor Modeling

Whereas bubbling beds may be described fairly accurately as a stirred reactor, the transport-type reactors are more a combination of a stirred reactor and a plug flow reactor.

Most fluid-bed gasifiers use a mixture of oxygen and steam as blast. The oxygen consumption per unit product is higher than those of moving-bed processes and lower than for entrained-flow gasifiers. Some data for typical feedstocks calculated for 1100°C are given in Table 5-6.

5.2.2 The Winkler Process

The Winkler atmospheric fluid-bed process was the first modern continuous gasification process using oxygen rather than air as blast. The process was patented in 1922 and the first plant built in 1925. Since then some 70 reactors have been built and brought into commercial service with a total capacity of about 20 million Nm³/d (Bögner and Wintrup 1984). This process is now, however, only of historic interest, since all but one of these plants has been shut down almost entirely for economic reasons. (See Figure 5-8.)

Table 5-6
Typical Performance for Fluid-Bed Gasifiers for Different Feedstocks and Blasts

Type of Coal		Biomass	Lignite	Bituminous	Bituminous
Blast		Air	O_2/steam	O_2/steam	Air
Temperature	°C	900	1000	1000	1000
Pressure	bar	30	30	30	30
Components (maf)					
C, wt%		50.45	66.66	81.65	81.65
H, wt%		5.62	4.87	5.68	5.68
S, wt%		0.10	0.41	1.13	1.13
N, wt%		0.10	1.14	1.71	1.71
O, wt%		43.73	26.92	9.83	9.83
Raw gas composition(dry)					
CO_2, mol%		6.7	6.2	5.3	1.9
CO, mol%		31.0	56.7	52.0	30.7
H_2, mol%		18.9	32.8	37.3	18.7
CH_4, mol%		2.1	2.6	3.5	0.9
A, mol%		0.5	0.6	0.6	0.6
N_2, mol%		40.8	0.9	1.0	47.0
H_2S, mol%		0.03	0.2	0.3	0.2
Feed components per 1000 Nm^3 H$_2$ + CO					
Fuel maf, kg		893	777	517	516
Steam, kg		0	1	213	112
Air or Oxygen, Nm^3		1358	339	324	1581

Note: Biomass is dry wood. Oxygen purity is 95 mol%. Air preheat is 400°C, Oxygen preheat is 200°C.

The Winkler process is operable with practically any fuel. Commercial plants have operated on browncoal coke, as well as on sub-bituminous and bituminous coals. Coal preparation requires milling to a particle size below 10 mm but does not require drying if the moisture content is below 10%. The feed is conveyed into the gasifier or generator by a screw conveyor. The fluid bed is maintained by the blast, which enters the reactor via a conical grate area at the base. An additional amount of blast is fed in above the bed to assist gasification of small, entrained coal particles. This also raises the temperature above that of the bed itself, thus reducing the tar content of the syngas. The reactor itself is refractory lined. Operation temperature is maintained below the ash melting point. Most commercial plants have operated between 950 and 1050°C. At maximum load the gas velocity in a Winkler generator

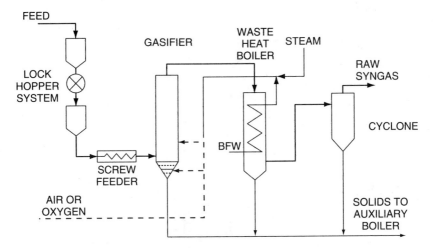

Figure 5-8. Winkler Atmospheric Fluid-Bed Gasification

is about 5 m/s. The flow sheet incorporates a radiant waste heat boiler and a cyclone to remove the ash. The ash contains a considerable amount of unreacted carbon—over 20% loss on feed (Kunii and Levenspiel 1991)—which can be burnt in an auxiliary boiler. Final solids removal is effected with a water wash (not shown in Figure 5-8).

5.2.3 The High-Temperature Winkler (HTW) Process

The name "high-temperature Winkler" for the process developed by Rheinbraun primarily for lignite gasification is to some extent a misnomer. The most important development vis-à-vis the original Winkler process is the increase of pressure, which has now been demonstrated at 30 bar.

Rheinbraun, an important lignite producer in Germany, began work on the process in the 1970s. A Rheinbraun subsidiary, Union Kraftstoff, Wesseling, had operated atmospheric Winkler generators between 1956 and 1964, and Rheinbraun was able to build on the experience gained.

An important motivation for the initial development was to ensure the availability of a suitable process to utilize existing lignite reserves, should economic conditions (i.e., the price of oil) justify it. The focus was on methanol syngas and hydrogen generation for the parallel development of a hydrogenating gasifier for SNG production.

Development goals included raising the pressure so as to increase output and reduce compression energy, raising operating temperatures so as to improve gas quality and carbon conversion, and to include a solids recycle from the cyclone to the fluid-bed as a further measure to increase carbon conversion (Teggers and Theis 1980).

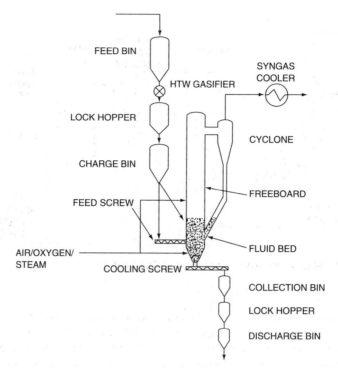

Figure 5-9. HTW Gasifier

The feed system comprises a lock hopper for pressurizing and a screw feeder for the transport of coal from the high pressure hopper to the gasifier.

The HTW process includes heat recovery in a syngas cooler in which the raw synthesis gas is cooled from 900°C to about 300°C. Both fire tube and water tube concepts have been used in the demonstration plants, and selection is based on project specific criteria, such as desired steam pressure (Renzenbrink et al. 1998).

A ceramic candle filter is used downstream of the syngas cooler for particulate removal.

The 600 t/d, 10 bar demonstration unit in Berrenrath, which was operated over 12 years and achieved an availability of 84%, was used to supply gas to a commercial methanol plant. This provided the basis for a plant of similar size operating at 13.5 bar and gasifying mainly peat in Oulu, Finland. A further 160 t/d pilot plant was built in Wesseling to prove various aspects connected with IGCC applications, such as a higher pressure (25 bar). These three plants have now been shut down. Two 980 t/d units are currently under consideration for an IGCC facility at Vřesová in the Czech Republic, where they will replace twenty-six existing fixed-bed gasifiers (Bucko et al. 2000).

In addition to the above, the HTW process can be applied to waste gasification (Adlhoch et al. 2000). A 20 t/d atmospheric demonstration unit has been built in Japan.

5.2.4 Circulating Fluid-Bed (CFB) Processes

The characteristics of a circulating fluid bed combine many advantages of the stationary fluid bed and the transport reactor. The high-slip velocities ensure good mixing of gas and solids, and thus promote excellent heat and mass transfer. Small particles are converted in one pass, or are entrained, separated from the gas, and returned via an external recycle. Larger particles are consumed more slowly and are recycled internally inside the bed until they are small enough for external recycling. The CFB operates with a much higher circulation rate than a classical stationary bed, thus creating a specific advantage in the higher heating rate experienced by the incoming feed particles. This reduces significantly the tar formation during the heating up process.

More recently the focus of new designs for fluid-bed gasifiers has shifted from the lower-velocity bubbling beds to higher-velocity circulating or transport-type designs, which feature higher char circulation rates with consequent improvements to the overall carbon conversion. Another advantage of the circulating fluid beds is that the size and shape of the particles is less important. This is one of the reasons why this type of gasifier is eminently suitable for the gasification of biomass and wastes of which the size, shape, and hence the fluidizing characteristics are even more difficult to control than of coal.

There are two companies offering CFB gasification systems, Lurgi and Foster Wheeler. The fundamental technologies supplied by these two companies are similar, although there are naturally a number of differences in detail.

Lurgi's atmospheric CFB technology (Figure 5-10) was originally developed for alumina calcination and later, during the 1980s, it was adapted for the combustion of coal. It has since been applied to the gasification of biomass.

The CFB system comprises the reactor, an integral recycle cyclone, and a seal pot. The high gas velocities (5–8 m/s) ensure that most of the larger particles are entrained and leave the reactor overhead. The solids separated from the gas in the cyclone are returned to the reactor via the seal pot. The gasifying agent, usually air, is fed as primary air through the nozzle grate and as secondary air at a level above the fuel supply point. For biomass applications the fuel must undergo size reduction to 25–50 mm (Greil et al. 2002).

The Foster Wheeler circulating fluid-bed technology (originally Ahlstrom) was developed in the environment of the Scandinavian forestry industry and is discussed further in Section 5.5.

5.2.5 The KBR Transport Gasifier

Fluid-bed gasification is also being developed in the high-velocity regime. Such a gasifier is the Kellogg Brown and Root (KBR) transport gasifier, for which a gas velocity in the riser of 11–18 m/s is reported (Smith et al. 2002). The objective of this development is to demonstrate higher circulation rates, velocities, and riser

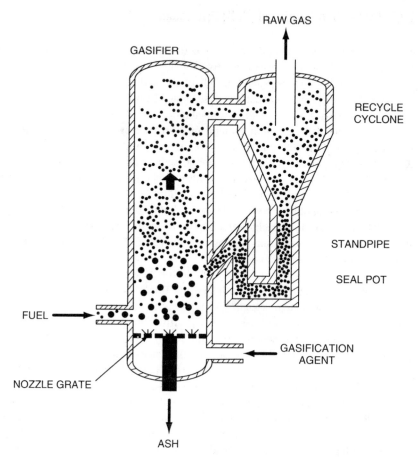

RAW GAS

GASIFIER

RECYCLE
CYCLONE

STANDPIPE

SEAL POT

FUEL

GASIFICATION
AGENT

NOZZLE GRATE

ASH

Figure 5-10. Lurgi Circulating Fluid-Bed Gasifier (*Source: Greil and Hirschfelder 1998*)

densities than in conventional circulating beds, resulting in higher throughput and better mixing and heat transfer rates.

The fuel and sorbent (limestone for sulfur removal) are fed to the reactor through separate lock hoppers. They are mixed in the mixing zone with oxidant and steam, and with recirculated solids from the standpipe. The gas with entrained solids moves up from the mixing zone into the riser. The riser outlet makes two turns before entering the disengager, where larger particles are removed by gravity separation. Smaller particles are largely removed from the gas in the cyclone. The solids collected by the disengager and cyclone are recycled to the mixing zone via the standpipe and J-leg.

The gas is cooled in a syngas cooler prior to fine particulate removal in a candle filter. In the demonstration facility, both ceramic and sintered metal candles have

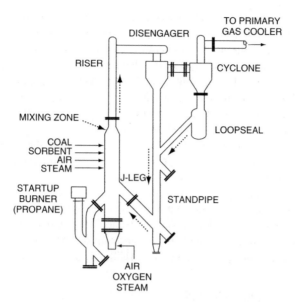

Figure 5-11. KBR Transport Gasifier (*Source: Smith et al. 2002*)

been tested. The temperature of the test filter can be varied between 370 and 870°C by bypassing the syngas cooler.

The sorbent added to the fuel reacts with the sulfur present to form CaS. Together with a char-ash mixture, this leaves the reactor from the standpipe via a screw cooler. These solids and the fines from the candle filter are combusted in an atmospheric fluid-bed combustor.

The transport reactor was operated in combustion mode from 1997 to 1999. Between September 1999 and 2002 the plant was operated over 3000 hours in gasification mode, most of this with air as oxidant. Average carbon conversion rates are about 95%, and values of up to 98% have been achieved.

Gasification takes place at 900–1000°C and pressures between 11 and 18 bar. There has been little oxygen-blown operation to date. All oxygen data has been taken at lower pressures.

5.2.6 Agglomerating Fluid-Bed Processes

The idea behind agglomerating fluid-bed processes is to have a localized area of higher temperature where the ash reaches its softening point and can begin to fuse. The purpose of this concept is to allow a limited agglomeration of ash particles that, as they grow, become too heavy to remain in the bed and fall out at the bottom. This preferential separation of low-carbon ash particles is designed to permit a higher carbon conversion than conventional fluid-bed processes.

A potential advantage for such processes over conventional fluid beds is that the problem of a leachable ash is less serious because of the ash agglomeration step incorporated near the burner(s) in the bottom of the reactor. The burner(s) in these gasifiers are in fact oxygen/air lances that have two functions: that of introducing the fluidizing gas, and also creating a hot region where ash agglomeration occurs. As mentioned before, such two-in-one features are nice but always put restrictions on the operation as one tries to operate in the "no-go" temperature range between the ash softening point and the ash melting point.

Two processes have been developed using this principle: the U-gas technology developed by the Institute of Gas Technology (IGT), now offered by Carbona, and the Kellogg Rust Westinghouse (KRW) process.

Several U-gas gasifiers operating at about 4 bar have been installed in China. A description of the process is contained in Reimert and Schaub (1989).

The 100-MW IGCC Piñon Pine Plant near Reno, Nevada, in the United States uses the KRW process. This plant could not be started up successfully, primarily because of difficulties in the hot gas filtering section (U.S. Department of Energy 2002).

5.2.7   Development Potential

Many fluid-bed gasification processes have been and are being developed, but none of them incorporate the use of a heat carrier (sand and/or ash and/or char) in such a way that the tars are combusted and char reacts with the gasifying agent to produce synthesis gas and/or fuel gas. Such a system could produce pure syngas without the need for an air separation unit (ASU). Moreover, the gas is free of tars and the carbon conversion is virtually complete.

An example for a simplified process scheme in which the problems with tar are circumvented is shown in Figure 5-12 (Holt and van der Burgt 1997). The feed coal is fed to a bubbling fluid-bed pyrolyser, which is fluidized with a small amount of air and/or steam and uses a relatively small part of the hot heat carrier from the top of the riser (entrained bed) as a heat source. The complete or partial combustion of all gases and tars from the pyrolyser takes place with air in the bottom of the riser reactor. Moreover all residual carbon left on the heat carrier leaving the gasifier proper is (partially) combusted in the riser. The hot gases leaving the top of the riser via cyclones have the typical composition of a low Btu fuel gas or of a flue gas, depending on whether partial or complete combustion is used. Ash that is virtually free of carbon is removed from the system as a bleed from the cyclones in the top of the riser. In principle, two modes of operation are possible.

When synthesis gas is the required product, most of the hot heat carrier leaving the top of the riser is used in the endothermic gasifier section where the char leaving the pyrolyser reacts with steam according to the water gas reaction. When insufficient pyrolysis products are available for (partial) combustion in the riser, some additional coal and/or oil may be injected into the bottom of the riser or the char slip from the gasifier to the riser may be increased. Where fuel gas is the required product the gasifier reactor may be omitted.

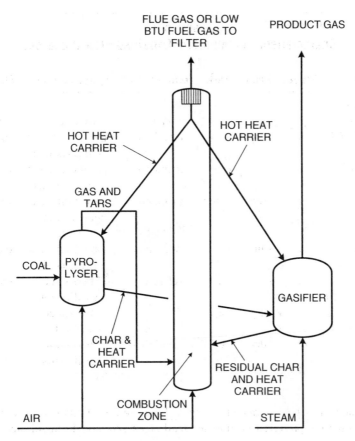

Figure 5-12. Modified Twin Reactor System

5.3 ENTRAINED-FLOW GASIFIERS

As discussed at the beginning of this chapter, the principal advantages of using entrained-flow are the ability to handle practically any coal as feedstock and to produce a clean, tar-free gas. Additionally, the ash is produced in the form of an inert slag or frit. This is achieved with the penalty of a high oxygen consumption, especially in the case of coal-water slurries or coals with a high moisture or ash content, as well as additional effort in coal preparation.

Nonetheless, even if entrained-flow has been selected as the means of contacting the fuel and gasification agent, this still leaves a considerable variety of alternatives open in the design approach, as can be judged from Table 5-7, which outlines characteristics of some important entrained-flow processes.

The majority of the most successful coal gasification processes that have been developed after 1950 are entrained-flow slagging gasifiers operating at pressures of

Table 5-7
Characteristics of Important Entrained-Flow Processes

Process	Stages	Feed	Flow	Reactor Wall	Syngas Cooling	Oxidant
Koppers-Totzek	1	dry	up	jacket	syngas cooler	oxygen
Shell SCGP	1	dry	up	membrane	gas quench and syngas cooler	oxygen
Prenflo	1	dry	up	membrane	gas quench and syngas cooler	oxygen
Noell	1	dry	down	membrane	water quench and/or syngas cooler	oxygen
Texaco	1	slurry	down	refractory	water quench or syngas cooler	oxygen
E-Gas	2	slurry	up	refractory	two-stage gasification	oxygen
CCP (Japan)	2	dry	up		two-stage gasification	air
Eagle	2	dry	up	membrane	two-stage gasification	oxygen

20–70 bar and at high temperatures of at least 1400°C. Entrained-flow gasifiers have become the preferred gasifier for hard coals and have been selected for the majority of commercial-sized IGCC applications.

In entrained-flow gasifiers the fine coal particles react with the concurrently flowing steam and oxygen. All entrained-flow gasifiers are of the slagging type, which implies that the operating temperature is above the ash melting point. This ensures the destruction of tars and oils and, if appropriately designed and operated, a high carbon conversion of over 99% although some water-slurry feed plants do not achieve this. Moreover, entrained-flow gasifiers produce the highest quality synthesis gas because of the low methane content. Entrained-flow gasifiers have relatively high oxygen requirements, and the raw gas has a high sensible heat content. The various designs of entrained-flow gasifiers differ in their feed systems (dry-coal feed in a high-density fluidized state or coal-water slurries), vessel containment for the hot conditions (refractory or membrane wall), configurations for introducing the reactants, and the ways in which sensible heat is recovered from the raw gas. The two best-known types of entrained-flow gasifiers are the top-fired coal-water-slurry feed gasifier, as used in the Texaco process and the dry coal feed side-fired gasifier as developed by Shell and Krupp-Koppers (Prenflo). Furthermore, there is the dry coal feed top-fired Noell gasifier. Simple sketches of these three reactor types are given together with temperature profiles for both the coal/char and the gas in Figures 5-13, 5-14, and 5-15. Some gasifiers (e.g., E-gas) use two stages to improve thermal efficiency and to reduce the sensible heat in the raw gas and to lower the oxygen requirements. In the present coal-water slurry-feed gasifiers, a substantial part of the reactor space is used to evaporate the water of the slurry. This is reflected in the temperature profile of this gasifier (see Figure 5-13).

Contrary to the moving-bed and fluid-bed processes, virtually all types of coals can be used in these processes, provided they are ground to the correct small size. The coals may be heavily caking and may range from sub-bituminous coals to anthracite. Browncoal and lignite can in principle be gasified, but for economic reasons this is not very attractive because of the ballast of inherent moisture that has to be evaporated and heated to the high temperatures required. High-ash coals are also not selected by preference, because all the ash has to be melted and that also constitutes thermal ballast to the gasifier. Coals with very high ash melting points are generally fluxed with limestone in order to lower the ash melting point and hence the operating temperature. This improves the process efficiency, reduces the oxygen consumption, and enables the use of a refractory-lined reactor.

Currently, most entrained-flow gasifiers are single-stage gasifiers. The fuel is introduced together with the blast via one or more burners. The blast is always pure oxygen or a mixture of oxygen and steam.

The ash produced in entrained-flow gasifiers is similar to that of the slagging moving-bed gasifiers and consists of the same fine, black, inert, gritty material.

Reactor Modeling

The upflow reactors, as employed in the Shell Coal Gasification Process (SCGP) and Prenflo reactors, can be essentially considered as continuously stirred tank reactors

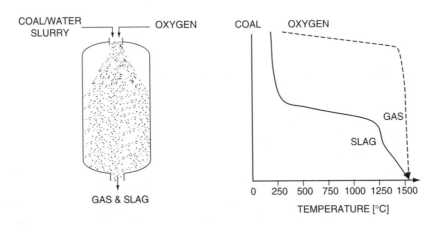

Figure 5-13. Top-Fired Coal-Water Slurry Feed Slagging Entrained-Flow Gasifier

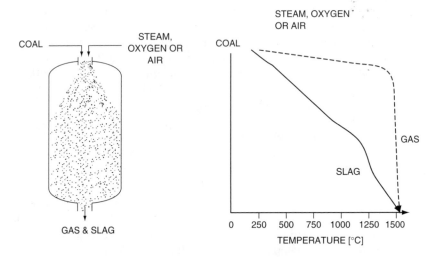

Figure 5-14. Top-Fired Dry-Coal Feed Slagging Entrained-Flow Gasifier

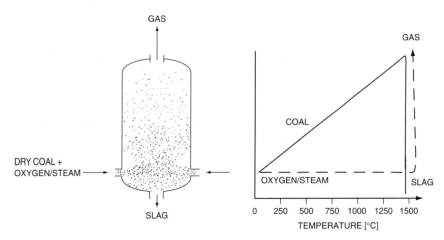

Figure 5-15. Side-Fired Dry-Coal Feed Slagging Entrained-Flow Gasifier

(CSTRs). The reason is the very large recirculation inside the reactor caused by small temperature differences and the way the reactants are introduced. Contrary to the fluid-bed reactors, the carbon conversion is almost 100% because most of the ash leaves the reactor as a slag that is very low in carbon. Moreover, measures can be taken to ensure that large carbon particles tend to remain longer in the reactor. This can be accomplished, as in the GSP gasifier, by introducing some swirl in the top burner or by tangential firing as, for example, in the EAGLE gasifier. The largest coal particles are thus preferentially deposited on the liquid slag flowing vertically downwards along the reactor wall. The coal particles having a lower density than the slag will float like "icebergs" on the slag. The velocity of this slag layer is much lower than of the gas in the reactor, and thus the ideal situation is obtained where the larger coal particles get the longest residence time. Careful design is important, as too much swirl causes reverse flow in the center of the reactor and can lead to unwanted situations.

Modeling of the second nonslagging stage of the E-Gas reactor is also not simple because of the evaporation of water and the pyrolysis reactions. The first slagging stage with the side introduction of coal-water slurry and recycled char in the E-Gas process can again be described as a CSTR.

CFD modeling of entrained-flow reactors has been initiated, and the first published results are encouraging (Bockelie et al. 2002).

5.3.1 General Considerations

Dry-Coal Feed Gasifiers

As discussed in Chapter 2, dry-coal feed gasifiers have the advantage over coal-water slurry feed gasifiers in that they can operate with almost the minimum amount of blast. This implies in practice that they have a 20–25% lower oxygen consumption than coal-water feed gasifiers. Also, as shown in Section 2.4, dry-coal feed entrained-flow gasifiers have in principal an additional degree of freedom that makes it possible to better optimize the synthesis gas production. Moreover, it is possible to adjust the H_2/CO ratio slightly. In practice, operation at a CO_2 content of the gas of 0.5–4 mol% and a temperature of 1500°C is generally adhered to.

Single-Stage Gasifiers. In particular, the single-stage entrained-flow gasifiers yield a high gas purity with only traces of hydrocarbons and with a CH_4 content of well below 0.1 mol%. Together with the low CO_2 and high carbon conversion, this ensures that almost all carbon in the feed is converted into CO, and hence a nonselective acid gas removal can be employed, as the H_2S/CO_2 is such that the combined acid gas may be routed directly to the Claus plant sulfur recovery. Details about the gas treating will be discussed in Chapter 8.

Examples of single-stage dry-coal feed gasifiers are the SCGP process, the Prenflo process, and the GSP process. The dry-coal-feed process SCGP is used in a 250 MW IGCC plant in Buggenum, The Netherlands, and the Prenflo process in a 300 MW IGCC plant in Puertollano, Spain. GSP has a 600 t/d plant operating on a variety of solid and liquid feedstocks in Schwarze Pumpe, Germany.

Process Performance. One of the most striking features of single stage dry-coal entrained-flow slagging gasifiers is that the gas composition is very insensitive to the coal quality. In the case of low-rank coals and high-ash coals, however, the gas yields suffer because of the ballast of water and ash, respectively. The performance of a variety of coals is given in Table 5-8. The coal analyses are those of Table 4-4.

Two-Stage Gasifiers. An improvement in the process efficiency can be obtained by adding a second nonslagging stage to the first slagging stage of an entrained-flow gasifier.

Based on the standardized, idealized conditions of Appendix E, present state of the art single-stage pressurized entrained-flow gasifiers have an efficiency of 50% LHV for IGCC applications (see data in Table 5-9). The process efficiency of these gasifiers can be increased by the introduction of a second nonslagging stage as applied in the CCP or EAGLE processes. The second stage results in a higher cold gas efficiency and lower oxygen consumption. Efficiencies are 50 and 50.9% for the one- and two-stage process-based IGCC, respectively, when operating with a dry-coal feed system (see Table 5-9). (The calculation basis for this data is given in Appendix E.)

This is achieved by operating the first stage under high temperature slagging conditions with only part of the reactants, and then adding the remainder in a second stage where the hot gas drives the endothermic reactions in the second, nonslagging stage. In practice, current processes all operate the first stage with a deficit of coal with a second stage coal feed. In principle, however, alternative staging concepts could be used, e.g. with steam addition as the second stage. Each staging concept has its advantages and disadvantages. With a coal feed second stage there is the fact that a certain amount of tars will inevitably leave the reactor with the syngas. The extent to which this constitutes a real problem depends on the feedstock quality and the actual outlet temperature of the reactor. With steam addition in the second stage, the fuel is present as unconverted char from the first stage so that the syngas would be tar free. On the other hand, in the context of a dry-feed system, this requires considerably more steam than for a single stage reactor. This increases the amount of process condensate to be handled and degrades some of the sensible heat in the gasifier exit gas to condensing temperature levels.

Irrespective of the choice staging arrangement, there are a number of interesting aspects to be considered. The lower temperature of the second stage will require more residence time than for a single stage gasifier and in practice one will have to reckon with some carbon carry over in the syngas. On the other hand precisely these lower temperatures would allow the use of a refractory wall in the second stage, which represents a significant cost saving compared with the membrane wall used in some single stage designs without the exposure to very high temperatures in other singlestage designs.

The second stage is nonslagging. The particulate matter in the syngas contains unreacted carbon and dry ash. This is removed from the gas downstream of the syngas cooler and recycled to the first stage. In this way almost all the ash is removed from the system as slag.

The overall effect is that a slagging gasifier is obtained in which the oxygen consumption is almost as low as for a gasifier operating at the temperature at which the gases leave the second nonslagging stage. This has the following process advantages:

Table 5-8
Performance of Various Types of Coals in Dry-Coal Entrained-Flow Gasifiers

Coal			Gas Analysis of Dry Gas, Mol%						Miscellaneous		
Country	Region	Classification	CO	H_2	CO_2	N_2	A	H_2S	Nm^3 CO+H_2/ ton maf coal	Nm^3 O_2/Nm^3 CO+H_2	kg steam/Nm^3 CO+H_2
Germany	Rhein	Browncoal	61	29	8	1	1	0.2	965	0.33	0
USA	North Dakota	Lignite	62	26	10	1	1	0.1	935	0.36	0
USA	Montana	Sub-bituminous	63	34	1	1	1	0.4	1950	0.26	0.06
USA	Illinois	Bituminous	61	35	1	1	1	1.5	2030	0.25	0.09
Poland	typical	Bituminous	58	39	1	1	1	0.2	2290	0.20	0.15
S. Africa	typical	Bituminous	64	33	1	1	1	0.3	2070	0.26	0.09
China	Datung	Bituminous	66	31	1	1	1	0.2	2060	0.27	0.09
India	typical	Bituminous	62	33	2	1	1	0.5	1730	0.31	0
Australia	typical	Bituminous	62	34	1	1	1	0.3	2100	0.26	0.07
Germany	Ruhr	Anthracite	65	31	1	1	1	0.2	2270	0.26	0.13

Table 5-9
IGCC Efficiencies for Various Entrained-Flow Gasifiers

Process Feed	Syngas Cooling	Gasifier Conditions	IGCC Efficiency, %LHV
Slurry feed	water quench	64 bar 1500°C	37.8
Slurry feed	gas quench	64 bar 1500°C	43.6
Slurry feed 320°C	gas quench	64 bar 1500°C	48.8
Dry feed	gas quench	32 bar 1500°C	50.0
Dry feed	two-stage gas quench	32 bar 1500/1100°C	50.9

Note: Efficiencies are based on the standardized, idealized conditions of Appendix E.

1. No gas quench and a lower syngas cooler duty.
2. A higher cold gas efficiency and a 20% lower oxygen consumption.

The methane content of the gas will slightly increase compared with the single stage gasifier and so will the CO_2 content.

A drawback of incorporating a second-stage to a dry-feed gasifier is the added complexity and the higher steam consumption. The extra efficiency is about one percentage point (see Table 5-8).

Another process advantage that can be incorporated in all slagging processes is a boiling water slag bath where the steam generated can be used for example as process steam or as a quench medium. Further, the slag bath could be used as a sour water stripper.

Coal-Water Slurry-Fed Gasifiers

The big advantage of coal-water slurry feed gasifiers is the more elegant method of pressurizing the coal. Lock hoppers as used in dry-coal feed gasifiers are costly and bulky equipment with complex valve systems that have to provide a gas-tight block in a dusty atmosphere. Pumping a coal-water slurry is not a simple operation either, but it is definitely less complex than lock-hoppering. In addition, the practical ultimate pressure for dry-pulverized-coal lock hoppers is about 50 bar, whereas for coal-water slurry pumps the pressure could, in principle, be as high as 200 bar.

As there is always a surplus of gasifying agent to be added to the gasifier, there is less benefit in adding a nonslagging second stage. Only when coal is added to the second stage, as is done in the E-Gas process, are part of the disadvantages of having so much water in the feed eliminated.

Single-Stage Gasifiers. In a single-stage coal-water slurry-feed gasifier all the water in the slurry must be evaporated and raised to the full outlet temperature of the slagging operation. This imposes a considerable penalty on the cold gas efficiency of the process to offset against the simplified feed system. This is also reflected in a higher oxygen consumption compared with a dry-feed system. Furthermore space in the reactor is required for the evaporation process.

The high water vapour content of the hot, raw syngas also influences its composition. The CO shift reaction is driven further to the right resulting in a higher H_2/CO ratio and higher CO_2 content than the equivalent dry-feed gasifier. Whether this is important or not will depend on the application under consideration. Additionally the methane content will be even lower although if the potential for higher operating pressures is exploited, this would tend to increase the methane content. However the methane content of all single-stage entrained-flow gasifiers is so small that this is unlikely to be of consequence.

Two-stage gasifiers. The main issues surrounding the staging of coal-water slurry-feed gasifiers are similar to those for dry feed gasifiers. The efficiency of a two-stage gasifier is higher than that of a single-stage process for the same reasons as for the dry feed case. The disadvantages of a higher steam consumption described for dry feed gasifiers are however not applicable, as no steam is required in coal-water slurry fed gasifiers.

5.3.2   The Koppers-Totzek Atmospheric Process

Just as with moving-bed and fluid-bed processes, the first entrained-flow slagging gasification process operated at atmospheric pressure. The atmospheric pressure Koppers-Totzek (KT) process was developed in the 1950s, and commercial units were built in Finland, Greece, Turkey, India, South Africa, Zambia, and elsewhere, mostly for ammonia manufacture. The South African unit has been reported as achieving a 95% availability (Krupp-Koppers 1996). In recent years no new units of this type have been built.

Process Description

The KT reactor features side-fired burners for the introduction of coal and oxygen, a top gas outlet, and a bottom outlet for the slag. The early units had a capacity of 5000 Nm³/h and featured two diametrically opposed burners that were situated in horizontal truncated cones (see Figure 5-16). Later units featured four burners that increased the maximum capacity to 32,000 Nm³/h. The gas leaving the top of the gasifier at about 1500°C is quenched with water near the top of the reactor to a temperature of about 900°C so as to render the slag nonsticky before it enters a water tube syngas cooler for the production of steam. The reactor has a steam jacket to protect the reactor shell from high temperatures. A significant portion of the sensible heat is transformed into low-pressure steam in the jacket, which represents a

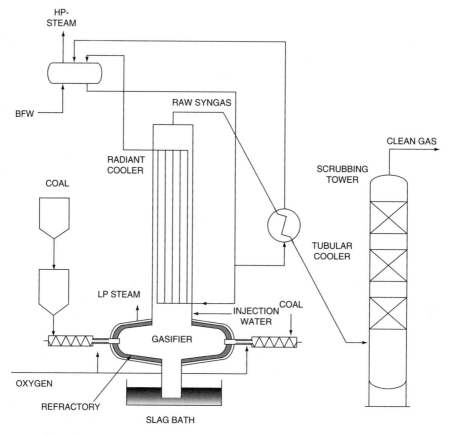

Figure 5-16. Koppers-Totzek Gasifier

considerable energy penalty for the process. The burners are of the premix type
which means that the velocity in the burners must be quite high in order to avoid
flash backs. For pressurized burners, premixing is considered too dangerous and has
therefore never been applied. The slag is quenched and granulated in a water bath
underneath the reactor. The water in the slag bath also provides a water seal to avoid
gas escaping via the bottom of the reactor.

5.3.3   Shell Coal Gasification Process (SCGP) and Prenflo Process

The origin of both these processes goes back to the Koppers-Totzek process. Shell
and Koppers jointly developed a pressurized version of the process, and in 1978
they started to operate a 150 t/d gasifier in Harburg, Germany. The main interest for
Koppers was to have a better process available for the production of syngas. For
Shell the main interest was at the time the production of synthetic fuels from coal by

the route gasification and Fischer-Tropsch synthesis. After the joint Harburg unit, Koppers and Shell decided to develop the process further along separate routes. Subsequently, Shell built a 250 t/d demonstration unit in Houston, and Krupp-Koppers a 48 t/d unit in Fürstenhausen, Germany. Based on the work in these units, two commercial plants were built as part of IGCC power station. In 1994 a 2000 t/d Shell gasification unit was built for Demkolec in Buggenum in The Netherlands using internationally traded coal as original feedstock, and in 1997 Krupp-Koppers built a 3000 t/d Prenflo unit for Elcogas in Puertollano in Spain using a blend of high-ash coal and petcoke as feedstock. Currently, Shell has various projects underway in China.

Process Description

The SCGP and Prenflo processes, which are very similar (Anon 1990), feature an even number (typically four, see Figure 5-17) of diametrically opposed burners in the side-wall at the bottom of the reactor through which the pulverized coal is introduced in a dense phase using an inert gas as carrier gas (van der Burgt and Naber 1983). The small niches in which the burners are placed are a remnant of the truncated

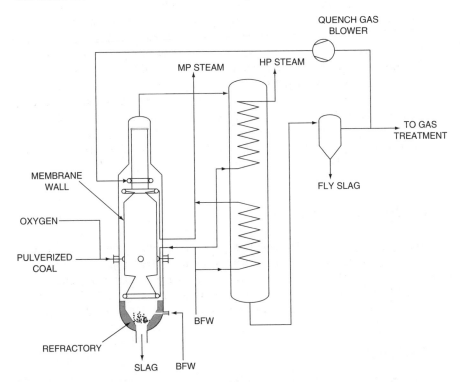

Figure 5-17. Shell Coal Gasification Process (*Source: Adapted from Koopman, Regenbogen, and Zuideveld. Used with permission from Shell. 1993*)

cones of the KT gasifiers. The main gasification volume constitutes the vertical cylindrical part of the gasifier. As a result the gasifiers have become upflow units as far as the gas flow is concerned. Coal is ground in a milling and drying unit to a size of 90% below 90 µm, pressurized in lock hoppers, transported as a dense phase in nitrogen, and mixed near the outlet nozzle of the burner with a mixture of oxygen and steam. The reactions are very fast, and after a residence time of 0.5–4 seconds the product gas leaves the reactor at the top, whereas the slag leaves through an opening in the bottom of the reactor where it is quenched in a water bath. The temperature in the gasifier is typically 1500°C and the pressure 30–40 bar. The sudden drop in temperature when entering the water bath solidifies the slag which breaks up into a fine, inert, glassy, black grit that may replace sand and aggregate in concrete. The granulated slag leaves the gasifier through a lock hopper system where water is the continuous phase.

The reactor wall is a membrane wall construction that is studded and covered with a castible refractory mix that is rammed in, in order to protect the metal wall from the direct radiation and the liquid slag. In the tubes, steam is generated that is used for additional power generation in the combined cycle (Anon. 1990). The heat loss through the wall is dependent on the quantity and quality of the slag and the size of the reactor, and generally lies between 2 and 4% of the heat of combustion of the coal feed. The gas produced consists roughly of two-thirds CO and one-third H_2. The hot gas leaving the slagging reactor is quenched to 900°C with cold recycle gas of 280°C. The quench is designed in such a way that the hot gases and the slag do not come into contact with the wall before they are cooled to a temperature where the slag becomes nonsticky. After the quench, the gas enters a syngas cooler where steam is raised for use in the combined cycle. The gas leaves the syngas cooler at a temperature of about 280°C and passes a candle filter unit where the solids in the gas are removed. About half the gas is then recycled via a recycle gas compressor to be used as quench gas. The other half, constituting the net production, is further cooled in a water scrubbing section.

5.3.4 The Noell Process

The Noell process was first developed by Deutsches Brennstoffinstitut Freiberg in 1975 for the gasification of the local browncoal and other solid fuels. It became known under the name GSP. After the Noell Group acquired the technology in 1991 it was further developed to gasify waste materials and liquid residues. This technology is now owned by Future Energy GmbH, which has revived the old GSP name.

Process Description

The GSP process features a top fired reactor where the reactants are introduced into the reactor through a single centrally mounted burner [Lorson, Schingnitz and Leipnitz 1995]. This concept has a number of special advantages in addition to those generic

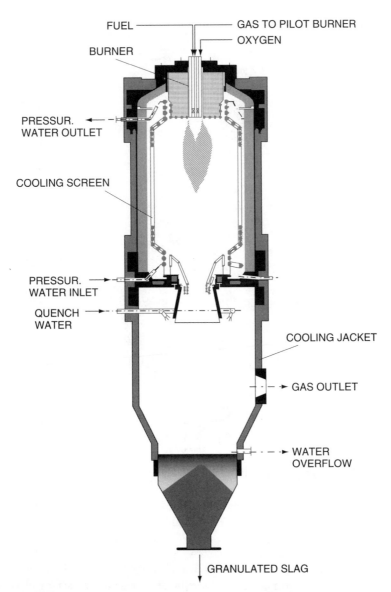

Figure 5-18. (a) GSP Gasifier with Cooling Screen (*Source: Future Energy GmbH*)

to all dry-feed systems. These include the simple rotational-symmetrical construction without penetrations through the cylinder wall, which reduces equipment costs. Secondly, the use of a single burner reduces the number of flows to be controlled to three (coal, oxygen and steam). Thirdly, slag and hot gas leave the gasification section of the reactor together, which reduces any potential for blockages in the slag tap as well as allowing for both partial and total water quenches, depending on application.

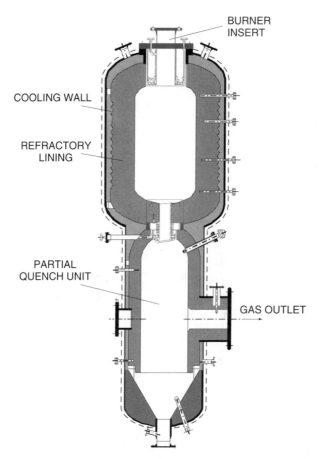

BURNER
INSERT

COOLING WALL

REFRACTORY
LINING

PARTIAL
QUENCH UNIT

GAS OUTLET

Figure 5-18. (b) GSP Gasifier with Cooling Wall (*Source: Future Energy GmbH*)

Within this overall concept there are a number of different variations of reactor design for the GSP gasifier, which can be selected and optimised for different feedstocks. Figure 5-18a shows the reactor with a spirally wound cooling screen, typically used for ash-containing conventional fuels (coal, browncoal) and liquids (residual oils, tars, and sludges). The cooling screen is covered with a SiC castible and a layer of molten slag. A partial quench is included in the lower section of the reactor. A reactor with cooling wall is shown in Figure 5-18b, which is used for applications with low or zero-ash feeds such as gas or organic liquid wastes. A third reactor type, which also uses a cooling screen, but has a total quench has been developed for black liquor gasification.

5.3.5 The Texaco Process

The Texaco gasification process was developed in the late 1940s. Although the main focus at that time was on utilization of natural gas reserves, some work on coal gasification was also performed (Schlinger 1984). The process achieved commercialization initially with gas feed (1950) and later with liquids (1956). This technology is discussed in more detail in Section 5.4. Against the background of a perceived medium-term oil shortage at the beginning of the 1970s, the previous work on coal gasification was taken up again. Both its background from the previous work on coal as well as its decision to maintain many of the concepts already proven in commercial oil gasification service allowed Texaco to develop its coal-gasification technology in a relatively short space of time, despite the many differences in detail between coal and oil gasifiers.

Two demonstration scale projects were operated in the late 1970s (RAG, Holten in Germany and Cool Water, CA), and three commercial-scale facilities were started up between 1983 and 1985 (two for Ube in Japan, and the Eastman plant at Kingsport, Tennessee) (Curran and Tyree 1998).

Since 1990 nine commercial coal-based plants have been brought into service, five in China and four in the United Sates. These have been predominantly for ammonia and electricity production, although two of the Chinese plants are for methanol and town gas production. Three of the United States plants (Coffeyville, El Dorado, and Delaware) use petroleum coke as feed. The fourth, Polk Power Station in Florida, is a 250 MWe IGCC unit, which went on stream in 1996.

Process Description

The Texaco process for coal gasification uses a slurry-feed downflow entrained-flow gasifier. The reactor shell is an uncooled refractory-lined vessel. As with their oil and gas gasification processes, Texaco maintains flexibility in syngas cooling concepts, offering both a radiant boiler and a total quench. The selection between these two alternatives is a matter of economics for the specific application.

The coal or petcoke feedstock is wet milled to a particle size of about 100 μm and slurried in essentially conventional equipment. The slurry is charged to the reactor with a membrane pump. The reactor pressure is typically about 30 bar for IGCC applications, where no gas expander is included in the scheme. For chemical applications it may be as much as 70 to 80 bar. The slurry feed is introduced into the reactor with the oxidant (usually oxygen) through the feed-injector (burner), which is located centrally on the top of the gasifier. The gasification takes place at slagging temperatures, typically about 1500°C, depending on the ash quality of the feed (Figure 5-19).

In the quench configuration, the hot syngas leaves the reactor at the bottom together with the liquid ash and enters the quench chamber. Texaco's quench system provides a total quench, so that the gas leaves the quench chamber fully water-saturated at a temperature of between 200 and 300°C. For chemical applications such as hydrogen or ammonia manufacture, these are suitable conditions for direct CO shift conversion. Particulates and hydrogen chloride are removed from the gas in a hot scrubber before it enters the catalyst bed.

The ash solidifies to a slag in the quench vessel and leaves it via a lock hopper. The water leaving the lock hopper is separated from the slag and recycled for slurry preparation.

In the radiant cooler configuration (Figure 5-20), which was used in the Cool Water and Polk IGCC plants, full use is made of the potential for heat recovery for maximum efficiency. The feed preparation and gasifier are identical to the quench configuration.

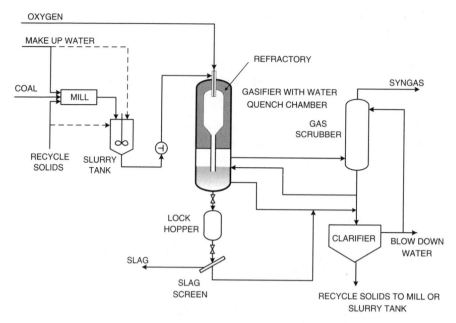

Figure 5-19. Texaco Quench Gasifier (*With permission: ChevronTexaco*)

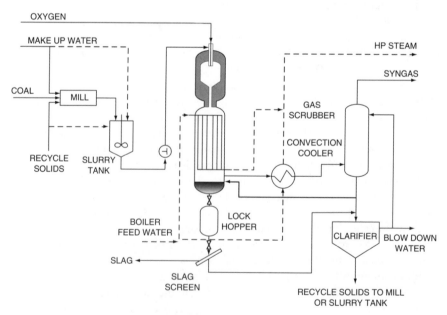

Figure 5-20. Texaco Radiant Cooler Configuration (*With permission: ChevronTexaco*)

The hot syngas leaves the gasifier at the bottom and enters the radiant cooler where it is cooled to about 760°C. The molten slag falls to the quench bath at the bottom of the cooler where it solidifies. As with quench configuration, the slag is removed through a lock hopper arrangement. The gas leaving the radiant cooler is then cooled further in a horizontal fire-tube convection cooler to a temperature of about 425°C. Both coolers are used to raise high pressure steam. In the Polk plant the steam pressure is 115 bar.

As in the quench configuration, there is a final hot-gas scrubber to remove hydrogen chloride and particulates.

Equipment Issues

The Texaco quench gasifier is definitely the most inexpensive design on the market. On the other hand, it is maintenance-intensive. To achieve the greater than 97% availability quoted by Eastman (Moock and Trapp 2002), it is necessary to have an installed standby reactor, which negates the low capital expenditure to a large extent.

5.3.6   The E-Gas Process

The E-Gas process utilizes a two-stage gasifier with a coal-slurry feed and is currently the only two-stage process with an operating commercial-scale demonstration plant.

The E-Gas process was developed by Dow, which started in 1978 with a 12 t/d pilot plant operating in Plaquemine, Louisiana. This was followed by a 550 t/d demonstration plant (in 1983) and a 1600 t/d 165 MW IGCC production facility (1987), both at the same site. Based on these results a 2500 t/d coal (2100 t/d petcoke) commercial unit was built at the Wabash River site in Terra Haute, Indiana, as part of a repowering project. This Wabash River Plant began operations in 1996. The unit is equipped with a spare gasifier. The reactor has an insulated brick lining similar to Texaco gasifiers. The overall efficiency is about 40% HHV.

Process Description

The E-Gas gasifier is a two-stage coal-water slurry-feed entrained slagging gasifier. It was originally designed for the gasification of sub-bituminous coal, although more recently high-sulfur (up to 5.9 wt% on a dry basis) Midwestern bituminous coal has been used. The combination of a coal-water slurry and a low-rank coal in a single-stage gasifier would result in a low-efficiency and high-oxygen consumption. By adding a second nonslagging stage this problem was avoided. In the process scheme (see Figure 5-21) the sub-bituminous coal-water slurry is injected into the hot gases from the first slagging stage resulting in a much cooler exit gas which contains some char. This mixture, with a temperature of about 1040°C, passes through a fire-tube syngas cooler, after which the char is separated from the gas in a particulate-removal unit

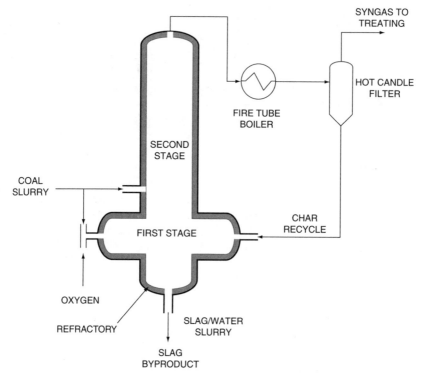

Figure 5-21. The E-Gas Gasifier

featuring metal candle filters. The char is then injected together with oxygen and/or steam into the first slagging stage with a temperature of about 1400°C. The advantage of this process is that although a sub-bituminous coal is used and introduced into the gasifier as a coal-water slurry, the slagging part of the gasifier sees a feed upgraded by a dry char stream that requires relatively little oxygen to be gasified. The waste heat from this stage is then used in the nonslagging stage to free the feed of all the water, as well to supply the heat for some pyrolysis reactions.

The slag is quenched in a water bath in the bottom of the slagging reactor. It is then crushed and, via a continuous pressure let-down system, brought to ambient pressure. The E-Gas process is the only process where no lock hoppers are used for this purpose.

5.3.7 The CCP Gasifier

Recently, a consortium of Japanese utilities led by Toyo Electric and The Central Research Institute of Electric Power Industry (CRIEPI) formed the Clean Coal Power R&D Company (CCP), which announced the construction of a commercial-scale

250 MW (1700 t/d) IGCC demonstration plant at Nakoso, Japan. Start-up is planned for 2007 (Kaneko, Ishibashi, and Wada 2002). The CCP gasifier features a dry feed with two-stage operation, but it uses air as the oxidant. The technology was initially developed by CRIEPI and Mitsubishi Heavy Industries (MHI) and tested in a 200 t/d pilot plant, also at Nakoso.

While pressurized dry-feed entrained-flow gasifiers can be considered to be proven technology (e.g., SCGP, GSP), and the same can be said of two-stage feeding (E-Gas), this is the first attempt to combine these two attractive features in a single gasifier.

Operating the first, "combustor," stage in a combustion mode promotes very high temperatures and simplifies separation of the liquid slag from the gas. The oxidant, although stated as being air, is in fact enriched slightly with surplus oxygen from the nitrogen plant, which supplies inert gas for feed transport.

At the second, "reductor," stage only coal is introduced, without any further oxidant. In the endothermic reaction with the gas from the first stage the coal is devolatilized and tars are cracked sufficiently that no problems occur in the downstream convective cooler. Most of the char is also gasified. Remaining char is separated from the gas in a cyclone and candle filter for recycle to the first stage. The

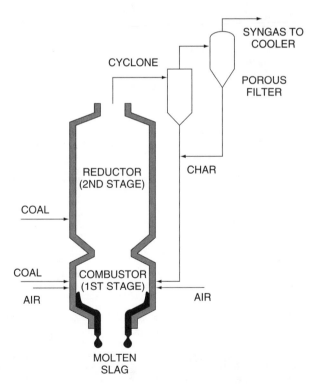

Figure 5-22. The CCP Gasifier (*Source: Kaneko, Ishibashi, and Wada 2002*)

temperature drop over the reductor stage is 700°C with a reactor outlet temperature of around 1000°C.

Carbon conversion rates of 99.8% and more have been achieved regularly with a variety of coals.

5.3.8 The EAGLE Gasifier

The EAGLE Gasifier is an oxygen-blown two-stage dry-feed reactor currently under development by the Electric Power Development Company in Japan. A 150 t/d pilot plant has been built and commenced trials in March 2002 (Tajima and Tsunoda 2002).

The first stage operates in an oxygen rich mode at temperatures of around 1600°C. The outlet temperature from the second stage, which operates oxygen lean with coal and recycled char, is of the order of 1150°C. The reactor uses tangential firing to promote a longer residence time for the coal particles. Coal is supplied in about equal quantities to the two stages, and the reactor is controlled by adjusting the oxygen rates.

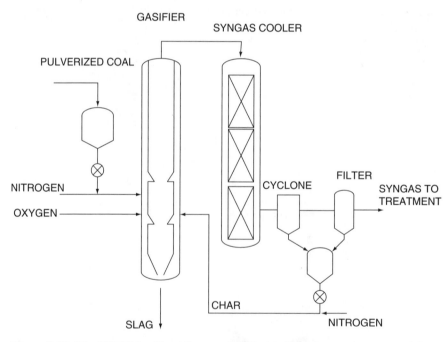

Figure 5-23. The EAGLE Gasifier (*Source: Adapted from Tajima and Tsunoda 2002*)

5.4   OIL GASIFICATION AND PARTIAL OXIDATION OF NATURAL GAS

Technologies for the gasification of liquid and gaseous feeds were developed at the end of the 1940s by Texaco and in the early 1950s by Shell. These two technologies have dominated this segment of the market since that time. In recent years Lurgi has begun marketing a third technology, known as multipurpose gasification (MPG), which was originally developed out of its coal gasification process specifically to handle the tars produced there. Montecatini and GIAP also developed technologies, but neither achieved commercial success.

Certain key features of all three processes are similar. All use entrained-flow reactors. The burners are top-mounted in the downflow, refractory-lined reactor vessels. Operation temperatures are similar (in the range 1250–1450°C). When operating on liquid feed, all three processes produce a small amount of residual carbon, which is necessary to sequester the ash from the reactor.

The important differences between the processes are in the details of burner design, in the method of syngas cooling, and in soot handling.

Partial Oxidation of Gaseous Feeds

Processes suitable for the gasification of liquid feeds can be used with very little modification for the partial oxidation of natural gas or other gaseous feedstocks. Typical differences include the design of the feed-preheat train and the burner. The main process difference is that very little carbon is formed (a few hundred ppm mass instead of values of about 0.5–1% mass) and that the carbon is free of metals, both of which simplify the soot capture and management substantially. And, of course, the gas quality is different, reflecting the C/H ratio of the feed. In the case of sulfur-free feeds, it may also be necessary to review special corrosion issues such as metal dusting.

For this reason, no specific, detailed description of gaseous feed processes is made. Where differences from oil gasification, such as those described above, are worthy of note, these are discussed as part of the relevant oil gasification technology.

5.4.1   The Texaco Gasification Process

The Texaco Gasification Process was developed in the late 1940s. Early research efforts focused on producing syngas from natural gas to produce liquid hydrocarbons via Fischer-Tropsch technology. The first commercial-scale plant based on natural gas as a feedstock was commissioned in 1950 for the production of ammonia. The first commercial-scale use of oil feedstocks occurred in 1956, and early coal work began at about the same time. In the 1970s research efforts were then focused on coal gasification (Weissman and Thone 1995).

During the succeeding 50 or more years over 100 reactors have been licensed for oil or gas service to produce nearly 100 million Nm^3/d syngas. One typical reference plant was commissioned in a German chemical plant in the 1960s; with two further expansions it still operates today with a modified product slate for the synthesis gas. Another has been producing 70,000 Nm^3/h hydrogen for refinery purposes since the mid-1980s.

Commercial plants have been built at pressures of up to 80 bar and experience with unit reactor sizes of up to 3.5 million Nm^3/d synthesis gas is now available from the ISAB installation in Sicily.

Process Description

The oil feedstock is mixed with the moderating steam and preheated in a fired heater. The Texaco burner (Figure 5-24) is of a water-cooled design in which steam and oil are fed together through an annular slit surrounding the central oxygen pipe. The process steam is used to atomize the oil, and mixing is ensured by imparting a counter-rotating vortex motion to the two streams (Pelofsky 1977; Brejc 1989).

The reactor itself is an empty, refractory-lined vessel. The soot make is 1–2 wt% based on feed flow (Appl 1999).

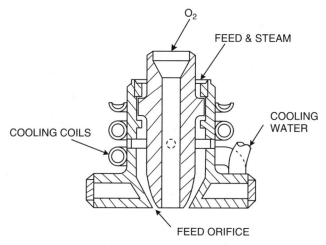

Figure 5-24. Texaco Oil Burner (*With permission: ChevronTexaco*)

Syngas Cooling

Texaco offers two different syngas cooling options: one by direct quenching with water, and another by using a syngas cooler to generate steam (see Figure 5-25).

In the quench mode the hot, raw syngas leaves the bottom of the reactor via a dip-tube into the quench section. The quenched syngas is saturated with water and leaves the quench section with a temperature of about 250°C. At an operating pressure of, say, 80 bar, this corresponds to water loading in the gas of about 2 kg H_2O per Nm^3 of gas. This high water loading makes the quenched gas suitable for CO shift conversion without further steam addition. The quench mode of syngas cooling is, therefore, Texaco's preferred mode for hydrogen and ammonia manufacture.

The quench removes the bulk of the solids in the gas, and these are extracted from the quench vessel as a soot-water slurry or "black water."

Texaco usually uses the syngas cooler mode in applications where a high CO content is required (e.g., oxo synthesis gas) and where the high steam loading of a quenched gas is of no advantage. For intermediate requirements in the H_2/CO ratio, such as methanol synthesis gas, a combination of quench and waste-heat boiler cooling is possible (Jungfer 1985).

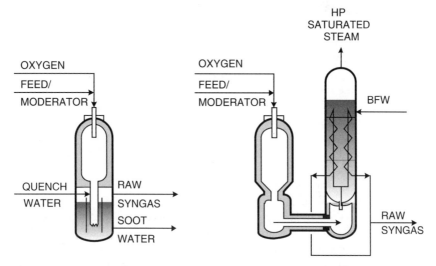

Figure 5-25. Texaco Reactors with Quench Cooling and Syngas Cooler (*With permission: ChevronTexaco*)

Carbon Removal

Following the flowsheet in Figure 5-26, which shows the quench configuration, the gas leaves the quench vessel and is then scrubbed with water twice, first in a Venturi scrubber and then in a packed column, to remove final traces of soot. The raw gas is then suitable for subsequent treatment in downstream units, such as CO shift and acid gas removal.

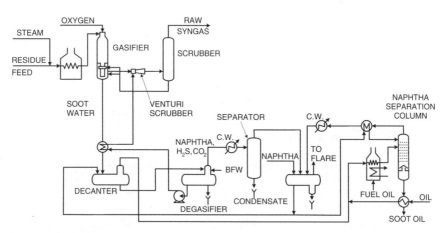

Figure 5-26. Typical Texaco Oil Gasification Flow Scheme (*With permission: ChevronTexaco*)

Carbon Management

In the Texaco process, soot is extracted from the carbon-water mixture with naphtha and recycled with the feedstock to the reactor where it is gasified to extinction. The black water from the quench and the scrubbing section is cooled and contacted with the naphtha in the decanter. In this vessel the naphtha extracts the soot from the water leaving much (but not all) of the ash present in the water phase (gray water). The soot-naphtha mixture is drawn off the top of the decanter and mixed with fresh feed oil. The naphtha is recovered in a distillation tower and recycled to the decanter, leaving the soot-oil mixture as a bottoms product for feeding to the gasifier. Traces of naphtha remain in the tower bottoms and are gasified as well. This naphtha slip has to be made up with fresh naphtha to the system.

The gray water is degassed to recover naphtha and recycled for use in the quench and scrubbing sections. When operating in quench mode the overall water balance is negative because of the large amount carried out with the syngas. Nonetheless, a bleed stream of gray water is bled from the circuit to remove ash. This is necessary to limit the buildup of ash in the circuit. This is by no means trivial, and a soot-oil gasifier feed metals content of about ten times that of the fresh feedstock has been reported. A device was developed that reduced this buildup factor to about 2.5 (Czytko, Gaupp, and Müller 1983).

When operating in the syngas cooler mode there is little water in the raw syngas, so that the bleed stream is necessary in any case to maintain the water balance.

Equipment Performance

The equipment for the process has proved reliable in service, and in a study on operation and maintenance aspects of the process, the data in Table 5-10 have been published.

Process Performance

Typical process performance for different feedstocks is shown in Table 5-11.

Table 5-10
Maintenance Intervals for Texaco Oil Gasifier

Component	Frequency	Time required for intervention
Burner	every six months	6–10 hrs
Quench	every two years	156–180 hrs
Refractory	every three years	680–760 hrs

Source: Bressan and Curcio 1997

Table 5-11
Performance Data for the Texaco Oil Gasification Process

Feedstock Type	Natural Gas	Naphtha	Heavy Fuel Oil	Tar (from Bituminous Coal)
Feedstock composition				
C, wt%	73.41	83.8	87.2	88.1
H, wt%	22.8	16.2	9.9	5.7
O, wt%	0.8		0.8	4.4
N, wt%	3.0		0.7	0.9
S, wt%			1.4	0.8
Ash, wt%				0.1
Raw gas composition				
Product gas (25 bar, quench)				
Carbon Dioxide, mol%	2.6	2.7	5.7	5.7
Carbon Monoxide, mol%	35.0	45.3	47.4	54.3
Hydrogen, mol%	61.1	51.2	45.8	38.9
Methane, mol%	0.3	0.7	0.5	0.1
Nitrogen + Argon, mol%	1.0	0.1	0.3	0.8
Hydrogen Sulfide, mol%			0.3	0.2
Soot, kg /1000 Nm^2		1.8	10	6.1
Consumption figures per				
1000 Nm^3 CO + H$_2$				
Feedstock, kg	262	297	323	356
Oxygen, Nm^3	248	239	240	243
Steam, kg		74	148	186

With permission: ChevronTexaco

5.4.2 The Shell Gasification Process (SGP)

The Shell gasification process (SGP) was developed in Shell's research center in Amsterdam during the early 1950s primarily as a means of manufacturing synthesis gas from fuel oil. The first gasifier, using heavy fuel oil as feedstock, was brought on stream in 1956.

Some 140–150 units have been installed worldwide with a processing capacity of some 7 million t/y of residue. One typical reference plant processes about 240,000 t/y of residues of varying quality, which are bought on the open market, for the production of ammonia. Another, which was started up in 1972, produces a mixed product slate of ammonia, methanol, and hydrogen and is fed with about 350,000 t/y residue directly out of a visbreaker. An interesting reference includes a reduction gas plant for nickel furnaces, one of the few air-blown units in commercial operation. Operating capability covers pressures up to about 65 bar and unit reactor sizes up to 1.8 million Nm^3/d syngas capacity.

Process Description

The noncatalytic partial oxidation of hydrocarbons by the Shell gasification process (Figure 5-28) takes place in a refractory-lined reactor (Figure 5-27) that is fitted with a specially designed burner. The oxidant is preheated and mixed with steam prior to being fed to the burner. The burner and reactor geometry are so designed that this mixture of oxidant and steam is intimately mixed with the preheated feedstock. Originally, a pressure atomizing burner was used, but during the mid-1980s an improved co-annular design using blast atomizing was developed. This burner is capable of handling residues of up to 300 cSt at the burner (Weigner et al. 2002).

Waste Heat Recovery

The product of the partial oxidation reaction is a raw synthesis gas at a temperature of about 1300°C that contains particles of residual carbon and ash. The recovery of the sensible heat in this gas is an integral feature of the SGP process.

Primary heat recovery takes place in a syngas cooler generating high-pressure steam (up to 120 bar) steam in which the reactor effluent is cooled to about 340°C. The syngas cooler is of a Shell proprietary design discussed in more detail in Section 6.6.

Secondary heat recovery takes place in a boiler feed-water economizer immediately downstream of the syngas cooler.

Carbon Removal

The partial-oxidation reactor-outlet gas contains a small amount of free carbon. The carbon particles are removed from the gas together with the ash in a two-stage water wash. The carbon formed in the partial oxidation reactor is removed from the system as a carbon slurry together with the ash and the process condensate. This slurry is subsequently processed in the ash-removal unit described in the following. The product syngas leaves the scrubber with a temperature of about 40°C and is essentially free of carbon. It is then suitable for treatment with any commercial desulfurization solvent.

Carbon Management

Over the course of its development SGP has gone through three distinct stages in its approach to management of the carbon produced in the gasification section.

The early plants were equipped with the *Shell Pelletizing System*, an extraction process using fuel oil as extraction medium. The fuel oil was put in contact with the carbon slurry in the pelletizer where carbon pellets of about 5–8 mm were formed leaving a clear water phase. These pellets were separated from the water on a vibrating screen.

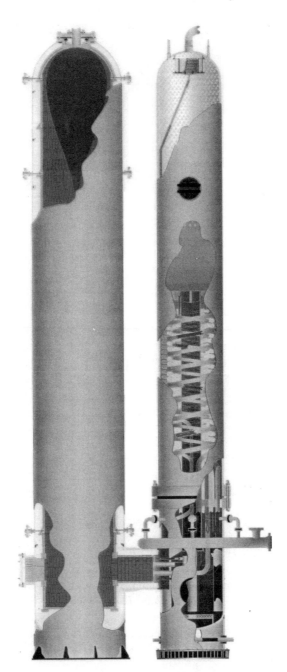

Figure 5-27. Shell Reactor and Syngas Cooler (*Source: de Graaf et al. 2000; With permission: Shell*)

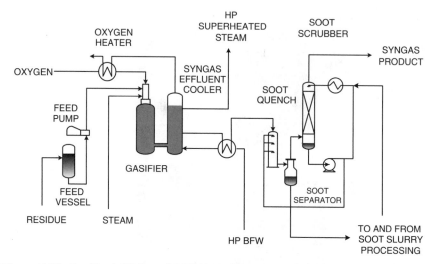

Figure 5-28. Residual Oil-Based SGP Units (*Source: de Graaf and Magri 2002; With permission: Shell*)

The pellets could be burned directly or mixed in with fuel oil to make a liquid fuel known as carbon oil. The carbon oil could in part be used as feedstock for the gasifier, thus providing partial recycle of the carbon. This process had the advantage of being cheap and simple to operate. However, in the extraction process with fuel oil, the separation of soot from the heavy metals (vanadium and nickel) from the gasifier feedstock was poor so that any attempt at 100% carbon recycle brought an unacceptable buildup of metals in the system. Furthermore, there was some water slip with the pellets in the carbon oil, so the carbon oil mixing process could not be operated above 100°C without causing foaming. This limitation meant that this process became unusable with the increasingly heavier feedstocks appearing on the market.

The next development was therefore to substitute the fuel oil with naphtha as extraction medium. This development was known as the *Naphtha Soot Carbon Recovery* process. The principle of extraction in a mixer to increase the size of the agglomerates as well as mechanical sieving was maintained so as to achieve a low naphtha/slurry ratio. The equipment was now operated under pressure, however. The naphtha-soot pellets are mixed with the main feedstock at whatever temperature is required to achieve the desired viscosity. The naphtha is then distilled off from the feed and recycled to the extraction stage leaving the soot behind in the feed (Brejc 1989). The use of naphtha as an intermediate allows the use of heavier, more viscous feedstocks than in the case of pelletizing with fuel oil. Also, an improvement in the separation between carbon and ash allows 100% carbon recycle. Nonetheless, an ash buildup factor of about 3:1 can be observed under 100% recycle conditions. These improvements are bought, however, at a cost in investment and operating expense. Furthermore, the ash buildup still places a limit on ash content in the feedstock.

The third generation of soot management now employed by Shell is based on filtration of the carbon slurry and subsequent handling of the soot-ash filter cake and goes under the name of *Soot-Ash Removal Unit* (SARU; Figure 5-29). The carbon slurry leaves the SGP under pressure at a temperature of some 125°C and is flashed into an intermediate slurry storage tank at atmospheric pressure. Thence it is cooled before water and filtrate are separated in a membrane filter press. The clear filtrate is mostly recycled to the SGP scrubber as wash water. The overall water balance produces a surplus, however, which is treated in a sour water stripper to remove dissolved gases such as H_2S, HCN, and ammonia before being sent to a biotreater.

The filter cake contains typically about 75–85% moisture, but nonetheless behaves for most purposes as a solid. It is then subjected to thermal treatment in a multiple hearth furnace (Figure 5-30). The carbon is burnt off under conditions that prevent the formation of liquid vanadium pentoxide, which has a melting point at about 700°C. In this type of furnace, which is used extensively in the vanadium industry, the filter cake is fed from the top of the furnace in counter-current to the combustion air/flue gas. Rakes, mounted to the central air-cooled shaft, rotate slowly drawing the solid material to downcomers, which are located on alternate hearths at the center and the periphery of the furnace. In the upper hearths the rising flue gas dries the filter cake. In the lower hearths the filter cake is gently burnt off. The bottom product has less than 2 wt% carbon and, depending on the metals in the SGP feedstock, can contain typically 75% V_2O_5. The soot combustion is under the prevailing conditions not quite complete, so that the off-gas contains not only the water vapor from the moisture in the filter cake but also carbon monoxide. In addition it contains traces of H_2S contained within the filter cake. This off-gas is incinerated either as part of the SARU facility or centrally depending on the site infrastructure.

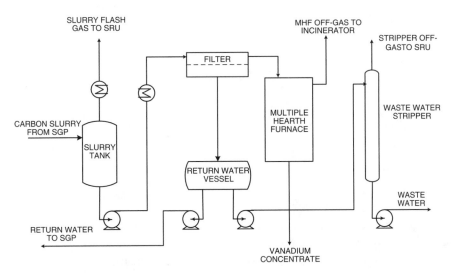

Figure 5-29. Shell Soot-Ash Removal Unit (SARU) (*Source: Higman 1993*)

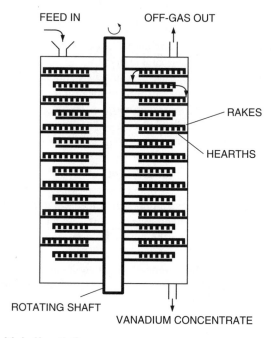

Figure 5-30. Multiple Hearth Furnace

Equipment Performance

SGP is a reliable process that has been proved in many applications worldwide. This reliability is based on the use of proven equipment in critical duties. Typical lifetimes are listed in Table 5-12 (Higman 1994).

Table 5-12 Typical SGP Equipment Lifetimes	
Burners (Co-annular Type)	
• Inspection intervals	~4000 hrs
• Repair intervals	8000–12,000 hrs
Refractory	
• dome repairs	~16,000 hrs
• wall	20,000–40,000 hrs
Syngas Cooler	
• coil inlet section	~60,000 hrs
Thermocouples	
• replacement intervals	2500–8000 hrs
Source: Higman 1994	

SGP employs a sophisticated automatic start-up and shutdown system. Since But-zert's description of the main characteristics (Butzert 1976), further developments include, for example, automated reactor heat-up and a system for minimizing flaring of sulfur-containing gases during start-up.

Process Performance

Table 5-13 provides some information on typical process performance with different feedstocks.

Table 5-13
SGP Process Performance with Different Feedstocks

Feedstock Type	Natural Gas	Heavy Fuel Oil	Vacuum Flash Cracked Residue
Feedstock properties			
Specific gravity (15/4)		0.99	1.10
C/H ratio, wt.	3.17	7.90	9.50
Sulfur, %wt		3.50	4.50
Ash, %wt		0.10	0.15
Feedstock preheat, °C	400	290	290
Oxygen, (t, 99:5%, 260°C)	1154	1103	954
Process steam, t(380°C)	–	350	350
Naphtha, t	–	4	4
Product gas (40°C, 56 bar, dry)			
Carbon Dioxide, mol%	1.71	2.75	2.30
Carbon Monoxide, mol%	34.89	49.52	52.27
Hydrogen, mol%	61.40	46.40	43.80
Methane, mol%	1.00	0.30	0.30
Nitrogen + Argon, mol%	1.00	0.23	0.25
Hydrogen Sulfide, mol%	–	0.77	1.04
Carbonyl Sulfide, mol%	–	0.03	0.04
Quantity, tmol	158	134	128
H_2/CO ratio, mol/mol	1.76	0.95	0.84
Product steam (92 bar sat'd), gross t	2182	2358	2283

Note: The above data for heavy fuel oil and vacuum flash cracked residue is based on the use of naphtha soot carbon recovery. When using SARU minor changes will be observed. Quantities are based on 1000t feed.

5.4.3   Lurgi's Multipurpose Gasification Process (MPG)

Lurgi has maintained a leading position in coal gasification since the 1930s, but for many years worked as contractor and licensing agent for the Shell SGP process for partial oxidation of liquids and gases. In 1998 Lurgi announced that it would now be marketing its own technology under the name of LurgiSVZ multipurpose gasification (MPG; Figures 5-31 and 5-32). This technology had been in existence since 1969 at what is today SVZ Schwarze Pumpe (Hirschfelder, Buttker, and Steiner 1997). It was developed originally out of a Lurgi moving-bed gasifier to process tars produced in the other twenty-three Lurgi gasifiers at the location, which produced town gas from lignite.

Recently the start-up after the revamp of an existing 60 bar 16 t/h asphalt feed reactor has been reported (Erdmann, Liebner, and Schlichting 2002).

Process Description

The gasification reactor is a refractory-lined vessel with a top-mounted burner. The burner has a multiple-nozzle design that allows it to accept separate feed streams of otherwise incompatible materials.

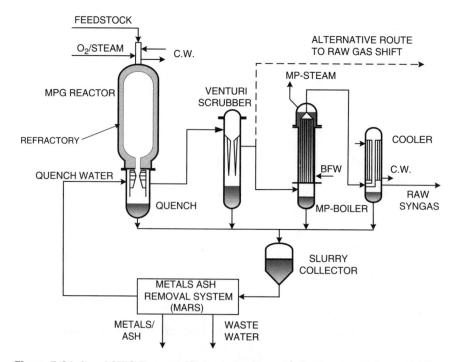

Figure 5-31. Lurgi MPG Process (Quench Configuration) (*Source: Liebner 1998*)

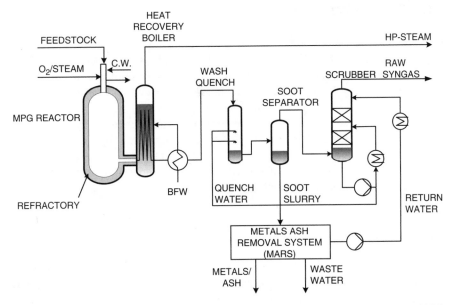

Figure 5-32. Lurgi MPG Process (Syngas Cooler Configuration) (*Source: Liebner 1998*)

Waste Heat Recovery

MPG is offered with two alternative syngas cooling configurations, quench and heat recovery. The criteria for selection of the cooling configuration are listed in Table 5-14.

Table 5-14
Selection Criteria for Quench versus Heat-Recovery Configuration

	Quench Configuration	Heat Recovery Configuration
Feedstocks: Gas, residue, wastes (sludges, coal, coke), extreme ash, and/or salt contents	Highest flexibility	Limited by possible salt precipitation
Product range: Syngas (H_2+CO, H_2, CO)	Fastest (cheapest) route to H_2	Syngas at high temperature H_2–CO equilibrium
Energy utilization	MP steam available Trade-off efficiency versus cost	HP-steam, heat recovery at highest efficiency for IGCC possible
Investment cost	Lowest cost for gasification unit	Boiler (i.e., a high efficiency) at extra cost
Source: Liebner 1998		

Carbon Management

The carbon is washed out of the gas with a conventional water wash. Lurgi's carbon management process for MPG, the metals ash removal system (MARS), is a filtration-multiple hearth furnace process. The flowsheet is very similar to that of Shell's SARU described in Section 5.4.2. Differences are a matter of detail in equipment design and selection. Lurgi uses its own proprietary design of multiple-hearth furnace, which already had a long track record in the vanadium industry before finding application in the field of residue gasification. Lurgi also propagates the use of belt filtration. This has the advantage of being a continuous process with easier operation and maintenance. In order to achieve the same dewatering performance as a membrane filter press, flocculants are required.

Process Performance

A particular feature of MPG is its multinozzle burner, allowing a wide range of feedstocks. The Table 5-15 lists operational ranges and maximum concentrations of base components and contaminants as experienced with MPG.

Typical product gas quality is in Table 5-16.

5.4.4 New Developments

Despite various improvements over the years, it is generally recognized that handling of the soot produced by partial oxidation of heavy residues places a considerable financial burden on the overall process. This has caused operating companies and others to investigate alternatives. In the 1970s one operating company was already using a toluene extraction process to recover the soot as saleable carbon black. A number of other companies made similar attempts, but the economics of these processes, together with the variable product quality depending on feed quality to the gasifier, have prevented commercialization beyond single demonstration plants. Nonetheless, two processes both based on filtration of soot slurry and subsequent treatment of the filter cake, have been reported on in recent years and may form the basis for further development.

Norsk Hydro Vanadium Recovery Process

Norsk Hydro developed its own process for its 65 t/h feed heavy oil gasification plant in Brunbüttel. This plant has been in operation now for several years and has definitely provided operating benefits compared with the original 1975-designed pelletizing plant (Maule and Kohnke 1999).

Table 5-15
MPG Feedstock Flexibility (Liquids and Slurries)

| Component | Actual Operating Ranges and Maximum Concentrations | |
	"Normal" Feeds	Waste Feeds
C, wt%	65–90	90
H, wt%	9–14	14
S, wt%	6	6
Cl, wt%	2	8
LHV, MJ/kg	35–42	5–330
Toluene insolubles, wt%	6	45
Ash, wt%	3	25
Water, wt%	2	5–100
Trace components (selection only)		
Al, ppmw	600	70,000
Ag, ppmw	5	10
Ba, ppmw	500	2000
Ca, ppmw	3000	170,000
Cu, ppmw	200	800
Fe, ppmw	2000	40,000
Hg, ppmw	10	25
Na, ppmw	1200	8000
Ni, ppmw	50	500
Pb, ppmw	200	10,000
V, ppmw	10	100
Zn, ppmw	1200	10,000
PCBs, ppmw	200	600
PAK, ppmw	20,000	40,000

Source: Liebner 1998

The Norsk Hydro VR (vanadium recovery) process (see Figure 5-33) is based on filtration of soot slurry and combustion of the filter cake. In this process the filter cake is first dried and pulverized before being burned in a special cyclone combustor in which part of the vanadium is combusted to a liquid V_2O_5 that is then scraped from the combustion chamber floor. The heat of combustion is used to generate steam, which in general is sufficient to provide the necessary heat for the drying stage. Fly ash from the combustion stage, which also contains V_2O_5, is collected in a bag filter and combined with that collected from the combustion chamber.

	Quench Mode (Coal Oil)	Heat-Recovery Mode (Heavy Residue)	
	Raw Gas	**Raw Gas**	**Clean Gas after Desulfurization**
CO_2, mol%	4.00	3.24	3.26
CO, mol%	53.03	48.25	48.63
H_2, mol%	40.80	46.02	46.39
CH_4, mol%	0.15	0.20	0.20
N_2, mol%	0.85	0.65	0.66
Ar, mol%	1.15	0.85	0.86
H_2S, mol%	0.02	0.79	≤10 ppmv
Total, mol%	100	100	100
HHV, MJ/Nm^3	11.96	12.24	12.13
LHV, MJ/Nm^3	11.15	11.31	11.22

Table 5-16
MPG Product Gases (IGCC Application)

Note: Gasification with oxygen (95%v) at about 30 bar
Source: Liebner 1998

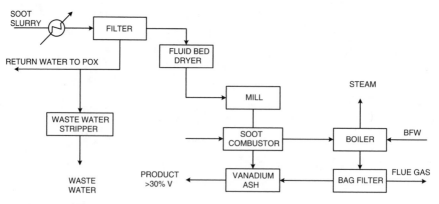

Figure 5-33. Norsk Hydro VR Process

In the published flow diagram, which is more complex than what is shown in Figure 5-33, there is no specific provision for sulfur recovery from the flue gas. Depending on the circumstances, this may have to be included. A fuller description of the process is available in the literature.

In the meantime, Texaco has acquired the rights to this process, and it remains to be seen just what further development the process will experience in the hands of a licensor. It is worth noting that Krupp Uhde reported a similar development under the name of CASH (Keller et al. 1997). No commercial application is known, however.

Soot Gasification

A totally new approach to handling filter cake is under development at the Engler-Bunte-Institut (EBI) of Universität Karlsruhe (Figure 5-34) (Higman 2002). The development of the filtration-based processes was driven by the recognition that the behavior of the vanadium in the ash is crucial to the oxidizing treatment of the filter cake. In particular, the MHF concepts operate at a low temperature specifically to prevent exceeding the melting temperature of the V_2O_5 formed, which is about 700°C. This low operating temperature, for all its benefits, has the disadvantage of a low reaction rate, and thus high residence times and large equipment.

Soot gasification retains the vanadium in the trioxide state and, as with the main gasifier, can operate at high temperatures without creating liquid vanadium pentoxide. The gasification is optimized to achieve maximum carbon conversion, whereby a higher level of CO_2 is tolerated than in most gasification processes, that is, minimizing the residual carbon in the ash is more important than $H_2 + CO$ yield. The process exploits the existing gasification infrastructure by using oxygen and can thus produce a low-pressure synthesis gas with a fuel value. This is in contrast to the large waste gas flow of a multiple hearth furnace, which contains carbon monoxide and requires incineration.

The process line-up includes a typical filtration step followed by fluid-bed drying and milling of the filter cake to <500 µm. Gasification takes place with oxygen and steam in an atmospheric entrained-flow reactor. The ash is removed from the product gas in a dry candle filter and meets the requirements of the metallurgical industry.

This process, which is still under development, has the potential to reduce the costs of carbon management significantly. Further possibilities include the development of a pressurized version, that could handle soot filtered dry directly out of the main

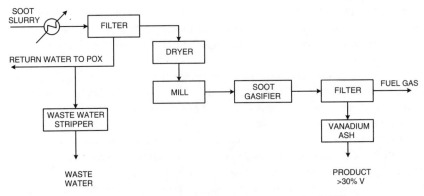

Figure 5-34. EBI Soot Gasification Process (*Source: Higman 2002*)

syngas stream. This would decrease equipment size further, reduce the cost of the wash water circuit, and offer an economic possibility to recycle the fuel gas into the main stream, thus increasing the syngas yield. Clearly, developments in the field of oil gasification are by no means at an end.

5.4.5   Process Safety

Process Automation

Partial oxidation is a process that requires careful control and monitoring so as to ensure that accidental temperature runaways are avoided, especially during transient conditions such as start-up and shutdown. Process licensors have developed sophisticated control systems together with integrated automated start-up and shutdown procedures, which play an important role in safe operation.

Butzert described the most important features of such systems in 1976. Each licensor includes details specific to his own process and experience. The fundamentals described by Butzert are still valid today, although the radical change in instrumentation and control hardware since then has allowed the incorporation of many additional functions. For instance, de Graaf and colleagues (1998; 2000) report on advanced features of the system in Pernis incorporating reactor preheat, panel restart after a spurious trip, and gas transfer to the turbines controlled so that flaring of sulfur-containing gas could be eliminated. Further refinements are reported by Weigner and colleagues (2002). Plants in a predominately power production application include a load following function to comply with the demands of the grid.

Reactor Shell Monitoring

Another safety aspect typical of all partial oxidation processes is the necessity of controlling and monitoring the reactor shell temperature in case of damage to the refractory lining. Early plants had to rely on a multiplicity of point thermocouples around the reactor, usually chosen to correspond with potential weak points in the refractory brickwork. This was often complemented by the use of thermosensitive paint and regular thermography. Later, special coaxial cables with a temperature-sensitive resistance between core and sheath became available. With these coils a continuity of coverage became possible far in advance of the discrete point measurements previously used. On the other hand, location of the hot spot from these measurements was only possible with a grid of such coils that noted which two were showing the high temperature.

Fiber-optic systems are now available that provide a combination of continuity and localization. With appropriate software this can be converted to a screen visualization identifying hot spots on the control panel screen (see Figure 5-35) (Nicholls 2001). This fiber-optic system has been used on oxygen-fired secondary reformers, and recently two Texaco gasifiers have been equipped in this manner.

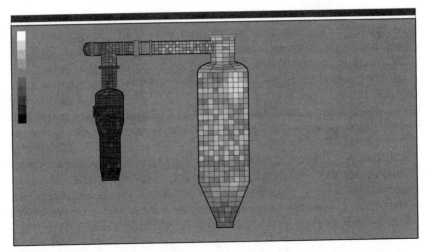

Figure 5-35. Screen Shot of Fiber-Optic Reactor Shell Temperature Monitor (*Source: Nicholls 2001*)

5.5   BIOMASS GASIFICATION

When looking at biomass conversion it is instructive to look at coal conversion, as there are many similarities. This is not so surprising since biomass is nothing but young coal. For coal gasification the minimum temperature required is about 900°C as is demonstrated in the old water-gas process in which the temperature during the steam run was allowed to drop from the maximum of 1300°C to 900°C. About the same maximum temperature of 800–900°C is required to gasify the most refractory part of almost any biomass. In other words, the temperature required for the complete thermal gasification of biomass is of the same order of magnitude as for coal. This high temperature in combination with the impurities, whether sulphur or ash components, is why indirectly heated coal and biomass gasification processes in which external heat has to be transferred via a metal surface have not yet achieved any commercial success.

On the other hand there are a number of significant differences between coal gasification and biomass gasification, which are directly attributable to the nature of the feedstock. Firstly there is the quality of biomass ash, which has a comparatively low melting point but in the molten state is very aggressive. Secondly, there is the generally high reactivity (see Figure 3-3) of biomass. Furthermore, particularly with vegetable biomass, there is its fibrous characteristic. Finally, there is the fact that, particularly in the lower temperature range biomass gasification has a very high tar make.

Although an entrained-flow process might have an apparent attraction in being able to generate a clean, tar-free gas as required for chemical applications, and the low melting point of the ash would keep the oxidant demand low, the aggressive quality of the molten slag speaks against such a solution, whether using a refractory or a cooling membrane for containment protection. Furthermore the short residence times of entrained-flow reactors require a small particle size, to ensure full gasification of the char. No method of size reduction has yet been found, which will perform satisfactorily on fibrous biomass.

A number of fixed-bed processes have been applied to lump wood, but they are limited to this material. They would not work on straw, miscanthus or other materials generally considered for large-scale biomass production unless these were previously bricketted. Furthermore in a counter-flow gasifier, the gas would be heavily laden with tar. The alternative of co-current flow could reduce the tar problem substantially, but the necessity to maintain good control over the blast distribution in the bed restricts this solution to units of very small size.

With this background it is probably not surprising that most processes for biomass gasification use fluid beds and aim at finding a solution to the tar problem outside the gasifier. In co-firing applications where the syngas is fired in an associated large-scale fossil fuel boiler, the problem can be circumvented by maintaining the gas at a temperature above the dewpoint of the tar. This has the added advantage of bringing the heating value of the tars and the sensible heat of the hot gas into the boiler.

There are many biomass processes at various stages of development. Summaries are given in, for example, Kwant (2001) and Ciferno and Marano (2002). The selection chosen here represents generally those that have reached some degree of commercialization.

5.5.1   Fluid-Bed Processes

Lurgi Circulating Fluid-Bed Process

The Lurgi CFB process is described in Section 5.2. Plants operating on biomass and or waste include those in Rüdersdorf in Germany (500 t/d waste) and Geertruidenberg in The Netherlands (400 t/d waste wood). In the latter plant the hot gas leaving the cyclone at a temperature of about 500°C is directly co-fired in a 600 MW$_e$ coal boiler. In Rüdersdorf, the gas is fired in a cement kiln (Greil et al. 2002).

Foster Wheeler Circulating Fluid-Bed Process

The Foster Wheeler (originally Ahlstrom) CFB process was developed to process waste biomass from the pulp and paper industry. The first unit was built in 1983, and the gas was used to replace oil firing of a lime kiln at a paper mill. Three further units were built in Sweden and Portugal for similar applications. The size range is between 17 and 35 MW$_{th}$.

The largest unit to date has a capacity of 40–70 MW$_{th}$ (depending on fuel) and operates co-firing the gas as a supplement fuel in an existing coal-fired boiler in Lahti, Finland (Anttikoski 2002). The feed is primarily biomass, but various refuse derived fuels are also used.

All these units operate at atmospheric pressure. In a different development, Foster Wheeler has also developed a pressurized version that formed the basis for the 6 MW$_e$ biomass IGCC at Värnamo in Sweden (see Figure 5-36). The gasifier operates at 20 bar and has a capacity of 18 MW$_{th}$.

The gasifier feed is pressurized in lock hoppers and a screw feeder is used for the transport from the high-pressure charge bin. Gasification takes place at 950–1000°C. Primary ash removal is via lock hoppers at the bottom of the gasifier. Fine particulate removal takes place in a hot gas filter with ceramic candles (later replaced by metal candles) at 350–400°C, at which temperature the gas enters the combustion chamber of the gas turbine. Tar production from the gasifier is reported to be less than 5 g/Nm^3 dry gas. The demonstration program was completed in 1999 after over 8500 hours operation. Technically, it has been a success. The economics are competitive vis-à-vis other biomass systems, but are still dependant on a general biomass-to-power support. (Sydkraft 1998).

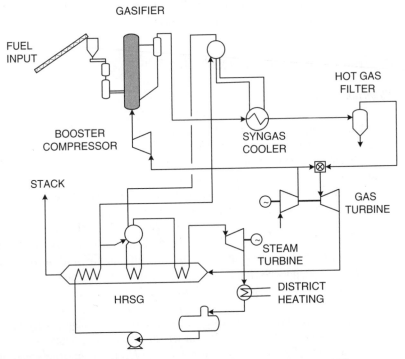

Figure 5-36. Flow Diagram of Värnamo IGCC (*Source: Sydkraft 1998*)

The TPS Process

The TPS process of TPS Termiska Processer AB is an atmospheric CFB that was, like the Foster Wheeler process described above, developed in the mid-1980s to provide energy from waste biomass in the pulp and paper industry. The first commercial application was at Greve-in-Chianti in Italy, where two 15 MW$_{th}$ units to process refuse-derived fuel went on stream in 1992 (Morris and Waldheim 2002). The process was selected and built for the 8 MW$_e$ ARBRE IGCC project in the United Kingdom (Morris and Waldheim 2002). A notable feature of the TPS process is the tar cracker, which uses a dolomite catalyst in a second CFB.

Pressurized Fluid-Bed Processes

In addition to the Foster Wheeler process applied in Värnamo and the HTW process described in Section 5.2, there are some other processes operating in a pressurized fluid bed. Of note are two processes both developed out of the IGT U-gas process and tailored for biomass application. One is the RENUGAS process that was applied in a 100 t/d bagasse fuelled unit in Hawaii, but which is no longer in operation (Ciferno and Marano 2002). Another is the Carbona process in Finland. A 20MW$_{th}$ pilot plant has been operated on various biomasses.

5.5.2   Twin Fluid-Bed Steam Gasification

The SilvaGas Process

This two-stage atmospheric biomass gasification process was developed by Battelle, and the first commercial demonstration unit with a feed capacity of 200 t/d was built in Burlington, Vermont. Commercialization of the process has been taken over by Future Energy Resources (FERCO), who market it under the name of SilvaGas. The medium Btu gas at the demonstration unit is fired in an existing biomass fired boiler and is planned to be used later in a combustion turbine (Paisley, and Overend 2002).

The principle of the SilvaGas process (see Figure 5-37) is similar to that of a catalytic cracker in an oil refinery or of the Exxon Flexicoker process. In all these processes two fluid-bed reactors are used. In one, an endothermic process takes place; in the SilvaGas process, for the gasification of biomass. The necessary heat for the reaction is supplied by a hot solid (sand, catalyst, or coke), which is heated by an exothermic reaction in the second reactor.

As in all biomass gasification processes, a feed preparation stage is necessary in which the biomass is reduced to 30–70 mm-length chips and oversize or foreign material such as metals are removed. The biomass is fed to the gasifier where it is mixed with hot sand (at about 980°C) and steam. During the ensuing endothermic cracking reaction, light gaseous hydrocarbons are formed together with hydrogen and carbon monoxide. After separating the heat carrier and the gas in cyclones, the relatively cold heat carrier and residual unreacted char are discharged to the combustor or regenerator. The sand is reheated in the combustor by burning the char with air. The reheated sand is removed from the flue gas by a cyclone separator and returned to the gasifier.

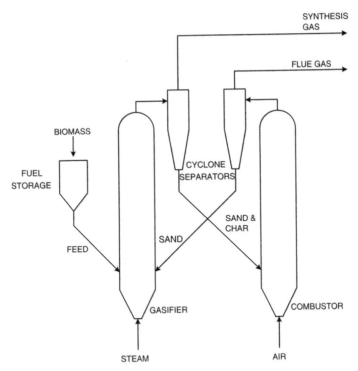

Figure 5-37. SilvaGas (Battelle) Process (*Source: Paisley, Irving, and Overend 2002*)

Table 5-17 Gas Composition of SilvaGas	
CO_2, mol%	12.2
CO, mol%	44.4
H_2, mol%	22.0
CH_4, mol%	15.6
C_2H_4, mol%	5.1
C_2H_6, mol%	0.7
HHV, MJ/Nm^3	17.3
Source: Paisley, Irving, and Overend 2002	

The syngas from the gasifier still contains typically about 16 g/m^3 tars. Depending on the application (e.g., for gas turbine fuel), these must be removed. Cracking catalysts, as used in the petroleum industry, are used to break down the heavy hydrocarbons. Work is continuing to find lower-cost disposable catalysts for this application. The syngas is cleaned up in a scrubber for alkali and particulate removal. A typical gas composition from the Burlington demonstration unit is shown in Table 5-17.

The flue gas is a valuable source of heat. Using it for pre-drying of the biomass feed helps increase the efficiency of the process, but alternative uses such as steam production may be applied if site-specific conditions favor this.

The FICFB Process

The FICFB (fast internal circulating fluid-bed) process developed by the Vienna University of Technology in Austria is another process that separates steam gasification of the biomass from combustion of char as a source of heat for the former (see Figure 5-38). A 42 t/d feed commercial demonstration combined heat and power (CHP) unit has been built in the town of Güssing, where it is integrated into the operations of the local district heating utility. The synthesis gas is fired in a gas motor generating 2 MW$_e$ and 4.5 MW heat is supplied to domestic and industrial consumers. The plant was taken on stream in December 2001.

The gasifier operates as a stationary fluid-bed reactor with sand as the fluidizing medium. The sand and ungasified char leave the reactor at the bottom and are transferred to the combustor where the char is burnt to heat the sand. The hot sand is separated from the flue gas in a cyclone and returned to the gasifier via a seal leg bringing in the necessary heat for the gasification reaction, which takes place at about 850°C. The synthesis gas is cooled and cleaned for use in a gas motor. Of note is the use of an oil wash to remove tars. In the demonstration unit in Güssing RME (Rape methylester) is used as washing oil (Hofbauer 2002). Gas compositions are given in Tables 5-18 and 5-19.

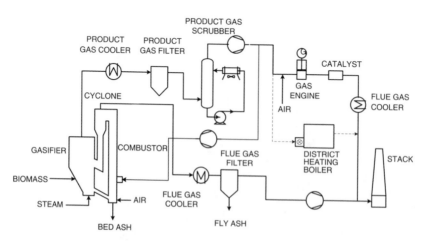

Figure 5-38. FICFB Process (*Source: Hofbauer 2002*)

Table 5-18
Gas Composition of FICFB Gas

CO_2, mol%	15–25
CO, mol%	20–30
H_2, mol%	30–45
CH_4, mol%	8–12
N_2, mol%	3–5
LHV, MJ/Nm^3	12–14

Source: Hofbauer 2002

Table 5-19
Impurities in FICFB Gas

		Raw Gas	Clean Gas
Tar	g/Nm^3	0.5–1.5	<0.020
Particulates	g/Nm^3	10–20	<0.010
NH_3	ppm	500–1000	<200
H_2S	ppm	20–50	

Source: Hofbauer 2002

5.5.3 Pyrolysis Processes

As discussed in Section 4.3, the logistics of biomass collection will in general limit biomass gasification facilities to a maximum of 30–40 MW$_{th}$. In order to overcome this limitation in benefiting from the economies of scale, the combination of decentralized pyrolysis plant and a central bio-oil gasifier has been proposed (e.g., Henrich, Dinjus, and Meier 2002).

Figure 5-39 shows a generic bio-oil plant in which typically about 75% of the dry feedstock is recovered as bio-oil. Char (10–15 wt%) and gas (15–20 wt%) are recovered and combusted to supply the heat required for drying the feedstock and heating the reactor. The pyrolysis takes place at about 450–475 °C with a residence time of the order of magnitude of 1 second.

Biomass pyrolysis processes are at this stage still in their infancy. There are a number of small-scale commercial and demonstration plants that have been built, the most important or representative of which are listed in Table 5-20.

The principle current use for bio-oil includes specialty chemicals, which are essential for economics at present (Freel 2002).

Various projects are in preparation for testing equipment with bio-oil feeds. Fortum have a burner-testing program with Oilon Oy. Test programs for slow-speed marine

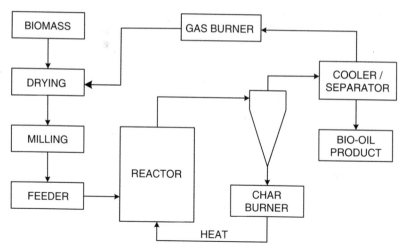

Figure 5-39. Block Flow Diagram for a Bio-Oil Plant (*Source: Meier 2002*)

Table 5-20
Bio-Oil Pilot and Demonstration Plants

Company, Country	Trade Name	Plant Size	Technology	Comments
Dynamotive, Canada	Biotherm™	10 t/d	Stationary fluid-bed	In planning 100 t/d UK, 200 t/d Canada
Wellman, UK		6 t/d	Stationary fluid-bed	Awaiting operation permit
ENSYN, Canada	RTP™	2*45 t/d	Circulating fluid-bed	40 t/d plant operating since 1996
ENEL, Italy			Circulating fluid-bed	
VTT, Finland		0.5 t/d	Circulating fluid-bed	
BTG, Netherlands		4 t/d	Rotating cone	In planning 10 t/d
Forschungszen-trum Karlsruhe, Germany			Double screw	Lurgi LR process
Pyrovac, Canada		35 t/d	Vacuum pyrolysis	
Fortum/Vapo, Finland	Forestera™	12 t/d	Vacuum pyrolysis	Start-up May 2002 (Gust, Nieminen, Nyrönrn 2002)

diesel engines (e.g., Omrod) are underway, and tests are also being conducted on gas turbines. The amount of bio-oil currently available is small, however, which limits the opportunities for such testing.

Proposals for testing the gasification characteristics of bio-oil have also been made (Henrich, Dinjus, and Meier 2002). Initial pilot testing has already been completed.

The European Union has a program to develop standards for bio-oil, considering end-user requirements. A guide to analysis and characterization methods specifically adapted to bio-oil products has been published (Oasmaa and Peacocke 2002).

5.5.4   Other Processes

Anaerobic Digestion of Biomass

One possible way to make use of a unique property of biomass is to convert it by means of biochemical reactions, but this subject, however interesting, falls outside the scope of the present book. Suffice it to mention that anaerobic digestion is the most elegant and efficient gasification process in which (dirty) liquid water can be used as a gasifying agent and a cold-gas efficiency of about 95% is obtained. Unfortunately, only the (hemi-)cellulose part of the biomass is converted in the presently available processes. The reaction for cellulose can be written as:

$$(C_6H_{10}O_5)_n + n\, H_2O = 3n\, CH_4 + 3n\, CO_2$$

Anaerobic digestion is only applied in small-scale units of below 5 MW to convert agricultural and liquid domestic waste (Krüger 1995). The use of thermophilic microorganisms has made this conversion more attractive, but for a reasonable conversion of the biomass a period of two to three weeks is still required, which makes this process less suitable for large-scale plants. It could be that with hyper-thermophilic microorganisms this period could be reduced such that it could also be applied for larger-scale plants. Moreover, by using hyper-thermophiles, no sterilization stage is required, which is necessary where the digestion waste will be recycled to farms and/or forests. Finally, hyper-thermophiles may eventually be found that also convert lignin. The gas produced is essentially a mixture of methane and CO_2 and can only be used advantageously for heating purposes and in combined heat and power (CHP) schemes using a gas motor or small gas turbine for power generation.

5.6   GASIFICATION OF WASTES

Because of its varied nature there are many different approaches to the gasification of waste. Some processes have already been described; in particular, but not only, fluid-bed processes have the possibility to be adapted to waste gasification. Others have been purpose developed for waste gasification. At present there are a large number of processes in various stages of development or demonstration. Schwager and Whiting (2002) report "some 71 novel thermal treatment plants that are already operating for waste applications including many that use gasification as their main conversion method"; they go on to list 26 of these, which are considered commercially available. It is noticeable that very few have more than one or two reference plants, and one must expect that over the course of time the market will show which of these are the most effective concepts.

The two most important aspects specific to the gasification of municipal solid wastes are first, the highly heterogeneous nature of the feed, and second, the extensive and stringent regulations on emissions.

5.6.1   Coal Gasifiers in Waste Service

One example of a process that was originally developed for coal feeds but that has been successfully adapted for municipal solid waste (MSW) or refuse derived fuel (RDF) service is the BGL moving-bed slagging gasifier described in Section 5.1. A 650 t/d unit is in service for waste gasification at the Schwarze Pumpe facility in Germany, where it is part of a larger complex producing electric power and methanol. The plant operates with a 75% waste/ 25% coal feedstock (Greil et al. 2002). Two other projects using this technology are in planning in the United States (Lockwood and Royer 2001).

A second technology with many references in coal combustion service that was later adapted for waste gasification is the circulating fluid-bed (see Section 5.2.3). Examples are a plant in Rüdersdorf near Berlin, which produces some 50,000 Nm³/h low Btu gas for a cement kiln from a wide variety of wastes (Greil et al. 2002), and one at Geertruidenberg in The Netherlands where waste wood is gasified to syngas that is fired in a 600 MW coal-fired power unit.

A third example is the HTW process (Section 5.2.4), for which a 20 t/d pilot MSW gasifier has been built by Sumitomo in Japan. Although the HTW process operates in a nonslagging mode, it is possible to add a separate slagging unit into the process, should this be appropriate (Adlhoch et al. 2000).

A final example is the liquid waste gasification of organic nitrogen compounds at Seal Sands using the GSP gasifier (Schingnitz et al. 2002).

Waste Addition to Coal Gasifier Feed

A number of other processes have demonstrated that they can accept small quantities of waste in the feed. For example, in the Shell Coal Gasification unit in Buggenum, 12% waste biomass has already been added to the fuel and plans are in progress to increase this amount (Hannemann et al. 2002).

5.6.2   Purpose Developed Processes

Pyrolysis Processes

One feature of many processes specific to waste gasification, is the use of a separate pyrolysis stage prior to partial oxidation. (In discussing waste gasification, it is important to keep a clear distinction, since the word gasification is often used indiscriminately for both.) Pyrolysis is sometimes used as a preliminary to partial oxidation of the tars and char in a separate reactor, as in, for example, the Thermoselect, Compact Power, Brightstar, PKA, and Alcyon processes. Others do not include a partial oxidation stage but have a more or less close-coupled combustion of the pyrolysis products, such as von Roll and Takuma. Also, where a partial oxidation follows the pyrolysis, there are different approaches. Thermoselect, for example, claims to operate the pyrolysis at 300°C, then gasify with oxygen and quench the syngas prior to cleaning, thus having the option to use the syngas for power or chemicals production (Calaminus and Stahlberg 1998). Compact Power, by contrast, operates with pyrolysis at 800°C, gasifies with air and burns the syngas directly in a close-coupled combustor (Cooper 2002).

Finally, there are some processes that only include a pyrolysis such as that of Thide, where the gas from the pyrolysis stage is used in a separate, not necessarily close-coupled thermal value-recovery stage.

The issue of close-coupling a combustion stage can be an important one, even if not only in the technical sense. Where a distinction is made in regulations between gasification (as a process that makes a synthesis gas) and incineration, the close-coupled combustor can be considered integral to the gasification stage and the whole unit is then classified as an incinerator. This can lead in some jurisdictions to unfortunate results, such as totally inappropriate personnel training requirements (Lockwood and Royer 2001).

Fluid-Bed Gasification

There are a number of processes that use fluid-bed gasification without a separate pyrolysis stage. The coal-derived HTW and CFB processes mentioned above are examples. Others have been developed primarily for waste feeds, such as automotive shredder residues. Such a process is that of Ebara, which in one variant close-couples an air-blown fluid-bed gasifier with a cyclonic combustion chamber. The latter operates at about 1400°C and produces a molten slag. The Ebara process, which in fact originated as an air-blown incineration process, has been developed via atmospheric gasification to include pressurized gasification with a chemicals-quality synthesis gas. The TwinRec variant of the Ebara process consists of a first stage fluid-bed air-blown gasifier operating under pyrolysis conditions at about 580°C, followed by a close-coupled downflow cyclonic combustion unit (Fujimura, Oshita, and Naruse 2001). The latter operates at 1350–1450°C and the slag is tapped at the bottom of this section. Six units are operational at the time of writing. A further fourteen are in various stages of design and construction.

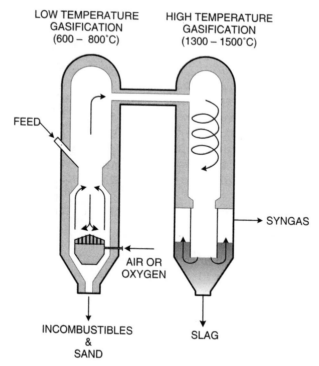

LOW TEMPERATURE
GASIFICATION
(600 – 800˚C)

HIGH TEMPERATURE
GASIFICATION
(1300 – 1500˚C)

FEED

SYNGAS

AIR OR
OXYGEN

INCOMBUSTIBLES
&
SAND

SLAG

Figure 5-40. Ebara-Ube Process (*Source: Steiner et al. 2002*)

A pressurized version of the fluid-bed pyrolysis unit, known as the Ebara-Ube process, has been developed in a 30 t/d pilot plant operating at about 10 bar and 600–800°C (see Figure 5-40). This is close-coupled to a high temperature gasifier operating at 1300–1500°C. The latter incorporates a water quench. An additional plant for 65 t/d is under construction at the same location, and the syngas produced will, after water scrubbing, be processed in a CO shift and PSA unit to provide hydrogen for an existing ammonia synthesis plant (Steiner et al. 2002).

Another process using fluid-bed gasification is that of Enerkem. The product syngas can be made available for separate use for powering a gas engine, for example.

Other Processes

Finally, there are a number of approaches that cannot be included in this summary classification. One such process is the Sauerstoff-Schmelz-Vergasung (2sv), which is one of the few current examples of a co-current moving-bed gasifier. The advantage of this concept lies in the much lower tar content in the gas compared with a counter-current moving-bed (Scheidig 2002).

5.7 BLACK LIQUOR GASIFICATION

Black liquor gasification is a specialized field (see also Section 4.3.2). Particular difficulties are the high inorganic load, particularly of sodium, and the requirement for recovering this for recycling. In addition, the sodium poses problems for conventional refractory solutions. Add to this the concern about the potential for tar formation from the lignin content, and it is understandable why only a few companies have attempted to realize this technology.

In comparison with conventional combustion technology (Tomlinson Boilers), pressurized black liquor gasification can increase the energy recovery in the pulping process from 65% using the most modern combustion equipment to about 75%. Compared with much of the existing installed equipment, of which the majority is 30 years or older, electric energy generation can increase by a factor of two to three. The quality of green liquor from a gasifier can provide process advantages, since sodium and sulfur are recovered separately for recycling to the digester, offering the opportunity for increased pulp yield and quality. The causticizing load does increase, however, which is a disadvantage.

In addition, the risk of a smelt-water explosion involved in conventional boiler technology is absent when gasifying black liquor due to the small smelt inventory in the process.

5.7.1 The Chemrec Process

Chemrec has built a number of small demonstration plants, including a 75 tDS/d (tons dry solids per day) as well as one commercial unit of 300 tDS/d (Chemrec, see Figure 5-41). These are all based on quench technology, and most use air as an oxidant. One of the pilot plants was converted to oxygen gasification, and a second oxygen-blown unit is under construction at the Energy Technology Centre (ETC) at Piteå, Sweden, close to the Kappa Kraftliner pulp and paper mill.

The Chemrec reactor is a refractory-lined entrained-flow quench reactor operating at a temperature of 950–1000°C. The organic material is gasified in the reaction zone. The inorganic material is decomposed into smelt droplets consisting of sodium and sulfur compounds. Carbon conversion is greater than 99.9%; tar formation is low.

The smelt droplets are separated from the gas phase in the quench zone, after which they are dissolved in the quench liquid to form a green liquor solution. The synthesis gas leaving the quench zone is scrubbed to remove particulate matter, primarily entrained alkaline particles in a countercurrent condensing tower.

In the booster configuration in which the gasifier is installed in parallel to an existing black liquor boiler as a de-bottlenecking measure, air is used as oxidant. The syngas is burnt untreated in a boiler to raise steam. Sulfur removal is effected by scrubbing the flue gas with oxidized white liquor.

Alternatively, a black liquor gasification combined cycle (BLGCC) can be used to replace the conventional black liquor boiler. In this configuration the gasifier is

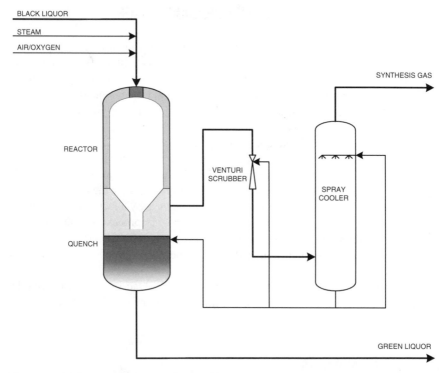

Figure 5-41. Chemrec Black Liquor Gasification Process (Booster Configuration) (*Source: Adapted from Chemrec website*)

oxygen blown at about 30 bar. A syngas cooler is installed for maximum heat recovery. Sulfur is then removed from the synthesis gas using conventional acid-gas removal technologies (see Chapter 8). A full BLGCC configuration can double the electric power production per ton of black liquor compared with a new recovery boiler.

The first commercial-sized booster unit was built for the Weyerhaeuser Company at New Bern, NC in 1996. Initial problems, particularly with respect to the refractory lining, were overcome, and in 1999 the plant achieved an availability of 85%. However, chloride-induced stress corrosion cracking has developed in the stainless steel reactor shell, and this will be replaced using a different metallurgy during 2003.

The use of a BLGCC can bring the same environmental advantages to pulp processing as other IGCC systems, as described in Chapter 7.

5.7.2 MTCI "PulseEnhanced" Steam Reformer

The MTCI process is based on steam gasification at about 600°C using indirect heat supply to the process via a heat exchanger built into the fluid-bed reactor. This has the advantage of operating at temperatures below the melting point of the smelt, but against this weighs the problem of tar production. The first 180 tDS/d demonstration plant is being built at Big Island, Virginia and is due to start up in 2004. A number of conceptual issues such as sulfur removal still have to be resolved during the detail execution of the project (DeCarrera 2001). Should the demonstration plant show this technology to be successful, it will share a number of the principle advantages of gasification over combustion of black liquor, such as elimination of the smelt-water explosion risk, superior thermal efficiency, and lower emissions to the environment.

5.8 MISCELLANEOUS PROCESSES

There are a number of gasification processes that for various reasons fall outside the categories that have been discussed here. These are in situ gasification of coal, also known as underground gasification, gasification in a molten iron bath, plasma gasification, and hydrogasification.

5.8.1 In situ Gasification

Gasification of coal in situ has a number of obvious attractions. Such a process has the potential to tap resources not otherwise readily or economically accessible. It would also eliminate the safety hazards and costs associated with underground mining. The ash would be left underground.

The first recorded proposal for underground coal gasification (UCG) was by Siemens in 1868, followed by Mendeleyev 20 years later. Initial experiments in the United Kingdom were broken off by the advent of World War I. No further work was done until the 1930s when an experimental station was started in the Donetsk coalfield in the then–Soviet Union, to be followed by a commercial installation in 1940 (Weil and Lane 1949). Underground gasification continued at a number of locations in the Soviet Union until the late 1970s, with a gas production of some 25,000 million Nm^3 of gas being produced from around 6.6 million tons of coal (Ökten 1994). This production came from seams of 50 to 300 m depth.

In the 1980s a number of small experimental units were operated in the United States. In Europe, tests have been conducted in Belgium (1986–1987), and then in Spain. The Spanish trial was aimed at demonstrating the feasibility of UCG using modern directional drilling techniques at a depth of 600 m. The trial operation lasted for a period of 300 hours, during which 290 tons of coal were gasified (Green and Armitage 2000).

The basic concept of UCG is to drill one or more wells into the coal seam where the blast is injected and others from which the fuel gas can be collected. There are a number of different methods that have been used to link injection and gas wells, some of which are depicted by Ökten (1994). Simple though the basic idea may be, there are many practical difficulties still to be overcome, and it its already clear that the technology can only be applied to certain types of coal seam. The hydrogeology of the seam is important, since excessive ingress of water would render the process uneconomic, and leakage of gas into underground water supplies could represent an environmental hazard. Both air and oxygen gasification have been tried. With air, a very low Btu gas is produced, whereas with oxygen the cost of the blast and the losses make the process very costly. For these and other reasons, no commercial development has yet emerged.

There is currently one active project in Australia (Walker, Blinderman, and Brun 2001). Operations using air injection commenced in December 1999, and to date (October 2002) some 32,000 tons of coal have been gasified. The peak gas production capacity is 80,000 Nm^3/h. Gas with a heating value of about 5 MJ/m^3 is produced at a pressure of 10 bar and a temperature of 300°C. During this demonstration phase the gas has been flared. The operation has been shut down pending the proposed installation of a gas turbine.

5.8.2 Molten Iron Processes

A number of process developments were started using technology from the steel industry, typically the Klöckner molten iron process (Reimert 1989) in which use is made of a molten metal. Coal is introduced through tuyères and blown on top of a bath with molten iron that acts as a promoter for heat and mass transfer. The development of the process as a gasification technology has stopped, although further development has resulted in the HIsmelt steelmaking process.

5.8.3 Plasma Gasification

A number of processes have been or are being developed that use plasma technology to input heat into the gasification process. One such technology is the Westinghouse plasma gasification process. A demonstration waste-to-energy plant has been operated in Japan, and the first commercial-scale 200 t/d plant has recently been completed, also in Japan (Dighe and Lazzara 2002). Quapp et al. (2002) describe another plasma gasification process.

These processes are in an early stage of development. They share some of the typical characteristics of other high-temperature processes, namely production of an inert, vitrified slag and a tar-free synthesis gas. Efforts appear to be concentrated particularly on the gasification of medical wastes, where sterility requirements demand high-temperature processing, and the nature of the material makes the size reduction required for an entrained-flow process difficult.

5.8.4 Hydrogasification

During the 1970s and 1980s there was considerable interest in developing a process to produce synthetic natural gas by direct hydrogenation of coal to methane. The principle reaction involved is that of methanation.

$$C + 2H_2 \leftrightarrows CH_4 \qquad\qquad -75\,MJ/kmol \qquad\qquad (2\text{-}6)$$

The thermodynamics of this reaction demand that it be operated at high pressure. Although the reaction is exothermic, the reactants must first be brought to a high temperature for the reaction to proceed at an economically acceptable rate. Furthermore, any development to exploit this reaction has to consider how to integrate the production of hydrogen into the process.

In many ways, the HKV (Hydrierende Kohlevergasung) process developed by Rheinbraun for the direct hydrogenation of browncoal to methane is typical of this class of gasification process (Speich 1981). The process development unit had a throughput of 0.4 t/d raw browncoal. A pilot plant was then built for 24 t/d and went into operation in 1983. The process operates in a fluid-bed at temperatures between 850 and 930°C and pressures between 60 and 95 bar. Pressures of up to 120 bar were foreseen. Carbon conversion rates of 50–70% were achieved. The residual char is then gasified in a High Temperature Winkler gasifier (HTW; see Section 5.2) for hydrogen manufacture.

Although the pilot plant achieved the goals that had been set for it, the increasing availability of natural gas rendered the whole direction of the development superfluous and all such developments were stopped long before 1990.

REFERENCES

Adlhoch, W., Sato, H., Wolff, J., and Radtke, K. "High Temperature Winkler Gasification of Municipal Solid Waste." Paper presented at Gasification Technologies Conference, San Francisco, October 2000.

Anon. "Coal Gasification, Clean Coal Technology." *International Power Generation*, May, 1990.

Anttikoski, T. "Circulating Fluidized-Bed Gasifier Offers Possibility for Biomass and Waste Utilization and for Substitution of Natural Gas by Syngas from Coal Gasification." Paper presented at IChemE Conference: "Gasification: The Clean Choice for Carbon Management" Noordwijk, The Netherlands, April 2002.

Appl, M. *Ammonia: Principles and Industrial Practice*. Weinheim: Wiley-VCH, 1999

Becker, J. "Examples for the Design of Heat Exchangers in Chemical Plants." *verfahrenstechnik*, 3(8) (1969):335–340.

Beránek, J., Rose, K., Winterstein, G. *Grundlagen der Wirbelschicht-Technik*. Leipzig: Krausskopf, 1975.

Bockelie, M. J., Denison, M. K., Chen, Z., Linjewile, T., Senior, C. L., Sarofim, A. F., and Holt, N. A."CFD Modeling for Entrained-Flow Gasifiers." Paper presented at Gasification Technologies Conference, San Francisco, October 2002.

Bögner, F., and Wintrup, K. "The Fluidized-Bed Coal Gasification Process (Winkler Type)." In *Handbook of Synfuels Technology*, ed. R. A. Meyers, New York: McGraw-Hill, 1984, pp. 3–111–3–125.

Brejc, M. "Gas Production." In *Ullmann's Encyclopedia of Industrial Chemistry*, 5th ed., vol. A12. Weinheim: VCH Verlagsgesellschaft, 1989, pp. 206–213.

Bressan, L., and Curcio, S. "A Key Aspect of Integrated Gasification Combined Cycle Plants Availability." Paper presented at IChemE Conference "Gasification Technology in Practice," Milan, February 1997.

Brooks, C. T., Stroud, H. J. F., and Tart, K. R. "British Gas/Lurgi Slagging Gasifier." In *Handbook of Synfuels Technology*, ed. R. A. Meyers. New York: McGraw-Hill, 1984, pp. 3–63–3–86.

Bucko, Z., Adlhoch, W., Mittelstädt, A., Vierrath, H., Wolff, J., and Ginsberg, D. "400 MW IGCC Power Plant Vřesová, Czech Republic: A New Application of the HTW-Gasification Technology." Paper presented at IChemE Conference, "Gasification for the Future," Noordwijk, April 2000.

Butzert, H. E. "Ammonia/Methanol Plant Startup." *Chemical Engineering Progress* 72 (January 1976):78–81.

Calaminus, B., and Stahlberg, R. "Continuous In-line Gasification/Vitrification Process for Thermal Waste Treatment: Process Technology and Current Status of Projects." *Waste Management* 18 (1998).

Chemrec. Available at: www.chemrec.se (2002).

Ciferno, J. P., and Marano, J. J. "Benchmarking Biomass Gasification Technologies for Fuels, Chemicals, and Hydrogen Production." U.S. Department of Energy/National Energy Technology Laboratory (NETL) Report, June 2002.

Clayton, S. J., Siegel, G. J., and Wimer, J. G. *Gasification Technologies: Gasification Markets and Technologies—Present and Future: An Industry Perspective.*" U.S. Department of Energy Report DOE/FE-0447, July 2002.

Cooper, N. "Compact Power Puts Pyrolysis at the Heart of Sustainable Developments Optimizing Value Recovery." Paper presented at Pyrolysis and Gasification of Biomass and Waste Expert Meeting, Strasbourg, September 2002.

Curran, P. F., and Tyree, R. F. "Feedstock Versatility for Texaco Gasifiers." Paper presented at IChemE Conference, "Gasification: The Gateway to a Cleaner Future," Dresden, September 1998.

Czytko, M. P., Gaupp, K., and Müller, R. "Syngas from Heavier Residues." *Hydrocarbon Processing* 62(9) (September 1983):115–119.

de Graaf, J. D., and Magri, A. "The Shell Gasification Process at the Agip Refinery in Sannazzaro." Paper presented at IChemE Conference, "Gasification: The Clean Choice for Carbon Management," Noordwijk, April 2002.

de Graaf, J. D., Zuideveld, P. L., Posthuma, S. A., and van Dongen, F. G. "Initial Operation of the Shell Pernis Residue Gasification Project." Paper presented at IChemE Conference, "Gasification: The Gateway to a Cleaner Future," Dresden, September 1998.

de Graaf, J. D., Zuideveld, P. L., van Dongen, F. G., and Hölscher, H. "Shell Pernis Netherlands Refinery Residue Gasification Project." Paper presented at IChemE Conference, "Gasification for the Future," Noordwijk, April 2000.

DeCarrera, R. "Demonstration of Black Liquor Gasification at Georgia-Pacific's Mill in Big Island, Virginia." Paper presented at Maine Energy and Technology Expo, September 2001.

Dighe, S. V., and Lazzara, D. "Westinghouse Plasma Gasification Technology for Application to Coal Gasification and Co-fired Coal/Biomass Gasification." Paper presented at 19th International Pittsburgh Coal Conference, Pittsburgh, September 2002.

Erasmus, H. B., and van Nierop, P. "Sasol: Fifty Years of Growth." Paper presented at IChemE Conference, "Gasification: The Clean Choice for Carbon Management," Noordwijk, April 2002.

Erdmann, C., Liebner, W., and Schlichting, H. "MPG Lurgi Multipurpose Gasification: Initial Operating Experience at a VR-based Ammonia Plant." Paper presented at IChemE Conference, "Gasification: The Clean Choice for Carbon Management," Noordwijk, April 2002.

Freel, B. "Ensyn Announces Commissioning of New Biomass Facility." *PyNe Newsletter* 14 (September 2002), Aston, UK.

Fujimura, H., Oshita, T., and Naruse, K. "Fluidized-Bed Gasification and Slagging Combustion System." Paper presented at IT3 Conference, Philadelphia, May 2001.

Garstang, J. H., and Vierrath, H. E. "BGL Gasification—Clean Energy from Coal." Paper presented at IChemE Gasification Conference, London, November 1995.

Green, M., and Armitage, M. "Underground Coal Gasification: What Next after the European Trial?" Paper presented at IChemE Conference, "Gasification for the Future," Noordwijk, April 2000.

Greil, C., and Hirschfelder, H. "Biomass Integrated CFB Gasification Combined Cycle Plants." Paper presented at IChemE Conference, "Gasification: Gateway to a Cleaner Future," Dresden, September 1998.

Greil, C., Hirschfelder, H., Turna, O., and Obermeier, T. "Operating Results from Gasification of Waste Material and Biomass in Fixed-Bed and Circulating Fluidized-Bed Gasifiers." Paper presented at Gasification Conference, Noordwijk, The Netherlands, April 2002. Also presented at IChemE Conference, "Gasification: The Clean Choice for Carbon Management." Noordwijk, April 2002.

Gumz, W. *Gas Producers and Blast Furnaces*. New York: John Wiley & Sons, 1950.

Gust, S., Nieminen, J.-P., and Nyrönrn, T. "Forestera: Liquefied Wood Fuel Pilot Plant." Paper presented at Pyrolysis and Gasification of Biomass and Waste Expert Meeting, Strasbourg, October 2002.

Hannemann, F., Kanaar, M., Karg, J., and Schiffers, U. "Buggenum Experience and Improved Concepts for Syngas Application." Paper presented at Gasification Conference, San Francisco, 2002.

Henrich, E., Dinjus, E., and Meier, D. "Flugstromvergasung von flüssigen Pyrolyseprodukten bei hohem Druck: ein Konzept zur Biomassevergasung." Paper presented at DGMK Conference "Energetische Nützung von Biomassen," velen: April 2002

Higman, C. A. A. "New Developments in Soot Management." Paper presented at IChemE Conference, "Gasification: The Clean Choice for Carbon Management," Noordwijk, 2002.

Higman, C. A. A. "Production-Integrated Environmental Protection." In *Ullmann's Encyclopedia of Industrial Chemistry*, 5th ed., vol. B8, p. 291. Weinheim: VCH Verlagsgesellschaft, 1995.

Higman, C. A. A. "Perspectives and Experience with Partial Oxidation of Heavy Residues." Paper presented at AFTP Conference, "L'Hydrogène, Maillon Essentiel du Raffinage de Demain," Paris, June 1994.

Higman, C. A. A., and Heurich, H. "Partial Oxidation in the Refinery Hydrogen Management Scheme." Paper presented at AIChE Spring Meeting, Houston, March 1993.

Hirschfelder, H., Buttker, B., and Steiner, G. "Concept and Realization of the Schwarze Pumpe 'Waste to Energy and Chemicals Centre.'" Paper presented at IChemE Conference, "Gasification Technology in Practice," Milan, February 1997.

Hofbauer, H. "Biomass CHP-Plant Güssing: A Success Story." Paper presented at Pyrolysis and Gasification of Biomass and Waste Expert Meeting, Strasbourg, October 2002.

Holt, N. A. "Coal Gasification Research, Development, and Demonstration: Needs and Opportunities." Paper presented at Gasification Technologies Conference, San Francisco, October 2001.

Holt, N. A., and van der Burgt, M. J. "Biomass Conversion: Prospects and Context." Paper presented at 16th EPRI Gasification Technology Conference, San Francisco, 1997.

Jungfer, H. "Synthesis Gas from Refinery Residues." *Linde Reports on Science and Technology* (40) (May 1985).

Kaneko, S., Ishibashi, Y., and Wada, J. "Project Status of 250 MW Air-Blown IGCC Demonstration Plant." Paper presented at Gasification Technologies Conference, San Francisco, October 2002.

Keller, H. J., Brandl, A., Buxel, M., and Klos, W. "Efficient and Reliable Oil-Based IGCC Plant Concept Combining Proven Technological Concepts: Advanced Burner Technology, Waste Heat Recovery, Soot Processing." Paper presented at IChemE Conference "Gasification Technology in Practice," Milan, February 1997.

Kersten, S. R. A., *Biomass Gasification in Circulating Fluidized Beds* Enschede: Twente University Press, 2002

Koopman, E., Regenbogen, R. W., and Zuideveld, P. L. "Experience with the Shell Coal Gasification Process." Paper presented at VGB Conference, "Buggenum IGCC Demonstration Plant," November 1993.

Krüger A/S. "Hashøj Biogas Plant: Combined Heat and Power for Two Urban Districts," Denmark: Rosenberg Bogtryk, June 1995.

Krupp-Koppers. "Prenflo: Clean Power Generation from Coal." Company brochure, June 1996.

Kunii, D., and Levenspiel, O. *Fluidization Engineering*, 2nd ed. New York: John Wiley & Sons, 1991.

Kwant, K. W. "Status of Gasification in Countries Participating in the IEA Bioenergy Gasification Activity." International Energy Agency Report, April 2001.

Landlälv, I. Private communication, November 2002.

Leonard, R., Rogers, L., Vimalchand, P., Liu, G., Smith, P. V., and Longanbach, J. "Development Status of the Transport Gasifier at the PSDF." Paper presented at Gasification Technologies Conference, San Francisco, 2001.

Liebner, W. "MPG—LurgiSVZ Multipurpose Gasification: A New and Proven Technology." Paper presented at 10th Refinery Technology Meet, Mumbai, February 1998.

Lockwood, D. N., and Royer, P. R. "Permitting and Regulatory Issues Associated with Development of an IGCC Project." Paper presented at Gasification Technologies Conference, San Francisco, October 2001.

Lohmann, C., and Langhoff, J. "The Development Project 'Ruhr 100'—an Advanced Lurgi Gasifier." Paper presented at 15th World Gas Conference, Lausanne, 1982.

Lorson, H., Schingnitz, M., and Leipnitz, Y. "The Thermal Treatment of Wastes and Sludges with the Noell Entrained-Flow Gasifier." Paper presented at IChemE Conference, "Gasification: An Alternative to Natural Gas," London, November 1995.

Lurgi. *Lurgi, der technologie-orientierte Anlagenbauer, 1897–1997*. Frankfurt: Lurgi, 1997.

Lurgi. "Lurgi Gas Production Technology: The Shell Gasification Process." Lurgi Brochure 189e/6.92/2.20. 1992.

Lurgi. *Lurgi Handbuch*. Frankfurt: Lurgi Gesellschaften, 1970.

Marklund, M. "Black Liquor Recovery: How Does It Work?" Available at: www.etcpitea.se/blg/document/PBLG:or_RB.pdf (December 2001).

Maule, K., and Kohnke, S. "Technical, Economic. and Environmental Improvements at Hydro Agri Brunsbüttel: The Solution to the Soot Problem in an HVR Gasification Plant." Paper presented at Gasification Technologies Conference, San Francisco, October 1999.

McDaniel, J. E., and Hornik, M. J. "Polk Power Station IGCC: Fourth Year of Commercial Operation." Paper presented at Gasification Technologies Conference, San Francisco, October 2000.

Meier, D., "Flash-Pyrolyse zur Verflüssigung von Biomasse—Stand der Technik." Paper presented at DGMK Conference, "Energetische Nutzungvan Biomassen." Velen 2002.

Milne, T. A., Evans, R. J., and Abatzoglou, N., "Biomass Gasifier 'Tars': Their Nature, Formation and Conversion" Golden, CO: NREL Report NREL/TP-570-25357, November 1998.

Moock, N., and Trapp, W. "Gasification Plant Design and Operation Considerations from an Operator's Perspective." Paper presented at Gasification Technologies Conference, San Francisco, October 2002.

Morris, M., and Waldheim, L. "Efficient Power Generation from Wood Gasification." Paper presented at IChemE Conference: "Gasification for the Future", Noordwijk, The Netherlands, April 2000.

Morris, M., and Waldheim, L. "Update on Project ARBRE, UK." Paper presented at IChemE Gasification Conference Gasification: The Clean Choice for Carbon Management", Noordwijk, The Netherlands, April 2002.

Nicholls, P. "Monitoring for Hot Spots and Leaks Using Fiber-Optic Technology." *Hydrocarbon Processing* 80(11) (November 2001):101–107.

Oasmaa, A., and Peacocke, C. "A Guide to Physical Property Characterization of Biomass-Derived Fast Pyrolysis Liquids." *VTT Publications* (2002).

Ökten, G. "Underground Gasification of Coal." In *Coal*, ed. O. Kural. Istanbul: Istanbul Technical University, 1994, pp. 371–378.

Paisley, M. A., and Overend, R. P. "Verification of the Performance of Future Energy Resources SilverGas Biomass Gasifier—Operating Experience in the Vermont Gasifier." Paper presented at 19th International Pittsburgh Coal Conference, Pittsburgh, September 2002.

Pelofsky, A. H. *Heavy Oil Gasification*. New York: Marcel Dekker, 1977.

Perry, J. H. *Chemical Engineers' Handbook*, 5th ed. New York: McGraw-Hill, 1973.

Quapp, W. J., Cohn, D. R., McDonald, T., Surma, J. E., and Lamar, D. E. "Steam Reforming Gasification of Coal, Wastes, and Other Renewable Feedstocks to Produce Hydrogen and Power Using the Plasma-Enhanced Melter." Paper presented at 19th International Pittsburgh Coal Conference, Pittsburgh, September 2002.

Reimert, R., and Schaub, G. "Gas Production." In *Ullmann's Encyclopedia of Industrial Chemistry*, 5th ed., vol. A12. Weinheim: VCH Verlagsgesellschaft, 1989.

Renzenbrink, W., Wischnewski, R., Engelhard, J., and Mittelstädt, A. "High Temperature Winkler Coal Gasification: A Fully Developed Process for Methanol and Electricity Production." Paper presented at Gasification Technologies Conference, San Francisco, 1998.

Rudolf, P. "Lurgi Coal Gasification (Moving-bed Gasifier)." In *Handbook of Synfuels Technology*, ed. R. A. Meyers. New York: McGraw-Hill, 1984, pp. 3–127–3–148.

Samant, G., Higman, C., Krishnan, V., Sturm, P., Heidsick, K., and Kowallik, W. "Verfahren zur thermischen Behandlung von Vanadium enthaltenden Rückständen." European patent EP 0 542 322 B1, November 1991.

Scheidig, K. "In 3. Generation." *UmweltMagazin* (May 2002):80–83.

Schingnitz, M., Gaudig, U., McVey, I., and Wood, K. "Gasifier to Convert Nitrogen Waste Organics at Seal Sands, UK." Paper presented at IChemE Conference, "Gasification: The Clean Choice for Carbon Management," Noordwijk, April 2002.

Schlinger, W. G. "The Texaco Coal Gasification Process." In *Handbook of Synfuels Technology*, ed. R. A. Meyers, New York: McGraw-Hill, 1984, pp. 3–148.

Schwager, J., and Whiting, K. J. "European Waste Gasification: Technical and Public Policy Trends and Deveiopments." Paper presented at Gasification Technologies Conference, San Francisco, October 2002.

Simbeck, D. R., Korens, D. R., Biasca, F. E., Vejtasa, S., and Dickenson, R. L. *Coal Gasification Guidebook: Status, Applications, and Technologies.* Palo Alto, Calif.: Electric Power Research Institute (EPRI), 1993.

Smith, P. V., Davis, B. M., Vimalchand, P., Liu, G., and Longanbach, J. "Operation of the PDSF Transport Gasifier." Paper presented at Gasification Technologies Conference, San Francisco, 2002.

Speich, P. "Braunkohle: auf dem Weg zur großtechnischen Veredlung." *VIK-Mitteilungen* 3/4 (1981).

Steiner, C., Kameda, O., Oshita, T., and Sato, T. "EBARA's Fluidized-Bed Gasification: Atmospheric 2×225 t/d for Shredding Residues Recycling and Two-Stage Pressurized 30 t/d for Ammonia Synthesis from Plastic Wastes." Paper presented at the 2nd International Symposium on Recycling of Plastics, Oostende, September 2002.

Stönner, H.-M. "Gas Production." In *Ullmann's Encylopedia of Industrial Chemistry*, 5th ed. Weinheim: VCH Verlagsgesellschaft, 1989.

Supp, E. *How to Produce Methanol from Coal.* Berlin: Springer, 1990.

Sydkraft, "Värnamo Demonstration Project—Construction and Commissioning", Malmö, Sydkraft AB, 1998.

Tajima, M., and Tsunoda, J. "Development Status of the EAGLE Gasification Pilot Plant." Paper presented at Gasification Technologies Conference, San Francisco, October 2002.

Teggers, H., and Theis, K. A. "The Rheinbraun High-Temperature Winkler and Hydrogasification Processes." Paper presented at First International Gas Research Conference, June 1980, Chicago.

Tils, H. M. G. C. European patent EP 0606957, "Carbon Burn-Off Process." January 1994.

Trapp, W. "Eastman and Gasification: The Next Step, Building on Past Success." Paper presented at Gasification Technologies Conference, San Francisco, October 2001.

U.S. Department of Energy (DoE). "Piñon Pine IGCC Power Project: A DoE Assessment." DoE/NETL report 2003/1183, December 2002.

U.S. Department of Energy (DoE). "Tampa Electric Polk Power Station Integrated Gasification Combined Cycle Project, Final Technical Report", August 2002.

U.S. Department of Energy (DoE). "The Wabash River Coal Gasificaiton Repowering Project, Final Technical Report", August 2000.

U.S. Department of Energy (DoE). "The Wabash River Coal Gasification Repowering Project" Topical report No. 20, September, 2000.

U.S. Department of Energy (DoE) "The Wabash River Coal Gasificaiton Repowering Project: A DoE Assessment" Report DoE/NETL-2002/1164, January 2002.

van der Burgt, M. J., Hope, T., Malchavek, R. V., and Perry, R. T. "Integrated Gasification and Combined Cycle Power Generation." Paper presented at Sixth Advanced Coal Gasification Symposium, Hangzhou, China, 1988.

van der Burgt, M. J., and Naber, J. E. "Development of the Shell Coal Gasification Process (SCGP)." Paper presented at Advanced Gasification Symposium, Shanghai, 1983. Also presented at Third BOC Priestley Conference, London, September, 1983.

Walker, L. K., Blinderman, M. S., and Brun, K. "An IGCC Project at Chinchilla, Australia Based on Underground Gasification." Paper presented at Gasification Technologies Conference, San Francisco, 2001.

Weigner, P., Martens, F., Uhlenberg, J., and Wolff, J. "Increased Flexibility of Shell Gasification Plant." Paper presented at IChemE Conference, "Gasification: The Clean Choice for Carbon Management," Noordwijk, April 2002.

Weil, B. H., and Lane, J. C. *The Technology of the Fischer-Tropsch Process.* London: Constable, 1949.

Weissman, R., and Thone, P. "Gasification of Solid, Liquid and Gaseous Feedstocks: Commercial Portfolio of Texaco Technology." Paper presented at IChemE Conference, "Gasification: An Alternative to Natural Gas," London, November 1995.

Wellman. "Gasification." Company brochure. Oldbury, England: Wellman Process Engineering, (undated).

Chapter 6
Practical Issues

6.1 EFFECT OF PRESSURE

The effect of gasification pressure and temperature on gas composition, yield, and cold gas efficiency was discussed in Section 2.3.1. There are other aspects to consider, however, when deciding on the values of these parameters in a process, and these are discussed here.

Pressure

The pressure in a gasifier is generally based on the requirements of processes downstream of the gasifier. This requirement is easily met when the downstream process is a combined cycle (CC) that typically requires a pressure in the gasifier of 20–40 bar. Other processes such as methanol or ammonia synthesis require much higher pressures of 50–200 bar and thus compression of the synthesis gas.

In principle, it looks more attractive to pressurize the feed to a gasifier than to pressurize the gas. However, it should be realized that most of the advantages in terms of equipment compactness and lower compression energy are already obtained when gasifying at a pressure of 15–25 bar. Moreover, where the feedstock is a solid such as coal or biomass, pressurizing becomes more and more complicated at higher pressures. In the case of air gasification, there is in principle less reason to prefer pressurization of the blast, since the savings on syngas compression are much less due to the large percentage of inerts in both the blast and the product gas.

For high-temperature entrained-flow gasifiers, this theoretical argument of pressurizing the blast components remains valid for quite high pressures of 100–150 bar because of the low methane content in the gas. For fluid-bed gasifiers that operate at much lower temperatures, the higher methane contents in the gas at such high pressures would be unfavorable for nonfuel gas applications. In moving-bed gasifiers, the methane content is already high owing to the pyrolysis reactions. High pressures raise the methane content further, even to the extent of almost doubling it as was demonstrated in the Ruhr 100 plant (see Section 5.1.3). This may not be desirable for syngas applications, but for SNG production it reduces the load on the downstream plant considerably.

171

There are also a number of practical aspects to be reviewed when considering gasification at very high pressures, which sometimes reduce the attractiveness of such a solution.

Compression of Reactants

Large oxygen compressors are available for pressures up to 70 bar. Above this pressure oxygen is mostly pressurized by pumping liquid oxygen. This facilitates the pressurizing and reduces the energy for syngas compression. However, in the ASU more energy is required for compression as the cold from the evaporation of the liquid oxygen now comes available at a somewhat higher temperature. Overall, there may be still an advantage to gasify at a pressure of, say, 100 bar.

Raising the pressure of heavy oil residues for gasification at 80 bar with plunger pumps is normal commercial practice in Texaco plants, and pilot plants have operated at 100 bar. Coal-water slurries are also relatively easy to pump, although more difficult than a pure liquid. Gravity feed of lump coal through lock hoppers to a moving-bed gasifier has been demonstrated at 100 bar (Lohmann and Langhoff 1982). The situation is different for dry-coal feed systems relying on pneumatic conveying, as in entrained-flow systems or screw conveyors. For such systems the maximum practical pressure is about 50 bar (see also Section 6.2.1).

Compression of the moderator, which in virtually all cases is steam, is not a problem, as pumping water requires relatively little energy.

Equipment

All gasification reactors require some form of protection between the high-temperature reaction space and the outer pressure shell, which must be maintained at moderate temperatures of 200°C to 300°C. This protection either takes the form of a thick (50–70 cm) insulating refractory wall, or a membrane wall that in current designs is at least as thick. One of the potential advantages of gasifying at higher pressures is that the reactor volume and thus cost required for a given throughput decreases, particularly in fluid-bed and entrained-flow reactors, where the volume is determined by the gas phase. Since the volume taken up by the pressure shell protection system is virtually independent of the pressure, the economics of designs much above 30 to 40 bar tends to be confronted with diminishing returns in IGCC applications using coal as a feedstock. When the downstream application of the gas requires very high pressures, as has been the case in most heavy oil gasifiers where the gas is mostly used for ammonia or methanol synthesis, the combination of the savings in compression cost and the fact that oil is easy to pressurize outweigh the disadvantages of a somewhat higher cost reactor.

Side Reactions

When looking at the possibilities of high-pressure gasification, one should not forget the effect of pressure on side reactions. When the feedstock contains iron or nickel (the latter being typical for refinery residues), the formation of carbonyls is favored by higher pressures and becomes significant at pressures over about 30 bar (see Section 6.9). Although this is not an argument against higher pressure per se, it will cause additional expense in the gas clean up.

Similarly, formation of formic acid in the liquid phase is favored by higher partial pressures of carbon monoxide. This will tend to lower the pH in water washes or process condensate and at high pressures will need to be considered in the material selection.

6.2 PRESSURIZATION OF COAL

Pressurizing a solid material as coal is a bit of a misnomer. It is in fact the transport of a solid from one environment to another environment with a higher pressure. Feeding coal or any other solid material into a space with a pressure of more than 5–10 bar is not an easy matter. Reimert (1981 and 1986) has presented systematic reviews of systems that are available or under development. It has been identified as an area requiring additional research and development (Holt 2001; Clayton, Stigel, and Wimer 2002). For coal gasification two different approaches have been followed for pressurizing pulverized coal:

- Lock hoppering, or a sluicing system using an inert gas as the continuous phase.
- Pumping as a slurry with a piston pump water as the continuous phase.

6.2.1 Dry-Coal Feeding with Lock Hoppers

Lock hoppers have been used for over a century in water gas reactors and in blast furnaces for sluicing lump coal, coke, and iron ore into vessels that operated under a slight overpressure. They were developed further in the 1930s for operation at 25 to 30 bar in connection with the Lurgi pressurized moving-bed gasifier. In the Ruhr 100 pilot plant they have been demonstrated at 100 bar (Reimert 1986).

In general, a lock hopper system consists of three vessels that are situated on top of each other and separated from each other by valves (see Figure 6-1).

The top hopper is at atmospheric pressure, and the middle one is the actual lock hopper. The bottom hopper can be a storage vessel that is at an elevated pressure, but it can also be the gasifier itself, as is the case with moving-bed gasifiers. The principle is that of any sluicing system. During loading, the valve between the atmospheric hopper and the lock hopper is open, and the valve between the lock hopper and the bottom hopper is closed. After the lock hopper has been filled, the first valve is closed and the second opened, after which the pressure in the lock hopper increases from atmospheric to the elevated pressure, and the solid material will drop into the bottom hopper. The valve between the lock hopper and the bottom hopper is then closed, the gas in the lock hopper depressurized and the valve between the top hopper and the lock hopper opened, after which the cycle can be repeated.

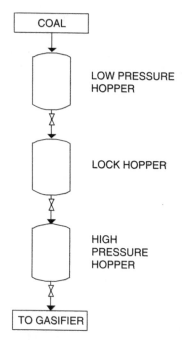

Figure 6-1. Lock Hopper for Dry Feed

Adopting the lock-hopper system for pressurizing pulverized materials such as coal requires major modifications to the lock-hopper system. The solids, for example, have to be kept fluidized during transport and in the hoppers. This requires providing the hoppers with spargers or other means for the introduction of the fluidizing gas. The dusty gases leaving the hoppers on depressurizing have to be cleaned and sometimes repressurized, which further complicates the lock-hopper system and makes it a costly piece of equipment. Finally, there is the drawback that lock hoppers are discontinuous. This is not a problem for processes, which have long residence times such as a blast furnace or a moving-bed coal gasifier, but it is more problematic for entrained-flow gasifiers, which have residence times in the order of seconds. In the latter case the pressurized hopper must be sized such that it is filled during the whole lock-hopper cycle so as to ensure a continuous flow of solids to the downstream equipment.

The use of lock hoppers for coal pressurization presents a problem for dry-coal feed entrained-flow slagging gasifiers when pressures higher than 30–40 bar are required. The problem is not limited to more complex equipment such as valves, fluidizing systems, and the compression of fluidizing gases. More important is that the gas consumption for fluidizing the pulverized coal in the pressurized hoppers becomes higher at higher pressures. Furthermore, the amount of gas required for the transport of the coal to the burners increases, creating a burden for the gasifier, as this gas has to be heated to the high gasification temperature.

Transport Gases

Nitrogen. Using nitrogen as transport gas has the drawback that the product gas becomes contaminated, which is particularly relevant when the gas is to be used for chemical synthesis or for the production of hydrogen. The only chemical application where the presence of nitrogen does not pose a problem is ammonia synthesis. In IGCC power stations the presence of nitrogen means that less nitrogen is available for quenching, for example. However, in IGCC applications the presence of some inert material in the gas has hardly any effect on the overall process efficiency.

In IGCC applications nitrogen is therefore the gas that is most commonly used in lock hoppers and for the subsequent dense phase transport to the burners. The nitrogen is available from the air separation unit (ASU), supplying the oxygen required for the gasification. It should be possible to get a loading during dense phase transport of 400 kg/actual m^3. In practice, the loading is about 300 kg/actual m^3 as then the coal flows more smoothly. This implies that, when operating at a pressure of 30 bar and a temperature of about 90°C, for every kg of coal 0.09 kg nitrogen is required for transport. At a pressure of 70 bar the latter figure would increase to 0.21 kg. The nitrogen (plus argon) percentages in the product gas correspond to 2.7 and 5.1 mol% for pressures of 30 and 70 bar, respectively (see Table 6-1). The same percentage of 5 mol% nitrogen is obtained at 30 bar when the oxygen purity is reduced from 99 to 95 mol%. Although in IGCC applications the higher nitrogen content in the gas has only a marginal effect on the overall process efficiency, it does slightly increase the duty of the syngas cooler and of the gas treating.

For chemical applications, the higher inert content of the gas will cause a subsequent synthesis to run under less favorable conditions. In such a situation, if nitrogen is to be used as transport gas it is often more attractive to run the gasifier at a lower pressure and to increase the duty of the syngas or hydrogen compressor, which is in any case required in most such applications. Examples where this applies are methanol synthesis and hydrocrackers.

Syngas. Using syngas for the high-density transport of pulverized coal to the gasifier instead of nitrogen largely reduces the problem of nitrogen contamination. In case a gas quench is used, as is the case currently in the SCGP gasifier, the syngas can best be taken from the discharge of the recycle gas compressor. Nevertheless, the use of syngas for transport of coal is in most cases not an attractive solution, although the nitrogen contamination of the gas is typically reduced from 3–5 mol% to less than 1 mol% (see Table 6-1). The problem with syngas as transport gas is that in the lock hoppers, the gas also has the function of providing a barrier between the oxidizing atmosphere of the atmospheric pressure coal and the reducing atmosphere of the gasifier, a function that syngas cannot fulfill. The obvious choice for the barrier function is nitrogen. It is inevitable, therefore, that the transport syngas will always be contaminated with some nitrogen. All in all, syngas is not an attractive option, and in practice the only practical alternatives are nitrogen and CO_2.

Table 6-1
Influence of the Coal Transport Medium, Pressure, and Oxygen Purity on the Syngas Purity

Process	Single-Stage Slurry Feed	Single-Stage Dry Feed						
Temp.,°C	1500	1500		1500		1500		
Press, bara	70	30	70	30	70	30	70	30
Coal transport medium	Water	CO_2	CO_2	Syngas	Syngas	N_2	N_2	N_2
CGE, %	65	82	82	82	82	82	82	82
IGCC eff., %	38	50	50	50	50	50	50	50
O_2, mol%	99	99	99	99	99	99	99	95
Wet raw product gas, mol%								
CO	37.4	64.5	62.9	63.2	63.5	61.9	60.6	60.8
H_2	15.4	31.9	30.3	32.8	32.4	32.2	31.0	31.3
CO_2	6.0	2.0	4.7	1.0	1.0	1.0	1.0	1.0
H_2O	40.3	0.2	0.4	1.9	1.9	1.9	1.9	1.9
CH_4	0	0.2	0.5	0	0.1	0	0.1	0
H_2S	0.2	0.3	0.3	0.3	0.3	0.3	0.3	0.3
$N_2 + A$	0.7	0.9	0.9	0.8	0.8	2.7	5.1	4.7
Dry raw product gas, mol%								
CO	62.6	64.6	63.2	64.4	64.8	63.2	61.9	62.1
H_2	25.8	32.0	30.4	33.5	33.0	32.8	31.6	31.9
CO_2	10.1	2.0	4.7	1.0	1.0	1.0	1.0	1.0
CH_4	0	0.2	0.5	0	0.1	0	0.1	0
H_2S	0.3	0.3	0.3	0.3	0.3	0.3	0.3	0.3
$N_2 + A$	1.2	0.9	0.9	0.8	0.8	2.7	5.1	4.7
H_2/CO molar ratio	0.41	0.50	0.48	0.52	0.51	0.52	0.51	0.51
CO_2/H_2S molar ratio	34	7	16	3.3	3.3	3.3	3.3	3.3

Note: The IGCC efficencies are calculated on the basis of the standardized, idealized conditions of Appendix E.

The effect of syngas on process efficiency and the syngas cooler duty is the same as for nitrogen, provided that the pressures and temperatures are similar (see Table 6-1).

Carbon Dioxide. The use of CO_2 as transport gas is only a serious option where it is available at no additional cost, that is, where a CO shift and subsequent CO_2 removal is already part of the downstream gas processing. For many chemical applications, such as hydrogen or methanol production, this is the case. If CO_2 capture and sequestration becomes a requirement for power production, it would also be the case for IGCC applications. The effect of CO_2 on process efficiency and the syngas cooler duty is only marginally different from nitrogen, provided the pressures and temperatures are the same (see Table 6-1). The H_2/CO ratio of the syngas may decrease slightly, but this generally would have little influence on subsequent gas processing. The effect of the H_2S/CO_2 ratio on the acid-gas removal system will be discussed in Chapter 8.

The major advantage of CO_2 over nitrogen as transport gas is that it does not dilute the gas with additional inerts. It has the advantage over syngas as transport gas in that it is not toxic and it slightly reduces the process steam requirements.

Although the most complex lock hoppers are required for pulverized coal, they are often also used for the discharge of fly slag that is separated in cyclones and or filters downstream of the gasifier. Lock hoppers in which the continuous phase is a liquid are used in some gasifiers for sluicing the slag out of the gasifier (see also Section 6.2.2).

6.2.2   Pumping Coal as a Coal-Water Slurry

Pumping coal as a slurry is in principle and in practice a more elegant route to coal pressurization than lock hoppering. In water, coal concentrations of 60–70 wt% can be used. A drawback is that only a small part of the water is required for the gasification, and the majority just constitutes a burden, as it has to be vaporized and heated to 1500°C. This in turn implies that the oxygen consumption is much higher than for dry-coal feed systems and that the CGE is substantially lower. For IGCC applications, this inevitably results in a lower station efficiency (50 and 38%; see Table 6-1).

In order to compensate for virtually all drawbacks of using water as a slurry medium, the merits of substantially preheating the slurry have been investigated.

Preheating has the following advantages:

1. The water has to be heated less in the reactor, and the heat of evaporation becomes lower at higher temperatures (see Figure 6-2B).
2. Atomization becomes better with hotter liquid containing feedstocks, increasing the accessibility of the coal for gaseous reactants, especially when the feedstock slurry is preheated such that the carrier flashes upon introduction into the gasification reactor. Also, the reduction in surface tension of the carrier liquid at higher

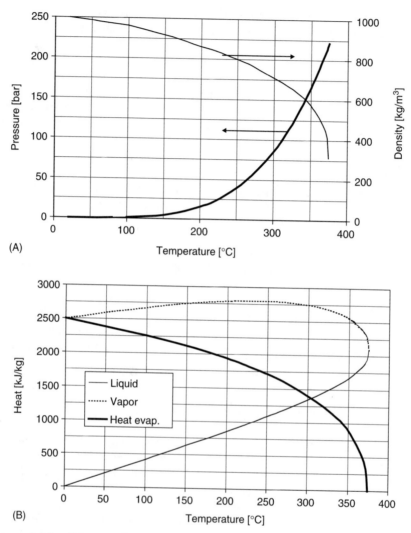

(A)

(B)

Figures 6-2A and B. Properties of Water for Extreme Preheat

temperatures enhances the atomization. The risk of rogue water droplets passing practically through the reactor completely is minimized (Bockelie et al. 2002).

3. More reactor space becomes available for the gasification per se, which will increase the carbon conversion.
4. The oxygen consumption will decrease and the cold gas efficiency will increase.
5. The water will expand (see Figure 6-2A), resulting in a lower water requirement to maintain good slurry conditions. To exploit this phenomenon to the fullest, use can be made of a circulating hot-coal slurry at high pressure, in which the relatively cold slurry leaving the slurry pump is blended in.

The water is preheated close to its critical point (see Figure 6-2B), the above effects will be most pronounced and the heat of evaporation is then minimum. Furthermore, not only the water but also the coal is preheated to above 300°C, a feature that is not possible with gasifiers using dry-coal feeding, because there the coal particles will become sticky and will interfere with the fluidization [Holt 2001(a)].

Where higher preheat temperatures can be accomplished in practice, the IGCC efficiency of a slurry feed process is almost equal to that of a dry-coal feed gasifier (49 and 50%, respectively; see Table 5-9).

As shown in Table 5-9, extreme coal-water slurry preheat is far more effective to increase the IGCC efficiency of a gasifier than adding a second stage to a dry-coal feed gasifier. The reason is that the efficiency of the state-of-the-art single-stage slurry-feed gasifier is lower to start with than that of a dry-coal feed gasifier (38 and 50%, respectively; see Table 5-9), and it is much easier to improve the efficiency from a low level of 38% than at a level of 50%. Adding a second stage to a dry-coal feed gasifier increases the efficiency by only 1–2%. The main reason for this low increase is that the steam that is injected into the second stage is for thermodynamic reasons only partly converted, and the remainder just adds burden to the gasifier. Moreover, the syngas cooler duty is decreased as the outlet of the gasifier drops from 1500 to 1100°C. These disadvantages in large part outweigh the advantage of the lower oxygen consumption. By combining the ability of a slurry-feed process to operate at higher pressures of, say, 70 to 100 bar with high-temperature slurry preheating, additional efficiency gains can be made by using a fuel gas expander in the solids free gas.

6.2.3 Wet Lock Hoppers

In principle it is possible to pressurize coal slurries making use of lock hoppers. It offers certain advantages over piston pumps for pressurizing coal-water slurries. Moving the coal-water slurry in a piston device is not simple. Valves for this service that are required in both lock hoppers and piston pumps are much less of a problem. Moreover, thicker slurries can be transported in lock hoppers. Before reaching the burners this thick slurry (paste) will become more dilute due to the expansion of the water.

The slurry is fed to the lock hopper with an open top valve and closed bottom valve (see Figure 6-3). Subsequently, the top valve is closed and the lock hopper is pressurized by opening a valve in a gas line between the high-pressure space and the top of the lock hopper. Then the bottom valve of the lock hopper is opened and the slurry leaves the lock hopper, flowing into the high-pressure system. The bottom valve of the lock hopper is then closed, and a second valve in the bottom of the lock hopper is opened through which water is admitted that pushes the gas in the lock hopper back into the high-pressure space. All valves are closed, and the top valve of the lock hopper is opened, thereby depressurizing the lock hopper. The water is then drained from the lock hopper through a valve in the bottom of the hopper, after which the cycle is repeated.

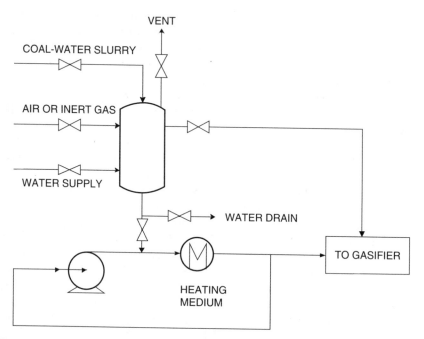

Figure 6-3. Wet Lock Hopper System with Heating Circuit

In principle, it is possible to pressurize the slurry to 200 bar, which enables subsequent preheating to about 350°C without steam formation. This requires a high-pressure lock hopper, but the temperatures in this hopper are below 100°C, so low-alloy steels can be used.

Almost the same system can be used for pressurizing dry lump coal. After the lock hopper has been filled with water to push out the syngas in the hopper, all valves are closed and the valve in the water drain is opened, thereby depressurizing the lock hopper. After draining the water, the valve in the drain is closed and the cycle is repeated.

The only place where in some gasifiers wet lock hoppering is currently practiced is in the discharge of the slag-water slurry from the gasifier.

6.2.4 Tall Hoppers for Pressurizing

Another option is to feed the coal via so-called dynamic hoppers (see Figure 6-4) (Visconty 1956). Pulverized coal is fed to the top of a high hopper and flows down through the hopper to a vessel that is pressurized. The height of the coal column in the hopper should be such that the pressure in the vessel is less or equal to the static height of the column. Moreover, the downward velocity of the coal should be higher or equal to the velocity of the gas from the pressurized vessel that wants to migrate through the interstices between the coal particles to the lower pressure space at the

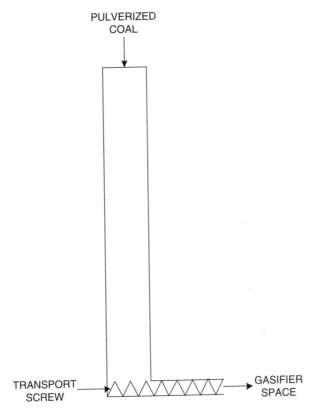

Figure 6-4. Tall Hopper Feeding Device

top of the column. When this is not the case, the gas has a chance to expand and the coal column would be blown out of the hopper. The relevant formulae are:

$$\Delta p < \rho h$$

and

$$v_{coal} \geq \bar{v} = \frac{\Delta p \cdot \varepsilon^2}{5 \cdot h \cdot \eta \cdot S^2}$$

where Δp is the pressure required to overcome the difference between, for example, the pressure in the gasifier and the atmospheric pressure. The bulk density of the coal is ρ, the height of the column filling the tall bunker is h, the average vertical velocity of the gas in the interstices is v, the dynamic viscosity of the gas η, the specific surface area of the pulverized coal is S, and the porosity of the coal/gas mixture in the column is ε. The "5" is an empirical constant.

Taking SI units and the bulk density of coal as $1000 \, kg/m^3$, the minimum height of the bunker to overcome a pressure difference of 10 bar ($= 10^6$ Pascal) is 100 m. Taking $h = 100 \, m$, η as $1.5 \times 10^{-6} \, kg/s.m$, S as $40000 \, m^2/m^3$, and ε as 0.4, this results in $v_{coal} = 0.13 \, m/s$. This means that for a $2000 \, t/d = 0.023 \, m^3/s$ coal gasifier, the bunker should have an internal cross-section of $0.023/0.13 = 0.18 \, m^2$, that is, 500 mm diameter.

The calculation shows that such bunkers are very thin and could advantageously be located near, or rather be incorporated in, a tall structure such as a stack. The coal can be either transported pneumatically to the top of the bunker or by mechanical means.

A problem with the use of tall hoppers is that the maximum pressure that can be obtained in a one-stage operation is about 20 bar, and even then a 200 m high structure is required. Proposals have been made to build them as a multistage machine in which much higher pressures can be reached. Were a 200 m structure to be built, a pressure of 40 bar could be reached with a two-stage operation. This would be sufficient for most applications (van der Burgt 1983).

Hang-ups in the bunker are not so likely to occur, as the bunker will operate near the point of incipient fluidization. Moreover, they can be avoided by building a bunker with an annular cross-section where the center part is turning very slowly to avoid any bridging or blocking.

The use of a small amount of inert gas in the column is probably mandatory.

The principle of these hoppers is the same as that of centrifugal devices that have been proposed (van der Burgt 1978, 1982). The difference is that instead of the centrifugal field, the gravitational field is used. In all these dynamic hoppers there will be hardly any contamination with the gas at the top of the hopper.

6.2.5   Miscellaneous Methods of Pressurizing

Coal Preheat with Hot Syngas

In various process schemes it has been proposed to inject coal-water slurry into the hot gas leaving the gasifier, thus cooling the gas, evaporating the water, and preheating the coal. This preheated coal is separated in cyclones and is fed to the gasifier as a dense phase in syngas (van der Burgt and van Liere 1994). Such a process scheme achieves a similar effect to the countercurrent flow in a Lurgi gasifier or the second stage of an E-Gas gasifier. It has also been proposed for feeding Victorian browncoal in Australia (Anderson et al. 1998).

Solids Pumping

Various other types of equipment have been developed for pressurizing pulverized coal, but thus far none have been very successful. Examples are piston machines and the Stamet pump (Chambert 1993; O'Keefe 1994). However, mostly the machines are either too complicated or vulnerable to fouling and erosion. Moreover, most of them are only suitable for pressurizing up to 5–10 bar.

Atmospheric Operation

One conclusion to be drawn from the whole discussion of pressurizing coal is that it is not easy, and there is a permanent concern for availability. This has led some to question the wisdom of trying and to propose an advanced atmospheric process as a preferred development route (Davey et al. 1998). The economics of such an approach are discussed in Chapter 9.

6.3   COAL SIZING AND DRYING

6.3.1   Coal Sizing

Coal preparation before gasification always involves some form of sizing of the coal. For moving-bed gasifiers this could be restricted to crushing of the coal followed by sieving out the lumps required for gasification. A separate drying step is not required for moving-bed gasifiers as the drying takes place in the gasifier itself, using the lowest level sensible heat in the product gas.

For the finer-sized coals or petcoke it is essential for dry-coal feed entrained-flow gasifiers that the feed is dry before entering the ring-roller mill that is usually used for the grinding. The drier is integrated with the mill in one recycle loop that is directly heated with a gas burner, as used in conventional pulverized coal (PC) boilers. For details, the reader is referred to the literature in this field (Perry and Chilton 1973, pp. 8–16; Stultz 1992). The gas that dries the coal is part of the same loop that also classifies the coal particles leaving the mill. For fluid-bed gasifiers, drying is not always essential, and more diverse grinding machines are used that depend on the feedstock and the particle size required.

For coal-water slurry gasifiers, driers are not required, and rod mills are mostly used for size reduction.

6.3.2   Coal Drying

Conventional drying processes are thermally based and typically use natural gas as a fuel. For coal gasifiers this is equivalent to using clean gas to dry the coal, and hence a prime product is used for drying the feed. Exergetically this is not attractive, as it lowers the overall energy efficiency. Therefore, for drying the coal it is better to use waste heat of an appropriate low temperature level.

In an IGCC scheme the most logical solution would be to dry it in direct contact with the warm exhaust gas from the heat recovery steam generator (HRSG) of the gas turbine. In this case, low-level heat is used for the drying. However, the exhaust gases from all present gas turbines contain about 15 mol% oxygen, which introduces the danger of spontaneous combustion in the dryer. Recycling flue gas back to the compressor inlet to replace part of the large excess air flow, as discussed in Section 7.3.4, would reduce the oxygen content to 3–5 mol% and would eliminate this problem. The dried coal would then still require a separate coal pressurizing system.

Using coal-water slurries eliminates the problem of drying the coal. By injection of the coal as a coal-water slurry into the hot gas leaving the gasifier, use can be made of waste heat in the fuel gas to evaporate the water from the coal, while still allowing use of the more elegant coal-slurry pump for pressurizing. This principle is used, for example, in the E-Gas process (Section 5.3.5). The problem remains, though, that drying with such hot gases always introduces the risk of some tar formation, and one of the advantages of entrained-flow slagging gasifiers is that no tars were formed in the gasifier proper. Although better than using clean syngas or natural gas for drying, using such high-temperature sensible heat for drying is also exergetically not attractive.

6.4   REACTOR DESIGN

6.4.1   Reactor Embodiment

Gasification reactors vary from process to process. At first glance, moving-bed processes may look like a simple stove, but they are in fact mechanically often the most complicated. This complexity is either due to the presence of stirrers in the top of the reactor or by the presence of a rotating ash grid in the bottom of the reactor as, for example, in the Lurgi dry-ash gasifier (see Figure 5-1). In the BGL-Lurgi slagger the rotating ash grid is absent and the ash is present as a molten slag.

With fluid-beds the complexity lies in the fluidization. For the remainder, the reactors are relatively simple. The main advantage is that temperatures are below the ash melting point, and pressures are also lower than in most other processes. Insulating refractory is used to protect the reactor pressure shell from high temperatures.

Most complex are the reactors of entrained-flow slagging gasifiers. The containment of both the high temperatures of 1500–1700°C in combination with pressures of 30–70 bar is not easy. The wall construction will be discussed in Section 6.4.2

The simplest and lowest cost construction is a refractory brick-lined reactor. In the case of top-fired single-stage gasifiers, it consists of a simple cylindrical vessel with, in principle, a hole in the top for the burner to supply the feedstocks, and a central outlet in the bottom for both the product gas and the slag. The reactor has only one burner, which has the advantage of a simple construction and a simple control system. Also, having one outlet for both the product gas and the slag is an advantage, as the chance of plugging the slag tap or the gas outlet is virtually absent even in small-capacity gasifiers. Last but not least, the construction is completely rotational symmetrical, which makes it elegant and relatively low cost.

An additional advantage of an insulating brick-wall reactor, which has been used for decades in all oil gasifiers, is that in contrast to a membrane wall the brick lining has a large heat capacity. Therefore, no extra wall penetrations are required for an ignition burner. The heat-up facility can be integrated into the main burner. The reactor and the hot face brickwork is heated first with a start-up burner, after which the start-up burner is replaced with the coal or oil burner and the reactor is (partly) pressurized. Ignition of the flame follows immediately after introduction of the reactants at the hot brick wall, after which the final reactor conditions are obtained.

Up-flow slagging entrained-flow processes provide separate outlets for the gas at the top of the reactor and for slag at the bottom. The burners are located in the cylindrical wall of the reactor near the bottom (see Figure 5-15). This construction is more complex than the simple cylindrical reactors. Wall penetrations are numerous as the gasifiers feature (at least) four burners, and a complex system is required for the ignition burner.

6.4.2 Reactor Containment and Heat Loss

Gasification is a process that is carried out under harsh conditions. Even in a steam methane reforming furnace where the feedstock, natural gas, is clean and the use of catalysts allows syngas generation at 850–900°C, special alloys are required to work at their limit. In a gasification reactor the temperatures can be much higher—up to 1500–1600°C—making the environment subject to chemical attack from slag and the pressures are often higher.

Refractory Linings

The simplest and lowest cost design of a reactor wall is a lining with a refractory capable of withstanding the prevailing temperature and chemical conditions.

Refractory lining is used universally for partial oxidation of petroleum residues and natural gas. Typically, the design consists of three layers, as shown in Figure 6-5. The inner "hot face" layer is a high-quality corundum brick (>99% Al_2O_3) suitable for temperatures up to about 1600°C. The intermediate layer is a castable bubble alumina, and the outer "cold face" is a silica firebrick with good insulation properties. This three-layer design combines the properties of high temperature resistance and good insulation. At the same time, it hinders the propagation of cracks, which may arise in the hot face through to the vessel shell. The design is selected to ensure that no condensation takes place on the inner wall of the steel shell, which would for a pressurized reactor typically have an operating temperature somewhere between 200 and 300°C. In locations with extremes of temperature, this requires care to ensure that the wall temperature does not fall below the dewpoint of the synthesis gas in a cold winter wind, but the maximum allowable wall temperature is not exceeded in summer.

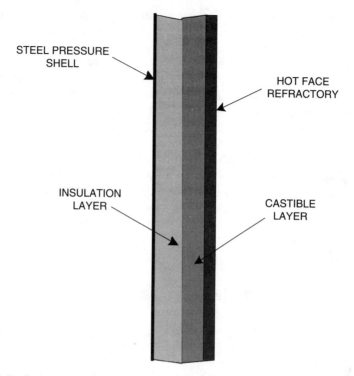

STEEL PRESSURE
SHELL

HOT FACE
REFRACTORY

INSULATION
LAYER

CASTIBLE
LAYER

Figure 6-5. Refractory Wall

In oil service the principle source of chemical attack is from the vanadium content in the feedstock. In normal operation this is not a major concern, since in the reducing atmosphere the vanadium is present as V_2O_3. Care must be exercised, however, during start-up or shutdown to avoid significant quantities of V_2O_5 being formed, since at temperatures above 700°C this is liquid and penetrates the refractory very quickly, causing a breakdown of the bonding matrix. With suitable operating procedures, refractory lifetimes of between 25,000 and 40,000 hours are possible.

For coal, however, the nature of the ash creates a very different situation. The large quantities of silica in most coal ashes would break down an Al_2O_3 hot face in a very short time (weeks rather than months). In addition to chemical attack, the refractory is also subject to erosion by the liquid slag flowing down the wall, although this only makes a minor contribution to the refractory wear. The solution currently employed is to use chromium oxide and/or zirconium oxide-based refractories, which have a better chemical resistance to the specific atmosphere prevailing. Nonetheless, this still cannot be considered satisfactory. Lifetimes are reported of 6–18 months (Clayton, Stiegel, and Wimer 2002), and considering that replacement of a refractory lining requires three to four weeks offline, a 25,000-hour life as with oil gasifiers must be achieved. A problem with all hot-face bricks is that for corrosion and erosion reasons these fusion cast materials should ideally be mono-crystalline, but because of thermal requirements during start-up and shutdown, poly-crystalline materials have to be used. The latter materials are less corrosion- and erosion-resistant at the crystal boundaries but have the advantage that they are more resistant to spalling.

Research into improved linings is being conducted as described by Dogan et al. (2002). In her paper Dogan describes the mechanism of liquid and vapor-phase penetration of silica, calcium oxide, and alumina into the matrix of the refractory. Subsurface swelling occurs and subsequent cracks develop parallel to and about 1 to 2 centimeters below the surface of the brick. While these cracks are developing, wear rates of about 0.003–0.005 mm/h can be expected, but when the cracks reach the edge of a brick, then there is a sudden loss of the whole material in front of the crack. Dogan then describes the development of a phosphated chromium oxide refractory with better resistance to liquid penetration. It is anticipated that test panels of this modified refractory may be installed into a commercial reactor late in 2003.

Fluid-bed coal gasifiers also have an insulating brick wall comprising dense erosion-resistant bricks and insulating bricks. Temperatures can be as high as 1150°C. Although there is no liquid slag present, there is mechanical erosion from the ash, limestone (for sulfur removal), and sometimes the heat carrier, that are circulated at high velocities in these gasifiers. The shape of the CFB and transport gasifiers is more complex (see Section 5.2); the construction in general and the domed and vaulted "roofs" in particular must be designed so as to keep their integrity over the whole temperature range, from ambient to the gasification temperature.

Biomass gasification, which is virtually always carried out in a fluid-bed, often in the presence of sand as a heat carrier, is performed at the lowest temperatures. This low temperature of 900–1050°C is often determined by the ash quality of the biomass rather than by the intrinsic reactivity towards gasification per se. Biomass ashes have relatively low softening and melting points and when molten are extremely aggressive in terms of corrosion owing to the high salt content.

Membrane Walls

The alternative to refractory linings is a water-cooled membrane wall construction such as that shown in Figure 6-6. The design shown is that of the GSP reactor, but it is typical also of other entrained-flow slagging gasifiers such as SCGP and Prenflo.

The membrane wall consists essentially of high-pressure tubes, in which steam is generated, connected by flat steel bridges of which the width is about equal to the outer diameter of the tubes. Tubes and bridges are welded together into a gas-tight wall. The tubes are provided with studs that act as anchors for a thin layer of castable refractory, usually silicon carbide. During operation of the gasifier the castable will ideally be covered by a layer of solid slag, over which the liquid slag will run to the bottom of the reactor (see Figure 6-6). In principle, the castable is not required, as it mainly acts as a "primer" to which the slag can adhere. There is a chance, though, that without this "primer" the coverage of the tube wall with slag may be erratic.

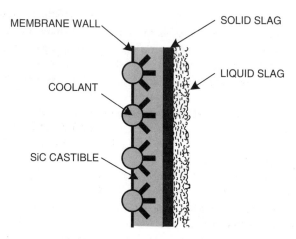

Figure 6-6. Membrane Wall *(Source: Lorson, Schingnitz, and Leipnitz 1995)*

In the ideal situation, the first liquid slag hitting the wall solidifies and forms a layer of solid slag on the castable. The elegance of a membrane wall is that the liquid slag then only comes in contact with solid slag, and hence no corrosion or erosion of the reactor wall takes place. Small temperature excursions will cause the boundary between solid and liquid slag will move a little, but in principle the solid slag layer seals the whole wall. Such a wall is self-repairing. Hence the membrane wall is very robust and has a long life. A service life of eight years or more can be achieved.

One drawback is that the heat loss through the reactor amounts to 2–4% of the heating value of the coal, whereas with an insulating brick wall this heat loss is less than 1%. In the case of a membrane wall, the heat loss is mainly determined by the radiant heat of the reactants and the total surface area of the gasifier reactor.

Another drawback is the high cost of a membrane wall. The wall itself is already expensive, but for constructional and maintenance reasons, these reactors are built with a space between the membrane and the steel outer shell of the reactor, and therefore the cost of penetrations for burners and instruments is also high. This space must also have an open connection with the gasification space, as the membrane wall is not built to stand pressure differences across the wall.

For control purposes, it is a great help if the steam production of the membrane wall can be accurately measured, as the heat loss through the wall will then be known. This heat loss is an important variable in reactor simulations and therefore for reactor control reasons. Moreover, the steam make is an important indicator for the reactor temperature. Its accuracy is however, heavily influenced by the state of the refractory/slag covering of the wall.

Jacket Construction

The use of a steam-generating water jacket is a well-proven solution in the context of the Lurgi dry-bottom gasifier and the Koppers-Totzek gasifier. The GSP technology also uses a jacket for applications with low ash feedstocks (Schingnitz et al. 2000). Internal jackets are an elegant and low-cost solution for protecting the pressure shell from high temperatures. The space within the jacket should be in open communication with the gasifier proper, as the internal wall of the jacket cannot withstand pressure differences. This (steam) connection may be located well downstream of the gasifier. Advantageously, the connection is, for example, made before a CO shift in case synthesis gas has to be produced or after the gas cleaning section in case the gasifier is part of an IGCC. The steam from the jacket can also be used as (part of) the quench gas. For slagging gasifiers, the hot inner wall of the jacket has to be protected on the gas side with a castable that must be anchored with studs.

The quality of the steam produced in a jacket is rather low, as the pressure of the saturated steam has to be equal to the pressure in the reactor, whereas in the tubes of a membrane wall, saturated steam of 100 bar can be produced. However, the jacket construction is not only lower in cost (not least because the vessel diameter can be up to a meter smaller), but also wall penetrations for such things as burners and instruments are simpler than for reactors with a membrane—or an insulating brick wall, because the wall is only 10–15 cm thick instead of 60–70 cm.

Heat Loss Calculations at a Membrane Wall

Heat losses through a brick lined reactor wall can be calculated easily. This is not the case for a membrane or jacket wall where liquid and solid slag layers cover the wall (Reid and Cohen 1944). The companion website includes a program to calculate the heat loss through all types of vertical reactor wall.

In the calculations, steady-state conditions have been assumed where the heat is flowing in a horizontal direction. The latter assumption is approximately correct, as the vertical heat flow is virtually limited to the flow of sensible heat contained in the liquid slag flowing down the vertical wall. In the calculations it has been further assumed that both the slag and the castable have a fixed melting point rather than a melting range. When the proper input data are used, good approximations can be obtained. For details about the calculation methods used in the program, the reader is referred to the help files associated with the program on the website.

Six different situations on a membrane wall may be distinguished, which are illustrated in Figure 6-7. Some of the conclusions of the calculations made with the computer program are discussed in the following.

1. Conditions where the refractory is at its melting point or can react with gaseous components should be avoided at all times, as the solid slag will then not adhere to it and its function as a "primer" for the subsequent slag coat is lost. Most important for a membrane wall is that the whole wall of the reactor is covered with solid slag. Without such a slag layer the wall will have a large heat loss, as the membrane is only protected by a thin layer of castable that has a relatively high heat conductivity owing to the use of SiC and the steel studs by which it is anchored to the membrane wall.

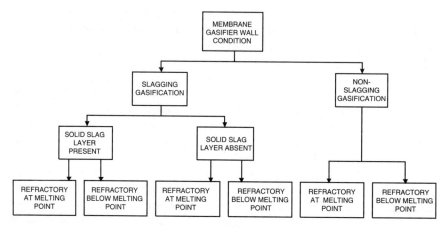

Figure 6-7. Conditions of a Membrane Wall

2. The above has consequences for the gasifier start-up procedure and operation if coal with a very high ash-melting point of, for example, 1650°C is used. The reactor must be started with a temperature well above the 1500°C required for gasification until the wall has been covered with a slag layer. This results temporarily in a less than favorable set of reaction conditions: a higher oxygen consumption and a lower CGE and gas make. Once the slag layer has built up, the temperature may be lowered and the operation will become nonslagging. Although this may result in a lower heat loss, the advantage of a slagging gasifier that most of the ash in the coal is turned into inert slag is lost. For this reason it is more attractive to add flux to the coal in order to lower the ash melting point and ensure a slagging operation.

The use of a thicker layer of castable on the membrane wall is not an alternative to using a slag layer for insulation under these circumstances, as the refractory may either melt or react with the gaseous reactants.

3. The design heat loss of the reactor will always be based on the ideal wall where a layer of solid slag covers the refractory, and this in turn is covered by a layer of liquid slag. Decreasing the melting point of the slag by adding fluxing material to the coal will always result in a lower heat loss and will make it possible to run with the ideal wall condition with almost any coal.

When processing feedstocks with a low ash content, the heat loss through the wall will only increase marginally although it will take longer to build up the layer of solid slag. The same holds for slags with a low viscosity. Low viscosities will result in a thicker layer of solid slag and a thinner layer of liquid slag.

4. The layer of liquid slag depends on how much slag reaches the wall. An increase in this slag flux can be accomplished by introducing the reactants in such a way into the cylindrically shaped reactor that the slag will preferentially be deposited on the wall, for example, by giving the reactants some swirl upon leaving the burner. Care should be taken to ensure that this cyclonic motion

does not result in a countercurrent flow in the center of the reactor, as this may have undesirable side-effects at the reactor outlets. Some swirl is also favorable for a good carbon conversion, since this will also increase the residence time (see Section 5.3.8).

5. The conditions in the reactor often make it ideal for reducing iron compounds present in the ash to liquid iron. For the vertical wall of the reactor this is not much of a problem, but in the bottom of the reactor this may lead to a layer of molten slag floating on top of molten iron. This situation is similar to that encountered in the bottom of a blast furnace. Generally, the geometry of the bottom of the reactor is such that the iron flows out of the reactor together with the molten slag. The iron is then found in the slag as small lumps that gives the slag particles a rusty appearance.

6.5 BURNERS

Most burner designs are confidential, although a fair amount of knowledge can be gleaned from the patent literature. Most of the burners (sometimes known as combustors or feed injectors) used are of the co-annular type where the reactants are fed through axisymmetrical annular openings at the front of the burner. The burners have to be protected from the hottest part of the reactor. Hence, having the oxygen come in contact with the hot syngas in the reactor near the burner opening must be avoided. Very large capacity burners are only possible when the coal also enters the reactor via an annular slit (van der Burgt 1990). In virtually all cases the reactants come into contact with each other inside the reactor. Premixing the reactants is not recommended because of the dangers associated with precombustion.

Special attention has to be paid to cooling the burner front. Water-cooling is applied in most cases. For safety reasons, the pressure of the water should be higher than in the gasifier so that the gas can never enter the cooling water system. This has the advantage that there is a large degree of freedom for selecting the cooling water temperature. In general, this temperature should not be too low, as then the steam present in the blast could condense inside the burner unless precautions are taken to insulate the steam tract in the burner.

For pressurizing gasifiers, special attention must be given to the heat-up procedure since the gas flows during this operation differ considerably from the normal design case. Where the reactor has an insulating refractory wall it is possible to heat up this wall with an atmospheric pressure gas or oil burner, exchange the burner for a coal burner, and ignite the coal/blast mixture on the hot brick wall. In the case of membrane walls with their very limited heat capacity, this is not possible as they will cool in less than a minute, and burner changes cannot be made in this short period. Therefore, in the case of a membrane wall gasifier, the ignition burner must remain lit until the first coal burner ignited. During this period there must be a continuous flame while the pressure is increased from atmospheric to operating pressure. This is particularly complicated for reactors with multiple burners or where the heat-up burner is not integrated into the main burner.

The relatively long time that is required for this operation is one of the reasons why IGCC power stations are not generally considered suitable for peak shaving duty. On the other hand, where a reactor is kept on hot stand-by, a quick start is possible, particularly since in most cases the procedure is at least semi- if not fully automatic.

Burners for oil service are also generally of a water-cooled co-annular design (Pelofsky 1977; Weigner et al. 2002). The design of such burners, which are centrally top-mounted, can include a removable gas-fired start-up burner with internal igniter. Weigner et al. (2002) describes an automatic temperature ramp system integrating firing of the start-up and main burners during reactor heat-up. The start-up burner is removed at 1100°C prior to ignition of the main feedstock. With such a burner, a turndown ratio of 60% is achievable.

A similar arrangement can be seen in the top-fired GSP reactor, which incorporates a central gas flow to the pilot burner surrounded by annular slits for oxygen that incorporate a swirler and an outer slit for fuel (Schingnitz et al. 2000).

Burner lifetime for coal service, particularly for slurry feeds, continues to be a source of concern. Typical lifetimes of between two and six months have been reported (Clayton, Stiegel, and Wimer 2002). Burners in oil service achieve a service life of over one year (Higman 1994; Weigner et al. 2002), which is generally considered acceptable even if a "long-term goal of two years" would be desirable (Clayton, Stiegel, and Wimer 2002).

6.6 SYNTHESIS GAS COOLING

Gases leave the gasifier reactor at high temperatures varying from 550°C for some dry-ash moving-bed gasifiers to 1600°C for dry-coal feed entrained-flow slagging gasifiers.

With the exception of natural gas feeds, the synthesis gas from a gasification reactor is contaminated with various components that must be removed before the syngas is suitable for its final use, whether as chemical feedstock or as fuel. These contaminants, which may be particulates, sulfur or chlorine compounds, tars, or others, must be removed, and all such removal processes, even the so-called hot-gas clean-up processes, operate at temperatures considerably lower than that of the gasifier itself. Thus there is always a necessity to cool the syngas. In most cases it will also be desirable to make good use of the sensible heat in the gas, for example, by raising steam. On the other hand, differences in the contaminants, which vary from feed to feed, as well as characteristics of the different gasification processes themselves, lead to a considerable variety of solutions to the syngas cooling task.

The highest temperature gasification processes are the entrained-flow slagging processes. On cooling the gas any entrained-flow ash particles will inevitably pass through the critical temperature range, where the ash becomes sticky. Every gas-cooling concept has to take this into account and quench the gas as quickly as possible to a temperature at which the ash becomes dry, typically about 900°C. There are a number of different methods for achieving this, which are described in more detail in what follows.

Fluid-bed coal gasifiers have typically outlet temperatures of 900–1000°C. Apart from some problems with tar, the hot gases can be used for the generation of a reasonable quality steam in a syngas cooler. Even after passing cyclones, the gas does contain some fly ash, and hence, in the design of the syngas cooler, the danger of potential erosion problems should be taken into account. When limestone is added to the feed to bind the sulfur, unconverted CaO may react back with CO_2 in the gas to $CaCO_3$ at temperatures below 950°C, which may enhance fouling in the syngas cooler.

Most biomass gasifiers operate around 900°C and the same syngas cooling issues apply as for other fluid-bed gasifiers. Moreover, as biomass ashes are rich in alkali carbonates, these may condense out or desublimate and cause additional fouling between 600 and 900°C. At lower temperatures the condensation point of volatiles in the gas is the main cause of fouling, as the condensate is ideal for catching ash particles.

Although in the case of moving-bed gasifiers the temperatures are low (300–550°C) the cooling of the gas is often complicated by the presence of tars in the gas that may foul heat exchangers when the temperatures drop below the condensation point of the heaviest tar components. In practice, this means that only low- or medium-pressure saturated steam can be produced in these syngas coolers.

In the Lurgi dry-ash and the BGL slagging gasifier the transition from slagging to nonslagging regime occurs within the coal bed and does not cause problems as the bed is continuously moving in a downward direction, which keeps the reactor wall free from slag deposits.

6.6.1   Quenching

As discussed above, the most demanding syngas cooling equipment is required for single-stage entrained-flow slagging gasifiers. The key problem is the transition stage between slagging and nonslagging conditions. This transition temperature range has to be crossed directly after leaving the slagging reactor, and ideally in such a way that the gas does not contact a wall before it is sufficiently cooled. One solution is to "sleeve" the inside of the quench section with clean gas (Staudinger and van der Burgt 1977).

For entrained-flow slagging gasifiers quenching can be accomplished in four different ways.

Radiant Syngas Cooler

Although attempts have been made at quenching by allowing the hot gas leaving the reactor flow into a radiant boiler, it appears to be difficult to ensure that liquid or sticky slag particles do not hit the wall and cause fouling. Moreover, radiant boilers have the disadvantage that they scale awkwardly. For a scale-up factor of say 2, keeping the gas velocities constant, the vessel diameter scales by $\sqrt{2}$, whereas the requirement of surface area increases by a factor of 2. The height therefore also has to be increased by $\sqrt{2}$. This is in contrast to other heat exchangers, for which the volume increases proportionally with the throughput of the gasifier without an increase in height. The reason is that only the surface of the vessel can be used for heat exchange. A solution has been sought in extending the wall surface of the radiant boiler by installing ribs perpendicular to the surface, but this further complicates the construction. Radiant coolers can be prone to fouling, and they are difficult to clean by rapping. Furthermore, the heat dissipated by the wall can only be used for generating saturated steam.

All this makes radiant syngas coolers an expensive piece of equipment in practice. For the 250 MW$_e$ Polk Power Station IGCC in Florida, the radiant syngas cooler (RSC) is "about 16 feet in diameter and 100 feet long, and weighs about 900 tons" (U.S. Department of Energy 2000). On the other hand the reported reliability of this cooler is satisfactory (McDaniel and Hornik 2000).

Water Quench

Hot gas can be quenched by evaporation of water into the gas. It is necessary to distinguish between a partial quench, in which only just enough water is evaporated to reduce the gas temperature to 900°C, and a total quench, in which sufficient water in evaporated to saturate the gas with water vapor.

A partial quench is a well-proven quenching system that was already applied in the atmospheric pressure Koppers-Totzek gasifiers between the burners and the radiant syngas cooler above them. While effective in this configuration, replacement of nozzles as a result of wear does represent a maintenance cost, though not a limitation on reactor run-time length. GSP also offers a partial quench system. The advantage of a partial quench is that it allows the sensible heat in the syngas to be exploited for high-pressure steam raising from 900°C in a downstream syngas cooler.

A total quench has been a feature of Texaco's oil gasification process since its inception and has also been adopted in most of the Texaco coal gasifiers. It is a low-cost and effective solution but has as a disadvantage that exergetically it is not very elegant. High-level heat that potentially can be put to better use is degraded to water vapor in the still dirty gas. Upon condensation, which is in any case required for fuel gas treating as practiced in most present day plants, this water will become available as a contaminated condensate stream that requires extensive cleaning. For slurry-feed processes, this problem of a large waste water plant is diminished by using this waste water for making the coal-water slurry feed for the gasifier.

If the final product is ammonia or hydrogen, the water vapor in the gas from a total quench may prove to be advantageous, as no additional steam has to be generated for the subsequent CO shift process. The only clean-up of the gas that is then required between the gasifier and the CO shift is a thorough solids removal. Either a hot water wash or candle filters will do this job.

One point to notice about water quenches, whether partial or total, is that the introduction of the water drives the shift reaction (2–7) to the right, and thus the CO_2 content and the H_2/CO ratio of the gas are increased to some extent.

Gas Quench

The gas quench is used in the SCGP and the Prenflo processes. The raw synthesis gas, which has been cooled in the syngas cooler and freed of solids in a candle filter, is split into two approximately equal portions. One is recycled with a compressor and used to quench the gas leaving the gasifier from about 1500 to about 900°C, and the remaining net gas production is routed to further downstream processing.

With gas cooling it is also possible to cool the gas further, to below 900°C, but in this case the amount of recycle gas required for the cooling will increase substantially. Even for the cooling of the gas from 1500 to 900°C, the molar gas flow (although not the volumetric flow) doubles, and as the heat has to be removed eventually by indirect means to make the quench effective, this leads to voluminous heat exchangers. Therefore, in practice, quenching with gas is limited to temperatures of 900°C. This is about the same temperature obtained after a chemical quench or after passing a radiant boiler.

Chemical Quench

In a chemical quench, ideally the sensible heat in the gas leaving the first slagging stage of an entrained-flow gasifier is used in the endothermic water gas reaction to gasify a second-stage feed. The second stage may be a dry feed as in the Japanese EAGLE and CCP processes, or a slurry feed as in the E-Gas process. Where the quench medium is a coal-water slurry a significant percentage of the heat is used to heat up this medium, to evaporate the water and for pyrolysis reactions, so that at least part of the cooling is actually attributable to a partial water quench. Either way, the heat absorbed is sufficient to cool the gas such that the ash from the second stage is dry.

Injecting coal as such or as water slurry into the hot gas leaving the first slagging stage has the disadvantage that some tars may be formed. In practice, with the E-Gas process this does not happen in normal operation, although it has been known to occur during upsets or low load operation (U.S. Department of Energy August 2000).

By introducing the chemical quench or a second non-slagging stage to a dry-coal feed-entrained slagging gasifier, a gasifier is obtained that has an outlet temperature some 400–500°C lower, and thus has a lower oxygen consumption as well as a higher CGE. As a result, the duty of the costly syngas cooler is substantially reduced. This has a cost advantage, which is attributable not only to the heat transfer surface area requirement, which is reduced by some 30%, but also to the possibility of using lower cost concepts such as a fire tube boiler.

The efficiency gain for a dry-feed gasifier is limited (see Section 5.3, page 114) but offers the advantage over a single-stage gasifier with the same outlet temperature of 1000–1100°C is that the bulk of the ash in the feed be comes available as an inert slag. The second non-slagging stage can be a simple brick-lined pressure vessel.

6.6.2 Synthesis Gas Coolers

When the gas from a slagging gasifier is quenched to about 900°C, or where the gas is produced at temperatures in the 900°C to 1000°C range, it has to be cooled further before the gas can be treated for use. Two aspects of gas cleaning have to be considered carefully and intimately with the design of this section of the cooling system. These are particulate removal and condensation, whether it be condensation of tars from biomass gasification, for example, ammonium chloride from coal gasification, or simply water.

The first cleaning stage after the syngas cooler comprises the removal of any solids present in the gas. Effective solids removal is only possible at temperatures below 500°C, whereas for the removal of acid gases and ammonia the gas has to be further cooled to essentially ambient temperatures.

The lowest-cost method of cooling the gas is to continue quenching to the temperatures required for the gas cleaning. This practice is only possible with a water quench. It results in the gas being loaded with even more steam which then has to be condensed out when acid gases and ammonia are removed from the gas.

In a typical syngas cooler the gas is cooled from 900 to 300°C. At 900°C there are no sticky ash or slag particles left in the gas, and at 300°C there is as yet no chance of NH_4Cl deposits. This is the ideal temperature range for raising good quality steam and for preheating the clean gas that is obtained after the near ambient temperature gas treating. Such preheating is beneficial, for example, when the gas is later used as clean fuel gas in a combined cycle power plant. Whether such pre-heating is economic is another matter, as it requires an expensive gas-gas heat exchanger. For corrosion and other material reasons, the metal temperatures should not exceed 500–600°C. When steam is to be made, the highest temperature gas is therefore used evaporation, followed then by a superheating section, and finally for further evaporation and water preheat. The temperature range of 300 to about 40°C is used for water preheat.

Where a gas quench is used, all the sensible heat in the gas leaving the gasifier is used for raising additional steam, which results in high efficiencies of the IGCC power station. The drawback is that this also results in the highest cost syngas cooler.

The syngas cooler is often one of the most expensive items in a gasification complex. Expensive high alloy steels have to be used in many places, as all the contaminants are still present in the gas. There is fly slag, which leads to erosion. There are also sulfur compounds, chlorine compounds, and so on. Frequent rapping of the boiler internals may be required, for example, to remove deposits from the boiler tubes. In order to accomplish this, expensive penetrations have to be made through the pressure wall of the syngas cooler.

As in so many occasions in gasification, there is the classical trade-off between efficiency and capital cost. Water quenching is cheap, but then efficiency is reduced; whereas with a syngas cooler, especially in combination with a gas quench, the capital costs are high but so is the efficiency.

Syngas Cooler Designs

There are two principle designs for syngas coolers: water-tube boilers and fire-tube boilers. Both have been operated successfully in various plants. Fire-tube boilers are lower in cost but have certain limitations, particularly with high-pressure steam. In practically all applications, the steam pressure is greater than the gas pressure, so that the tubes are subjected to external pressure. Depending on the details of the individual design, maximum steam pressures for fire-tube boilers lie between 100 and 150 bar. An advantage of fire-tube boilers is the well-defined flow of the gas in the tubes, but the inlets need to be designed carefully in order to ensure that the dust-laden gas does not cause erosion. Another detail to which attention must be paid is the adequacy of the cooling at the inlet where the heat fluxes are very high. In the field of oil gasification, fire-tube boilers are used almost exclusively, and some examples are discussed in Section 6.6.3.

With water-tube boilers the local flow pattern around the tubes is less even than a fire-tube boiler, and there can be areas of almost stagnant gas with the attendant risk of dust accumulation. A number of designs include rappers to shake off any dust (Keintzel and Gawlowski 1993). On the other hand, at the HTW plant at Berrenrath, rappers originally included as part of the design were dispensed with after tests showed them to be unnecessary. In fact, at this plant both fire-tube and water-tube boilers were tested in parallel, and both were deemed to be satisfactory. The conclusion of these parallel tests was that economics would be the deciding factor in syngas cooler design selection (Gorges, Renzenbrink, and Wischnewski 1998).

6.6.3 Syngas Cooling in Oil Service

There are a number of syngas cooler designs available for oil gasification service that have given excellent service over many years and that, in contrast to many designs for coal service, are relatively inexpensive. They all have a number of common features, not least that all are fire-tube designs. As discussed in Section 5.4, oil gasifiers are operated so as to leave a certain amount of residual carbon in the gas, and this ensures that the ash passes through the syngas cooler as a dry particulate. The heat transfer in the cooler takes place through a number of coils, which are designed with a gas velocity of 25–35 m/s and arranged vertically in a water chamber. The selection of gas velocity, tube size, and helical arrangement are chosen to ensure that ash and soot particles are transported through the waste heat boiler with neither fouling nor abrasion of the tubes. The tubes have two or three reductions of diameter over their length, so as to ensure that the velocity is maintained in the design range over the length of the tube. The very high gas temperatures and heat fluxes at the inlet demand extreme attention to detail in the design, and the solutions of this issue represent an important differentiation between the various designs.

Borsig Design

The syngas cooler design, which over the years has established itself as the standard for Texaco plants, was originally developed by Steinmüller (Figure 6-8) and is now manufactured and marketed by Borsig. Borsig's reference list includes 24 units with steam pressures as high as 110 bar.

The salient features of this fire-tube boiler are the coils and the hot gas inlet zone. The coils are made as individual "candles" mounted vertically and in parallel in the water bath. Each coil has its own "tube sheet" in the form of a double-walled tube. Boiler water flows through the annular space between the two tubes by forced circulation to provide intensive cooling. Recent developments include the use of stiffening ribs on the water side and a ceramic coating on the gas side of the inlet zone.

Shell Design

Shell has its own proprietary design for its SGP oil gasification process, which is used in about 135 installations worldwide (see Figure 5-27). Generally, these coolers are designed for the production of saturated steam (up to 120 bar), but designs can include a superheater as was installed for the natural gas fired unit in Bintulu. A design for a residue-based application with superheater has been announced (de Graaf and Magri 2002).

The Shell syngas cooler addresses the same issues as those described above, namely safe passage of the solids through the boiler coils and intensive cooling of the gas inlet zone. The Shell coil design integrates the coils into a single helix rather than having each coil separate. This keeps the radii of curvature larger than in the Borsig design, and thus the potential ovality of the tubes after bending is less. Since the mechanical design of the coils is defined by the maximum external pressure from steam side to gas side, this makes design for very high pressures somewhat easier.

Shell employs a patented double-tube sheet, one to provide the mechanical rigidity required, and in front of it a thin heat shield supported by the tubes. The incoming fresh boiler feed water cools the intermediate space.

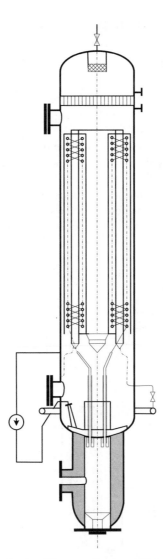

Figure 6-8. Borsig Syngas Cooler (*Source: Becker 1969*)

Alstom Design

A recent entry into the market for oil gasification syngas coolers is Alstom. For the tubes Alstom uses a multiple-candle concept similar to that of the original Steinmüller design. Alstom has its own solution to intensive cooling of the inlet zone (Alstom undated).

6.7   PARTICULATE REMOVAL

6.7.1   Dry Solids Removal

Only one step of the syngas treating can be carried out at an elevated temperature and that is filtering. The introduction of candle filters that can remove all solids from the gas at temperatures of up to 500°C was one of the most significant developments in gasification during the last quarter of the twentieth century. In Figure 6-9 a sketch is given of such a filter installation. The solids are deposited on the outside of the candles. Intermittently from the clean gas side the filters are blown back by a pulse of nitrogen or another gas that causes the solids, which have collected at the outside of the filters, to drop down to the bottom of the vessel, whence they can be removed via a lock hopper. The importance of this development for gasification-based power

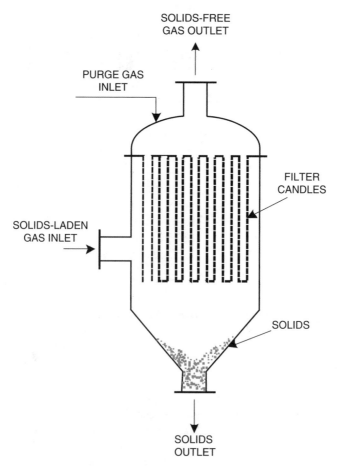

Figure 6-9. Candle Filter Vessel

stations is discussed in Section 7.3. Filtering has to be carried out at temperatures between 300 and 500°C. At about 250–300°C the filters may be blinded by deposits of NH_4Cl. Above 500°C the vapor pressure of alkali compounds may still be high, which means that significant amounts may pass the filters. Below about 500°C the amount of alkali compounds is negligible, provided they are properly filtered out. This is accomplished by removing them, together with the fly slag that acts as a substrate on which the alkali compounds are deposited. Operating in the higher temperature range near 500°C is also beneficial for avoiding problems with carbonyls. Carbonyls will hardly form under these thermodynamically unfavorable conditions. Any nickel or iron present in the gas will also be deposited on the fly ash at these high temperatures.

The candle filters are mostly made of ceramic material where a fine-grain ceramic layer is deposited on a wider pore support that gives strength to the filters. Special attention has to be given to the seal between the filters and the steel support plate. The filters may either rest on a steel plate or hang down from a steel plate, as shown in Figure 6-9.

Candle filter materials have also been developed on a metal basis. Although ceramic materials can be operated at higher temperatures than metals without risk of sintering, the latter are more robust and can resist localized damage without rupture. The selection depends on the location of the filter within any particular process and the situations to which it can be exposed.

6.7.2 Wet Solids Removal

In most existing plants the solids are washed out with water in Venturi scrubbers or wash towers. This scrubbing takes place below the dewpoint of the gas so that the finest solid particles can act as nuclei for condensation, thus ensuring that all solids are removed efficiently. The wet removal of solids causes them eventually to appear in the filter cake in the water treatment. The disadvantage of wet solids removal is that it is more difficult to separate the ash from compounds containing lead, zinc, cadmium, arsenic, and others, and thus increases the amount of chemical waste.

6.8 PROCESS MEASUREMENT

In this chapter the control of the gasifier itself will be discussed. This excludes the process control of, for example, an IGCC where the gasifier is often closely integrated with the ASU and the CC. However interesting the complex control of such systems is, this will not be discussed as it falls beyond the scope of this book.

6.8.1 Gasification Temperature Measurement

When operating a gasifier, one wants primarily to control the temperature. As discussed in Chapter 2, the temperature is the principle variable determining the gas composition.

The temperature is decisive in relation to the ash-rejection regime, whether slagging or nonslagging. Ultimately, too high a temperature will destroy the integrity of the reactor containment, be it refractory or membrane wall.

Unfortunately, any accurate measurement of a gasifier temperature is extremely difficult, both for a number of practical reasons and, surprisingly, also for one theoretical reason. Let us look at the latter first. The problem is that where there are still solid particles present that contain carbon, the measurements are influenced by the phenomena of the "chemical wet bulb temperature." This effect was first explained by van Loon (1952, pp. 17–34), who showed that at the surface of the solid carbon only endothermic reactions with H_2O and CO_2 occur, whereas in a sort of halo around the particles part of the CO and H_2 formed by these reactions, react exothermically with oxygen to form H_2O and CO_2 (see Figure 6-10). This mechanism renders pyrometers of little use for exact measurement, since one cannot establish whether the temperature measured is dominated by the relatively cool but more strongly radiating solid particles or by the hot gases in the halo.

The practical problems of temperature measurement are particularly relevant to entrained-flow slagging gasifiers. Partly this is caused by the harsh conditions of high temperatures per se, but also by the fact that slag can attack ceramic protective sheathing around the thermocouple, causing erosion damage to thermocouples by the ash and slag and allowing hydrogen to penetrate into the metals of the thermocouples and causing faulty readings. Where nitrogen or other purge gas is used to protect the thermocouple assembly, local cooling occurs, which gives rise to understated temperatures. A further disadvantage of thermocouples is that their exact location has a significant influence on the accuracy of measurement. In refractory-lined

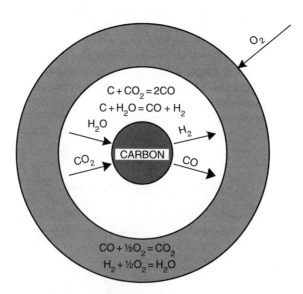

Figure 6-10. Van Loon's Gasification Model

gasifiers, the tip of the assembly is typically located slightly withdrawn into the wall, so as to protect it from slag or other erosion damage. The actual temperature measured is closer to that of the refractory rather than that of the reactor core and thus highly dependent on the extent of the depth of withdrawal from the reactor space.

Despite these disadvantages, platinum-rhodium thermocouples are still the most common device currently used for gasifier temperature measurement. It is accepted that real accuracy of temperature measurement is less important than consistency. For the gasifier operator, who has set his feed inputs on the basis of other parameters, a continuous and steady temperature reading is more important than the absolute value shown.

Nonetheless, investigations continue to develop alternative methods, not least because in oil gasification thermocouple life can be run-length determining. Systems under consideration or development include the following:

- *Pyrometers.* Texaco has used a pyrometer in its pilot unit for several years, and this will shortly be tested in a commercial environment (Leininger 2002). The principle advantage of such a system is that the sensor is located outside the reactor and not subject to the harsh environment. The necessity to ensure pressure integrity, including a high pressure nitrogen purge, does make it expensive, however. Interestingly, the actual temperature measured is dependant on the gasifier fuel. With gas firing, the visible path reaches to the opposite wall of the reactor, so that the temperature measured is that of the refractory. Depending on the degree of solids in the reactor, the visible path may reach only to the center, that is, the hottest location in the reactor, or even less where the temperature is cooler again. Furthermore, the nitrogen purge can cool the slag around the line of sight of the pyrometer leading to a loss of reading. Interruption of the nitrogen purge can solve this problem online, a distinct advantage over thermocouples, which generally require a reactor shutdown for replacement. Interpretation of such a loss of reading does require additional temperature measurements, so that any commercialization of pyrometry is likely to be in addition to rather than as a replacement of thermocouples.
- *Steam make in the membrane wall.* This measurement is of course limited to reactors having a membrane wall or water jacket. As was already mentioned in Section 6.4.2, the steam make in the membrane wall is a valuable indicator for both the heat loss through the wall and for the reactor temperature. It has the advantage of being an integral measurement of the temperature anywhere near the wall of the reactor. It is fast, with a response time of less than a minute, and very reliable. It does not, however, provide local temperature values.
- *Heat flux measurement.* This measurement comprises installing a small piece of membrane wall in the wall of a reactor and measuring the increase in water temperature of a known flow of water through the membrane wall. It can give a fast—10–30 second response time—indication of the local temperature.
- *Microwave-based measurements* have also been considered.
- *Other devices.* A new system based on temperature-dependant changes in the optical properties of single-crystal sapphire is under development. Despite some promising

results for other applications, this sensor will still have to survive the difficult reactor environment, and the fundamental uncertainties of temperature measurement in a gasifier will remain (Pickrell, Zhang, and Wang 2002).

- *Using the gas analysis.* Using the methane or the CO_2 content in the gas can give a valuable indication about the reactor temperature. The advantage of this method is that it gives an integral measurement of the temperature at the reactor outlet. But it does not give an indication about local hot spots. Moreover, the measurement has a certain time lag that with modern gas analyzers can be reduced to less than a minute. For the interpretation of the gas analysis, see Section 6.8.3.

6.8.2   Temperature Control

To use a temperature measurement for process control is especially difficult in entrained-flow gasifiers where the only option is an indirect temperature measurement based on the composition of the product gas. Moreover, the control is complicated by the very short residence time of 1–3 seconds. Operating for a short period at a temperature below the melting point of the ash immediately leads to the buildup of solid slag on the walls and on the bottom of the reactor. Membrane-wall reactors are worse in this respect than brick-lined reactors, as the membrane walls are relatively cool (250–300°C, depending on steam pressure) and keep their heat only for a very short period. Because most modern gasification plants have entrained-flow gasifiers, the discussion on process control will concentrate on this type of gasifier.

Fluid-beds are already less difficult to control because of the much lower temperatures of 900–1150°C and the longer residence time.

The easiest gasifiers in terms of control are moving-bed gasifiers because of their very long coal/char residence time of about one hour. On the other hand, the gases have a short residence time and leave the reactor at relatively low temperatures of below 600°C. Measuring the reactor outlet temperatures gives a fast indication about the gasification temperatures in the bottom of the reactor.

6.8.3   Gas Analysis

The coal footprint was already discussed in Section 2.3.3. This footprint is only relevant for fluid-bed and entrained-flow gasifiers, as it assumes chemical equilibrium between all major gas components under conditions where methane is the only hydrocarbon that is present. It was concluded that, in general, the methane content in the gas is the best indicator for monitoring the temperature of a nonwater-slurry feed gasifier. In fact, the only reason to calculate the methane content in entrained-flow gasifiers is for process control reasons, as the impact of the methane content (only a few hundred ppmv) on the total mass and energy balance is negligible.

Studying the coal footprint taught that the CO_2 content of a gas is a dangerous indicator, as the iso-CO_2 lines run almost perpendicular to the isotherms (see Figure 2-6). Those experienced in furnace control engaged in the start-up of a gasifier may have a tendency to look only at the CO_2 content in the gas. And there are entrained-flow gasifiers where this works very well, namely with coal-water slurry fed gasifiers. Control is relatively simple in this case, as the only variable that has to be watched is the ratio between the coal-water slurry and the oxygen. And if the reaction takes place in a brick-lined reactor where the heat loss is low, one neither has to worry about the heat loss in the reactor nor about the methane content in the gas. The coal footprint is then simple as it has only two dimensions (no heat loss as an additional variable), and the methane content in the gas is always very low because of the high water concentration, so there is no need to use it for control purposes.

With dry-coal feed entrained-flow gasifiers, life becomes more difficult, in comparison with coal-water slurry-fed systems, now three flows of reactants have to be monitored. Furthermore, dry-coal feed entrained-flow gasifiers so far always feature a membrane wall and hence also the heat loss plays a role. The methane content in the gas analysis must be checked with calculated values based on the mass and heat balance and chemical equilibria, as the methane value depends on variables as the feedstock, reactor geometry, and freezing-in temperatures of the various reactions.

Gasifier Outlet Temperature as a Function of the Gas Analysis

Calculating the gasifier outlet temperature from the gas analysis is not so simple. The reason is that it is not possible to measure the composition of the gas when it leaves the reactor. The best one can do is a "postmortem" when the gas has been cooled down. There have been attempts to devise a means to draw a gas sample directly after the gas has left the reactor through a cooled, high-alloy, thin tube and then perform the analysis. The idea is that by freezing-in the equilibria between the various possible reactions, a fair analysis will be obtained. However, the problem is that high-alloy steels mostly contain nickel or other metals, which may catalyze reactions between the various gas components. Moreover, by drawing the sample gas through a thin capillary tube, the exposed metal surface is relatively large and makes the situation only worse. Finally, it is difficult to avoid fouling of the entrance of the capillary tube and keep it open.

Because of these difficulties it is more practical to analyze the gas after it has been cooled down by quenching or by indirect cooling. In the case of single-stage entrained-flow gasifiers, the analysis has to be corrected for the fact that, for example, the CO shift reaction freezes in at a temperature below that prevailing at the outlet of the gasifier itself. For entrained-flow gasifiers the reactor outlet temperature is about 1500–1550°C. In most cases it may be assumed that the freezing-in of the CO shift occurs at a temperature of 1300°C. This causes a small part of the CO present in the gas leaving the reactor to be converted into H_2. This reaction is exothermic, and hence this heat effect increases slightly the duty of the subsequent gas-cooling train.

Cooling the gas by quenching with water will also result in some CO shift. In general, there is always some CO shift taking place until the point where the temperature has dropped to about 1300°C. Only below 1300°C the cooling will be purely physical.

The temperature at which the CO shift reaction freezes can be calculated from the gas analysis of the cooled gas. The first step is to correct the gas analysis for the fact that water has condensed out of the gas, as the gas analysis always takes place after the gas has been cooled below its dewpoint. From this corrected gas composition the k_p value of the CO-shift reaction can be calculated.

Certainly where the gas leaving the gasifier has been cooled by quenching with water, it will be found that the corresponding temperature T_{eq} associated with this k_p value is much higher than the outlet temperature of the gasifier. This is logical, as much of the water was added to the gas below the temperature at which the CO shift reaction freezes and has only increased the water content of the gas without having any chemical effect. In a trial-and-error calculation, it is now possible to subtract quench water from the gas in such a way that the mass and energy balance tally. The temperature of the water and of the gas after the gasifier reactor must of course be known before this exercise can be carried out. Subtracting the water results in a higher temperature of the gas in the now "shortened" quench and in a lower—but still too high—T_{eq} corresponding to the newly calculated k_p value. Hence the gas temperature and the equilibrium temperature come closer together. By repeating this procedure there comes a moment where both temperatures become equal, and this is the temperature at which the CO shift reaction was frozen in (see Figure 6-11).

With the gas analysis it is thus possible to calculate the freezing-in temperature of the CO shift reaction accurately, but it says nothing about the outlet temperature of the gasifier. It is possible, though, to make an element balance over the reactor, and then the outlet temperature can be calculated to within about 30°C. A computer program for calculating the freezing-in temperature of the CO shift equilibrium from the gas analysis is provided in the companion website. It is given for both water quenching and indirect cooling. The indirect cooling can be either a radiant boiler or a gas quench.

The freezing-in temperature of a suitable methane containing reaction can also be calculated.

The best reaction to use (see also Section 2.2.2) is:

$$CH_4 + CO_2 \leftrightarrows 2CO + 2H_2 \tag{2-10}$$

The advantage of this reaction is that there is no water present.

The data given for the coal footprint in Figures 2-5 and 2-6 have been obtained taking into account the effect of the adjustment of the CO shift reaction. These data show that, although it has been assumed that the freezing temperature of this reaction is the same in all calculations, the composition of the gas after cooling still clearly reflects the differences in the reactor outlet temperature.

Thus far one may wonder why the methane content in the product gas has been taken into account, as in entrained-flow slagging gasifiers it is only a few hundred

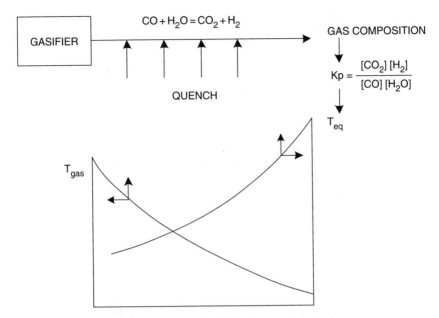

Figure 6-11. **Calculation of CO Shift Equilibrium Temperature**

ppmv and hardly plays a role in the overall mass and heat balance. Moreover, forgetting about methane greatly simplifies the calculations. The very reason is process control. In the preceding discussion on the coal footprint it was already mentioned that the iso-methane lines run more or less parallel to the isotherms and hence were a good indication of the outlet temperature of the gasifier. In fact, this is only partly true. They are a valuable indicator of this temperature, but when one calculates what the outlet temperature is on the basis of the gas analysis, one finds values that relate to different temperatures than can be reasonably expected. However, introducing correction factors with the calculated temperatures based on the other gas components, the methane content becomes an extremely valuable indicator of the reactor outlet temperature.

Overall Procedure for Process Control of Entrained-Flow Slagging Reactors

For the initial start up (or with a new feedstock) gasifier flows are set on the basis of values calculated for the coal footprint (van der Burgt 1992) from analyses of coal and blast and of a calculated heat loss from the gasifier. The reaction conditions will in general be set such that there is a surplus of the blast (in most cases is mostly oxygen). For dry-coal feed entrained-flow gasifiers this means that the reactor operates with a CO_2 content of, say, 3–4 mol% and a calculated temperature of well above the melting point of the ash components that may already contain fluxing material. The fastest indication for undesirable fluctuations in reactor temperature is reflected in the steam production in the wall of a reactor with a jacket or membrane wall. The next indication is the methane content of the gas. The CO_2 content can then be decreased to about 1 mol% by decreasing the oxygen/coal ratio and adjusting the steam/coal ratio. During this operation the theoretical coal footprint will be adjusted to the actual values found, for example, for the reactor heat loss and the CH_4 content of the gas, which data will then be later used for optimizing the gasification conditions and the control of the reactor. The coal footprint is a very valuable tool. For the CO_2 content and the percentages of the other major gas components, the theoretical figures and actual figures differ little provided the freezing-in of the CO shift equilibrium is taken into account. Only the actual CH_4 data can deviate considerably from the calculated data. Hence, calibrating the theoretical data with actual measurements becomes important.

6.9   TRACE COMPONENTS OF RAW SYNTHESIS GAS

In Chapter 2 we discussed the principal components of the raw synthesis gas from the gasification reactor, namely carbon, CO, CO_2, H_2, H_2O, and CH_4. A review of the raw gas composition would, however, be incomplete without a discussion of trace components, which can influence the selection of the downstream gas treating process, corrosion behavior, or the potential for fouling. In this respect it is, for instance, important to have an understanding of the fate of the sulfur and nitrogen in the feedstock. Similarly, one needs to be aware of those reactions that take place downstream of the reactor in the gas cooling and solids removal sections of the process, such as metal carbonyl or organic acid formation.

6.9.1   Sulfur Compounds

The existence of sulfur compounds in raw synthesis gas represents a poison for the catalysts of most chemical applications, including ammonia, methanol, Fischer-Tropsch, low temperature shift, and others. In power applications, if untreated they would be emitted with the flue gas as SO_2 and SO_3, major components of "acid rain."

In high-temperature processes all sulfur components in the feed are converted to H_2S or COS. Other compounds such as SO_x or CS_2 are essentially absent. This is not the case in low-temperature processes, where tars and other species have not been completely cracked. A detailed breakdown of the sulfur compounds in raw gas from a Lurgi dry-bottom gasifier is given in Table 6-2 as an example.

Table 6-2
Sulfur Compounds in Raw Gas from a Lurgi
Dry-Bottom Gasifier

Component	
COS, ppmv	180
H_2S, ppmv	15,300
Mercaptan S, ppmv	600
Thiophenes, ppmv	5
CS_2, ppmv	100
Source: Supp 1990	

The relationship between the H_2S and COS contents of a raw gas is determined by two reactions, the hydrogenation reaction

$$H_2 + COS \leftrightarrows H_2S + CO \qquad +7 \, MJ/kmol \qquad (6\text{-}1)$$

and the hydrolysis reaction

$$COS + H_2O \leftrightarrows H_2S + CO_2 \qquad -34 \, MJ/kmol. \qquad (6\text{-}2)$$

Equilibrium constants for these reactions can be found in the literature (Reimert and Schaub 1989). Under typical gasification conditions, H_2S is the dominant species, and approximately 93–96% of the sulfur is in this form, the rest being COS.

It is important to be aware of the COS content in the raw gas, since not all gas treatment systems will remove COS. In order to overcome this it may be necessary to perform a selective catalytic hydrolysis of COS to H_2S (reaction 6-2) prior to the acid gas removal. This is discussed in more detail in Chapter 8.

6.9.2 Nitrogen Compounds

Formation of HCN and NH₃

Nitrogen enters the gasifier in two forms, either as molecular nitrogen, generally in the gasification agent (but also as a component in gaseous feeds), or as organic nitrogen in the fuel. Although the bulk of the nitrogen in the synthesis gas is present as molecular nitrogen, most gasifiers produce small amounts of HCN and NH_3. There is little literature on the formation of nitrogen compounds in gasifiers. It is, however, possible to draw inferences from the well-researched mechanisms of NO_x formation in combustion flames.

Fuel-derived formation of HCN and NH_3 is far greater than that formed from molecular nitrogen, so that in most cases the latter can be neglected. Fuel nitrogen is often contained in structures with N–H and N–C bonds, which are much weaker than the triple bond in molecular nitrogen. The typical mechanism for NO formation during complete combustion can be depicted as follows:

$$\text{Fuel-N} \longrightarrow \text{HCN} \xrightarrow{\;H_2O\;} NH_i \overset{NO \rightarrow N_2}{\underset{O_x \rightarrow NO}{\diagdown}}$$

Initially, fuel nitrogen is converted to HCN, which rapidly decays to NH_i (i = 1,2,3), which under combustion conditions, where sufficient oxygen is present, reacts to form NO and N_2 (Smoot 1993). Under gasification conditions, the oxidation of NH_i radicals does not take place, and in the presence of a large hydrogen surplus, the nitrogen remains as HCN and NH_3. Research on NO_x formation indicates that HCN is the principle product when the nitrogen in the fuel is bound in aromatic rings, whereas NH_3 appears to be the principle product when the nitrogen is bound in amines. The proportions of HCN and NH_3 formed, therefore, vary in accordance with the fuel characteristics.

Only in the partial oxidation of natural gas, where no chemically bound fuel nitrogen is present, is it necessary to recognize that at least some thermal HCN and NH_3 formation does take place. Since thermal HCN and NH_3 formation is a function of the actual temperatures in the flame zone and thus of individual burner performance, one can only rely on the experience of licensors with their own burner designs to provide data on the expected HCN and NH_3 formation.

Typical Concentrations

Typical concentrations of nitrogen compounds in various syngases are shown in Table 6-3. It is unclear whether the figure of 0.05 ppmv NO_x given by Rowles for oil gasification (Slack and James 1974) was really measured or just represents the limit of detectability. For raw gas from a Koppers-Totzek gasifier Partridge (1978) provides a figure of 70 ppm NO. In the same source he gives a figure of 150 ppm for the oxygen content. These are both much higher than figures quoted for other entrained-flow processes and may well be due to the ingress of air and/or poor mixing of reactants in the gasifier.

Effects of Nitrogen Compound Impurities

Ammonia has a very high solubility in water (two orders of magnitude higher than CO_2). One effect of this is that ammonia is seldom removed from the wash or quench water of carbon removal systems. Sufficient ammonia is then recycled in the scrubber wash water and partially stripped out by the syngas in the scrubber such that the potential for full ammonia removal in the syngas water wash is seldom realized.

Table 6-3
Nitrogen Components in Synthesis Gas

Feed	Process	HCN	NH$_3$	NO/NO$_2$	Source
Coal	Lurgi dry bottom gasifier	22 ppmv	39 ppmv	NO$_x$ 0.02 ppmv	(Supp 1990, p. 23)
Coal	GSP	1.0 mg/Nm^3	0.24–0.4 mg/Nm^3	n.a.	(Lorson, Schingnitz, and Leipnitz 1995)
Oil		50 ppmv	1–20 ppmv	0.05 ppmv	(Weiss 1997; Slack and James 1974)
Gas		Traces	Traces	n.a.	
Biomass		<25 mg/Nm^3	2200 mg/Nm^3	n.a.	(Boerrigter, den Uil, and Calis 2002)

Where chlorine is present, typically when gasifying coal, ammonia will combine with the chlorides to form ammonium chloride (see Section 6.9.3).

In methanol plants, ammonia (and also nitrogen oxides) can contribute to the formation of amines on the methanol synthesis catalyst. The presence of amines is not permitted in internationally accepted methanol specifications (e.g., U.S. Federal Specification, Grade AA) and can only be removed from the raw methanol with an ion exchanger (Supp 1990). It is, therefore, preferable to ensure the absence of nitrogen compounds in the synthesis gas upstream of the synthesis itself.

Hydrogen cyanide also has a high solubility in water and other physical wash solutions. If the main acid gas removal (AGR) is based on a physical solvent, then an HCN pre-wash can be integrated with the main system. It can also be removed by a water wash, although it should be noted that the high solubility also has its downside, namely the cost of regeneration.

Care should be exercised when using an amine AGR on a gas with a high HCN content, since although amines will remove it satisfactorily the acidic cyanide will react with the amine and degrade it. This problem should be examined as part of the AGR selection process.

Any HCN or NO entering a raw gas shift will be hydrogenated to ammonia (BASF). For some catalytic processes, such as Fischer-Tropsch, HCN acts as a poison (Boerrigter, den Uil, and Calis 2002).

Nitrogen oxides require particular attention in ammonia plants. In the liquid nitrogen wash (LNW) of an ammonia plant they will form a resin with any unsaturated hydrocarbons in the gas, and this resin is "extremely susceptible to spontaneous detonation"

(Slack and James 1974). In most plants the molecular sieve immediately upstream of the liquid nitrogen wash (LNW) represents the last line of defense against ingress of both NO_x and unsaturated hydrocarbons into the cold box. If Rectisol is used as the acid-gas removal system, for instance, the unsaturated hydrocarbons would already be removed at this stage. Where a raw gas shift is installed, both nitrogen oxides and unsaturated hydrocarbons would be hydrogenated on the catalyst.

6.9.3   Chlorine Compounds

Chlorine compounds are present in most coals. They will react with ammonia in the raw gas to form ammonium chloride (NH_4Cl). At high temperature this is (dissociated) in the vapor phase, but below 250–280°C it becomes solid and presents a fouling risk to the gas cooling train. At lower temperatures still, below the water dewpoint of the gas, it goes into solution and is highly corrosive. These aspects have to be considered in the design of the cooling train.

Metals in the feedstock will also form chlorides (e.g., sodium chloride). Many of these have melting points in the range 350–800°C and represent a fouling risk in heat exchangers.

Note also that chlorine is a catalyst poison for ammonia and methanol syntheses as well as for the low temperature shift.

6.9.4   Unsaturated Hydrocarbons

The existence of unsaturated hydrocarbons in the raw synthesis gas varies very widely. In the Lurgi dry-bottom process, there will in general be large quantities of aromatics and other unsaturates in the volatiles and tars, though the exact amount also will depend heavily on the coal. For biomass gasification the presence of tars in the gas is also a problem. For high-temperature entrained-flow processes, including oil gasification, the presence of any hydrocarbon other than methane, whether saturated or unsaturated, is minimal.

Removal of hydrocarbons from the product gas from a Lurgi dry-bottom gasifier can be integrated into the design of a Rectisol wash. Kriebel (1989) provides a description of this.

The effect of unsaturated hydrocarbons entering a liquid nitrogen wash together with nitrogen oxides is described in Section 6.9.2.

Care should be taken with oil gasification using naphtha as a medium for soot extraction. Aromatic components in the naphtha dissolve in the water and can be introduced into the gas via the return water to the gas scrubber.

Unsaturated hydrocarbons will be hydrogenated when the gas is treated by a CoMo raw gas shift catalyst (BASF undated).

6.9.5   Oxygen

Oxygen is a catalyst poison for some catalysts. Some typical requirements limit the oxygen content in the syngas to ca. 5 ppmv. In high-temperature gasification processes, whether of coal or oil, the oxygen is completely consumed in the reaction and no oxygen is contained in the synthesis gas. One should, however, be aware of the danger of introducing small quantities of oxygen into the gas in the subsequent processing. Typically, if using a water wash for solids removal, it is possible to introduce oxygen via the water. It is therefore advisable in critical circumstances to eliminate such sources of accidental contamination by, for example, using only deaerated water for scrubbing. Atmospheric gasification processes also have a risk of introducing oxygen from unintended sources.

Any oxygen in gas subjected to CO shift will react with the hydrogen present (BASF undated). It is worth mentioning that oxygen is often measured together with argon, and occasionally an analysis result showing oxygen in syngas can cause some considerable concern before this "misleading message" is understood.

6.9.6   Formic Acid

At higher partial pressures carbon monoxide will react with water to form formic acid according to the equation

$$CO + H_2O \leftrightarrows HCOOH \qquad -34.6\,MJ/kmol \qquad (6\text{-}3)$$

The thermodynamics of the reaction favor formic-acid formation at lower temperatures so that this is particularly noticeable in the gas condensate.

At pressures up to about 60 bar there is usually sufficient ammonia formed to maintain a neutral pH in wash water. This is therefore seldom mentioned in connection with coal gasification because such pressures have only been seen in pilot plants operating under test conditions. It is, however, a phenomenon that has been observed in high-pressure oil gasification and requires consideration in material selection (Strelzoff 1974).

6.9.7   Carbon

It is necessary to distinguish between two very different sources of carbon, which can occur in raw synthesis gas.

Coal Gasification

In coal gasification there is always a certain amount of the initial carbon feedstock, which is carried over unconverted in the form of char as particulate matter into the gas. Typically, this can be extracted from the gas in a particulate filter and—in case of low carbon conversions— recycled to the gasifier. In the case of slagging gasifiers, this form of recycling has the added advantage that this carbon is usually intimately mixed with dry ash, which can also be recycled for slagging.

Oil Gasification

In contrast to coal gasification, the carbon in synthesis gas leaving an oil gasifier is actually formed in the gasifier itself. The soot leaving an oil gasifier has an extremely high surface area of 200–800 m²/g, depending on ash content (Higman 2002).

An oil gasifier is deliberately operated to maintain a small quantity of this soot in the raw gas as an aid to sequestration of the ash from the reactor, whether of the quench type or with a syngas cooler. Typically, the soot make in modern plants is about 0.5% to 1.0% of the initial feed, although it could be as much as 3% in older plants. Removal of this carbon with a water wash is an integral part of all commercial oil-gasification processes.

6.9.8 Metal Carbonyls

The steady increase in the metal content of liquid partial-oxidation feedstocks over the years has led to a developing awareness of the necessity to consider nickel and iron carbonyl formation in the raw synthesis gas. Nickel and iron carbonyl are toxic gaseous compounds that form during the cooling of the raw gas and pass on in the raw gas to the treating units. Depending on the treatment scheme, there may be a need for special handling to avoid problems.

Table 6-4 shows some of the principal chemical and physical data of these gases (IPCS 1995, 2001; Kerfoot 1991; Lascelles, Morgan, and Nicholls 1991; Wildermuth et al. 1990).

The formation of nickel and iron carbonyls can take place in the presence of gaseous carbon monoxide in contact with metallic nickel or iron or their sulfides. Industrially hydrogen sulfide or carbonyl sulfide are used as catalysts for the production of nickel carbonyl from active nickel. Ammonia has also been used as a catalyst. Given that all three of these gases are present in the raw synthesis gas, one needs to anticipate some carbonyl formation in a partial oxidation gas containing as much as 50 mol% CO if the feedstock contains significant quantities of nickel or iron.

The reactions leading to the formation of carbonyls in a partial oxidation unit are shown in Table 6-5 together with their equilibrium data.

Figure 6-12 shows a plot of the equilibrium concentrations of nickel and iron carbonyls against temperature for various CO partial pressures. From these plots one can see that carbonyl formation increases with increasing pressure and decreasing temperature, whereby nickel carbonyl formation takes place already at significantly higher temperatures than iron carbonyl formation. Based on this data and a plant pressure of 60 bar and 45 mol% CO in the raw gas, one could expect the formation of 1 ppmv $Ni\,(CO)_4$ from nickel sulfide below about 380°C and 1 ppm (v) $Fe\,(CO)_5$ from iron sulfide below 40°C. The corresponding temperatures for carbonyl formation from the metals are somewhat higher. Although the kinetics of the reactions, particularly at lower temperatures, may prevent equilibrium conditions arising in practice, these tendencies correspond with industrial practice (Soyez 1988; Beeg,

Table 6-4
Properties of Nickel and Iron Carbonyl

Name	Nickel Tetracarbonyl	Iron Pentacarbonyl
Formula	$Ni (CO)_4$	$Fe (CO)_5$
Molecular mass	170.7	195.9
Boiling point at 1.01 bar, °C	43	103
Melting point, °C	−19	−20
Vapor pressure, kPa	42 at 20°C	3.49 at 20°C
Vapor density (air = 1)	5.9	6.8
Explosive limits in air vol%	3–34	3.7–12.5
Auto-ignition Temperature, °C	60	
Flash point, °C	−24	−15
Solubility in water	None in water but soluble in many organic solvents	Contradictory 50–100 mg/l

Table 6-5
Formation of Nickel and Iron Carbonyl

Reaction	K_p	$Log\ K_p$	
$Ni + 4\,CO \leftrightarrows Ni\,(CO)_4$	$\dfrac{P_{Ni(CO)_4}}{P_{CO}^4}$	$\dfrac{8299}{21.11T}$	(6-4)
$Fe + 5\,CO \leftrightarrows Fe\,(CO)_5$	$\dfrac{P_{Fe(CO)_5}}{P_{CO}^5}$	$\dfrac{8852}{29.60T}$	(6-5)
$NiS + 4\,CO + H_2 \leftrightarrows Ni\,(CO)_4 + H_2S$	$\dfrac{P_{Ni(CO)_4} \cdot P_{H_2S}}{P_{CO}^5 \cdot P_{H_2}}$	$\dfrac{4903}{18.78T}$	(6-6)
$FeS + 5\,CO + H_2 \leftrightarrows Fe\,(CO)_5 + H_2S$	$\dfrac{P_{Fe(CO)_5} \cdot P_{H_2S}}{P_{CO}^5 \cdot P_{H_2}}$	$\dfrac{4875}{28.21T}$	(6-7)

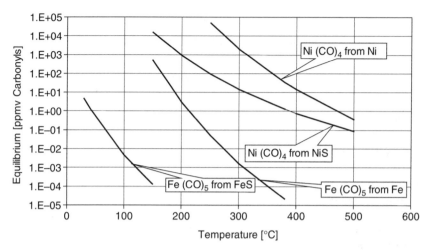

Figure 6-12. **Equilibrium Concentration of Carbonyls as a Function of the Temperature**

Schneider, and Sparing 1993). Carbonyl formation takes place in the cold section of the plant. And because of the lack of solubility of the carbonyls in water, they leave the partial oxidation unit with the raw gas.

Formation of carbonyls can be inhibited to some degree by the presence of free oxygen. There is, however, no recorded instance of such an approach being taken in any gasification unit.

The consequences of any metal carbonyl slip into the gas treatment units depend very much on the treatment scheme. Quench cooling leads to a lower carbonyl formation than the use of a syngas cooler, since much of the metals removal takes place at higher temperatures. This applies particularly to iron carbonyl formation. Nonetheless, in one plant with quench cooling and subsequent raw gas shift, significant depositing of nickel sulfide on the shift catalyst led to reduced catalyst life (BASF undated). This is caused by the reverse of reaction 6-6, decomposition of the carbonyls on heating in the shift unit.

As described in Table 6-4, the carbonyls are not soluble in water. They are not removed from the gas by amine washes. Most physical-chemical washing systems will also allow the carbonyls to pass through the absorber and appear in the clean gas, so that depending on the application, other problems may occur downstream.

Carbonyls are soluble in physical washes such as Rectisol and can be completely removed from the synthesis gas this way. It is, however, necessary to consider the subsequent fate of the metals. The relative partial pressures of carbon monoxide and hydrogen sulfide in liquor containing the dissolved carbonyls is substantially different to that of the raw gas, so that the reactions 6-6 and 6-7 are driven towards the left, particularly on heating the liquor for regeneration. The subsequent precipitation of the sulfides can cause problems, such as fouling of heat exchangers. If decomposition of the carbonyls is suppressed in the acid-gas removal unit, then they will appear in the sour-gas stream and may deposit on the Claus catalyst in the sulfur recovery unit. The various licensors of such physical wash processes have developed methods to control this phenomenon.

Effects of Carbonyls

Iron carbonyl can present problems in the methanol synthesis and was a regular difficulty in the older high-pressure processes because of its formation if CO came into contact with iron in the loop equipment. Irrespective of its origin, iron carbonyl will decompose at the conditions of the methanol synthesis (50–100 bar, 250°C) leaving iron deposits on the methanol catalyst. The iron will then catalyze Fischer-Tropsch reactions, contaminating the methanol with unwanted hydrocarbons (Supp 1990; Skrzypek, Sloczyński, and Ledakowicz 1994). Skrzypek and colleagues report that nickel carbonyl has the same effect. Carbonyls can act as a poison on other synthesis catalysts. This must be reviewed on a case-by-case basis.

In an IGCC situation, if carbonyls are permitted to enter the gas turbine, they will decompose at the high temperatures prevailing in the burners. There is a potential, then, for the metals to deposit on the turbine blades, causing imbalance. Care is generally exercised, therefore, to avoid this.

6.9.9 Mercury

Mercury can be present in both coal and natural gas, although the quantities vary widely from source to source. Mercury presents a potential hazard both for the integrity of the plant and as a toxic emission for the environment. Whether gasifying coal or partially oxidizing natural gas, mercury from the feed will appear at least in part in the synthesis gas, and so for these feeds it is necessary to address this feed contaminant.

Wilhelm describes a number of different mechanisms by which mercury degrades engineering materials (Wilhelm 1990). In particular, he mentions liquid metal embrittlement of high-strength steels. He also describes the formation of the highly explosive compound mercury nitride in the presence of ammonia.

Mercury is gaining increasing recognition as an important atmospheric pollutant from coal-fired power stations. The U.S. Department of Energy has reported that for conventional coal-fired power stations, there is "currently no single technology" available that can control mercury from all power plant flue gas emissions. The DoE has a major test and development program for processes to control mercury emission in flue gas (U.S. Department of Energy 2002).

The situation for gasification technologies is different. Proven and economic methods for mercury removal are available and have been practiced for many years. Mercury can be adsorbed onto sulfur-impregnated carbon, which can achieve an effluent concentration of less than $0.1 \mu g/m^3$ (Wilhelm 1990).

A prominent example of mercury removal in a coal gasification environment is provided by the Eastman Chemical operation in Kingsport, Tennessee. A sulfur-impregnated activated carbon bed was installed upstream of the acid-gas removal unit from the plant's inception in 1983 to protect downstream chemical processes from contamination and has operated successfully for nearly 20 years (Trapp 2002). Mercury capture is estimated to be between 90% and 95%. This experience was used as the basis for a cost-comparison study performed for the U.S. DoE showing that mercury removal from an IGCC plant could be as little as one-tenth of the cost of removal from a conventional PC power plant (Rutkowski, Klett, and Maxwell 2002).

Koss and Meyer (2002) report also on mercury removal from an existing coal gasification plant, in which metallic mercury removal is integrated into a Rectisol desulfurization unit operating at $-57°C$. Total mercury slip through the unit was measured at 1–2 ppbv.

In the case of a natural gas feed, Marsch (1990) has reported on the explosion of an ammonia separator after 10 years of operation that was attributed to the presence of mercury. The natural gas feed to the primary reformer of this 1000 t/d ammonia plant contained on average $150–180 \mu g/m^3$ mercury, amounting to an annual intake of 60–72 kg per year. Significant quantities of mercury passed through primary and secondary reformer (essentially a catalytic partial oxidation process), as well as a CO shift and acid-gas removal system, to enter the ammonia synthesis unit, where it caused the damage described. In evaluating the conclusions from this accident, Marsch recommends removal of any mercury in the feed to the lowest possible level. This message applies not only to ammonia manufacture but equally to any other application involving the partial oxidation of natural gas.

6.9.10 Arsenic

One of the problems associated with coal gasification is that in coal many of the elements of the periodic table can be found in minor concentrations. An element of emerging concern is arsenic, which may be present in concentrations in the order of 1–10 ppmw in coal (see Table 4-7). Toxic elements are of no concern when they end up bound in the slag or in stable chemical compounds. The problem with arsenic is that under reducing conditions it forms the volatile compound AsH_3. Arsenic is a known poison for ammonia catalysts, but recorded instances of this occurring in commercial plants have not been found.

Raw gas shift catalyst is reported as "removing arsenic very selectively," though arsenic deposits on the catalyst were low compared with those of nickel and carbon (BASF undated).

6.10   CHOICE OF OXIDANT

All gasification processes require an oxidant for the partial oxidation reaction. There are essentially two alternatives: air, which is available in unlimited quantities at the location of the gasifier; and oxygen, which has to be separated from the nitrogen in the air at considerable cost. A third alternative, oxygen-enriched air, is essentially a mixture of the two.

Historically, the first continuous partial oxidation systems, producer gas generators, operated with air. The idea of operating with pure oxygen was already developed in the 1890s, but it was only realized in the 1930s after the introduction of large-scale commercial cryogenic oxygen plants. Since then most gasification plants have operated with high purity (>90 mol% O_2) oxygen. To a large extent this has been dictated by the fact that in the period between 1935 and 1985, most gasifiers were built for chemical applications where the presence of large quantities of nitrogen originating from the air was detrimental to the downstream synthesis process. (Note that this also applies to ammonia, where only about 25–30% of the nitrogen associated with the oxygen used in the gasifier is required for the synthesis.)

These considerations of downstream chemistry do not apply to power applications, which have developed along with the increasing efficiencies of gas turbines, so that it was necessary to review the pros and cons of air versus oxygen for these applications. The result of such reviews in individual cases has been a decision in favor of oxygen in practically all large-scale projects. For small-scale projects (<50 MW$_e$), mostly operating with biomass or waste, the decisions have tended to favor air. It is therefore useful to understand the basic issues behind the choice of oxidant.

6.10.1   Effect of Oxidant on the Gasification Process

The most significant effects of varying oxidant composition can be seen in Figure 6-13. These results have been determined using a constant gasifier temperature of 1500°C (as determined by the ash characteristics of the coal) and constant preheat temperatures for the reactants.

Cold Gas Efficiency

The loss of cold gas efficiency with increasing nitrogen content of the oxidant is immediately noticeable. It falls off from 82% at 100% O_2 to 61% with air. The essential reason for this, and for the other effects visible in the Figure 6-13, is the amount of heat required to raise the nitrogen from its preheat temperature of 300°C up to the reactor outlet temperature of 1500°C. This can be partially compensated for by reducing the moderating steam, but this is only possible to the extent of reducing it to zero. For the chosen coal and conditions, this happens at about 26 mol% O_2 in the

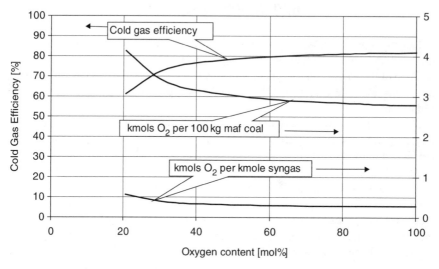

Figure 6-13. Gasification Parameters as a Function of Oxygen in Oxidant

oxidant. At this point more carbon is combusted to CO_2 to maintain the heat balance, which further reduces the efficiency. Should a particular process require a minimum amount of steam, for burner protection, for example, the point at which the CO_2 make begins to increase will be at a correspondingly higher oxygen content in the oxidant.

An alternative approach to compensating for the heat absorption by the nitrogen might appear to be an increase in preheat temperatures. Klosek, Sorensen, and Wong (1993) discuss this aspect in their paper but point out its limitations. The required preheat temperatures of about 1200°C become "excessively high and unacceptable."

Similarly, the drop in useful syngas ($H_2 + CO$) is visible as the oxygen content in the blast decreases.

Gasification Temperature

At lower gasification temperatures these effects are less but are nevertheless still present.

Use of Hot Syngas

Reviewing the implications of oxidant quality only on a cold-gas efficiency basis ignores the fate of the sensible heat contained in the syngas leaving the reactor. In an IGCC environment, the most efficient use of this heat would be to feed it uncooled to the gas turbine, since this has a fundamentally higher efficiency than the steam cycle (see Section 7.3). This theoretical possibility is, however, only realizable if the raw syngas quality is acceptable to the gas turbine, which in general is not the case. In particular, particulates and alkali metals need to be removed prior to the gas entering the gas turbine. These contaminants can be removed at 500°C using candle filters (see Section 6.7).

There is still an interest in "hot gas desulfurization," especially in connection with air-blown gasifiers. This is not because the gas turbine cannot cope with the sulfur once the alkali salts have been removed, but because of the environmental requirements. Improvements in flue gas desulphurization technology provide a continuously moving target. With double-scrubbing technology, sulfur removal from the flue gas can also be as high as 99%. However, it should be noted that any improvement obtained with such technology developments will benefit oxygen-blown systems as much as they will air-blown systems. So far all attempts at hot sulfur removal have failed, and there is no indication that a good solution will be found in the near future. Further details of gas cleaning technologies are described in Chapter 8.

An effective use of the hot gas, particularly where it is generated in a small-scale biomass gasifier, is to utilize it as a gaseous fuel in a large utility boiler. This is particularly effective, because the low efficiency inevitable with small-scale plants can be avoided by tying into the larger unit. A particularly interesting example is an air-blown CFB gasifier processing waste wood and firing the gas into a 600 MWe PC power boiler at Geetruidenberg in the Netherlands.

One other albeit rare case where the full sensible heat of the syngas from an air-blown gasifier is used in the downstream process is in the production of reducing gas, for example, for nickel reduction, where the gas is fed from an oil gasifier directly to the reduction furnaces without any cooling.

6.10.2 Equipment Sizing and Gas Flow Rate

The flow rate of synthesis gas when using air as an oxidant is approximately twice that of the equivalent oxygen-blown gasifier. This has consequences both for equipment sizing and for the gas-cleaning technologies.

An air-blown gasifier must either be twice the size of an oxygen-blown one, or two are required. Whether the design is refractory-lined (e.g., Texaco) or fitted with a cooling wall (e.g., Shell or GSP), the gasifier is an expensive piece of equipment. The same goes for the syngas cooler and other downstream equipment such as filters and gas-cleaning equipment.

Besides the effect on equipment, one also needs to look at the emissions. To achieve the same absolute sulfur emissions, the sulfur slip from the SRU for an air-blown unit must be half that of an oxygen-blown gasifier since there is twice as much gas. This will certainly require additional expenditure in the AGR unit.

6.10.3   Parasitic Power

Parasitic power is usually defined as the electrical energy required for drivers of auxiliary machines in an IGCC. In connection with an oxygen-blown gasifier, this includes the air compressor (if any) for the ASU, the oxygen compressor, and the nitrogen compressor (again, if any). For an air-blown gasifier, the parasitic power for the provision of oxidant is the air booster compressor (see Figure 6-14). Unfortunately, this definition very quickly creates misunderstandings or confusion, because it does not include the power used by the air compressor of the gas turbine. Any air integration included in the cycle removes mechanical energy in the shaft from the generator, where it can produce electric power. Since the compression energy internal to the gas turbine is not included in the normal definition of parasitic power, one needs to be very careful about drawing any conclusions from a single number provided for any particular project. In practice, finding an optimum integration between the gas-turbine air compressor and the provision of oxidant is highly dependant on the characteristics of the air compressor itself, and it should come as no surprise to find that different optima can be presented by different gas turbine suppliers for the same project.

Klosek, Sorensen, and Wong (1993) point out that the parasitic power demand for an oxygen-blown IGCC without nitrogen integration can be significantly lower than that of an air-blown unit. This initially surprising result can be attributed to two facts. First, the air-blown unit must handle much more air, and second, the air must all be compressed to the gasifier inlet pressure, which is in the current designs significantly higher than the turbine combustor pressure. Reducing the pressure drop over the gasifier and treating could well change some of the above conclusions.

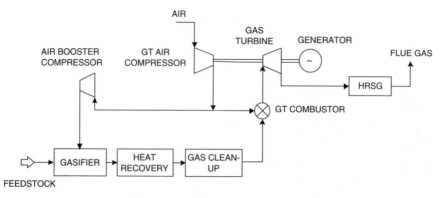

Figure 6-14. Typical Air-Blown IGCC

6.10.4   Deductions

Essentially, the reasons why most real IGCCs have been built using oxygen as oxidant are economic, and the technical background to these economics has been described in the preceding sections. These plants, whether demonstration (e.g., Buggenum, Puertollano, Wabash, and Tampa at 250–300 MW) or commercial (e.g., Sarlux at 500 MW), are, however, all large projects. There is no disguising the fact that the favorable economics of oxygen-blown technology has an entry price, which is the investment cost of the ASU. At this scale, the entry price is well worth paying.

For small plants, smaller than say 50 MW, which is largely the realm of biomass and waste fuels, the initial investment in an ASU is less attractive. Savings on the much smaller equipment and the improved efficiency of oxygen operation is no longer able to pay for this investment. This is why many gasification projects in this size range use air-blown gasifiers. It is, however, not possible to provide any hard-and-fast guidelines for determining the break-even point between the two technologies. The number of variables, which include feedstock pricing and supply, gas turbine characteristics as discussed, and the potential for synergies in oxygen supply (see Section 8.1), make a project-specific evaluation inevitable.

6.11   CORROSION ASPECTS

Gasification takes place at high temperatures, depending on the process, anywhere between about 900 and 1500°C. Processes have been developed with the capability of operating at pressures of over 100 bar, although commercial experience is limited to about 80 bar (oil) or 40 bar (coal). Other processes operate at atmospheric pressure. The synthesis gas produced contains CO_2, CO, H_2, CH_4, N_2, H_2O, and almost invariably sulfur compounds, mostly H_2S. This gas is usually cooled down to ambient temperatures, and in doing so the temperature range imposed on the containing equipment is very wide and includes sub-dewpoint temperatures. It is not surprising, therefore, that particular attention has to be paid to material selection in a potentially corrosive environment. Some of the corrosion mechanisms that can arise under these circumstances are well known from other circumstances. The purpose of this section is not to address these, however, but to highlight those potential problems that might be less familiar to those with little prior experience of gasification, and to point to suitable literature where more information may be obtained.

6.11.1   Sulfur

The metal temperatures of evaporator sections of syngas coolers is determined primarily by the temperature and thus by the pressure of the steam side. With typical high-pressure steam conditions of 100 bar, this implies metal temperatures between 300 and 350°C. EPRI and KEMA have conducted corrosion tests specifically to determine corrosion rates by sulfiding in this temperature range, and they found a clear dependency between corrosion rate and partial pressure of H_2S (van Liere, Bakker, and Bolt 1993). According to these results, in an atmosphere with 1 mol% H_2S at 40 bar and 300°C, a corrosion rate of 1.5 mm/y can be expected with low alloy steels, with which they specifically include alloy T22 (10CrMo910). The investigation was made primarily with coal gasification in mind, so it is unclear from the published results what other trace components were in the gas. In our experience with gasification of heavy petroleum residues at 60 bar where 1 mol% H_2S is by no means uncommon, any corrosion rate must be at least an order of magnitude lower than this, since evaporator coil lifetimes of 60,000 hours are common (Higman 1994). It is unclear what protection mechanism exists in the latter case to explain this apparent contradiction.

When superheating is included in the syngas cooler, then much higher metal temperatures must be expected, and they are likely to be in the range 300 to 600°C. Van Liere, Bakker, and Bolt's results indicate clearly that an austenitic steel is required for this service, typically SS310, which interestingly performs better than alloy 800.

Corrosion during shutdowns, particularly on the ingress of oxygen, requires careful attention. Appropriate operation and maintenance procedures have been published (National Association of Corrosion Engineers 1993).

6.11.2 Hydrogen

A typical oil gasifier may have a pressure of 60 to 80 bar and a hydrogen content of, say, 30% (wet basis). In other words, a hydrogen partial pressure of 20 bar or more may prevail. The equivalent figures for a coal gasifier will probably be lower. For partial oxidation of natural gas there may be over 50% hydrogen, although for most applications the absolute pressures may be somewhat lower than for oil. The shell of a refractory-lined gasifier will have a normal operating temperature of around 250°C, but it will probably have a design temperature of 350°C to allow for a small amount of bypass gas streams through cracks in the refractory, or in the case of coal gasifiers, simple refractory wear. Materials in the equipment immediately downstream the gasifier will be subjected to similar conditions, which are well into the range where hydrogen embrittlement can occur. Attention must be paid to the selection of hydrogen tolerant alloys, which typically may be chrome-molybdenum steels or higher alloys, depending on actual hydrogen partial pressure and temperature. The best guide to material selection for this service is API 941 (API 1990), which contains the well-known Nelson curves.

6.11.3 Chlorides

In the reducing environment of a gasifier, a significant portion of any chlorine in the feed is converted to HCl. The most important corrosion aspect to consider is that at temperatures below the water dewpoint the condensate can cause pitting or stress corrosion cracking of the commoner austenitic steels. Good plant design can take care of exchangers and piping containing process condensate. It is an important matter for operations and maintenance procedures to avoid cold bridges through the insulation of warmer equipment (e.g., at pressure gauges), which can cause local condensation. Another point where operational care is required is during shutdown, when it would be necessary to avoid condensate formation.

6.11.4 Metal Dusting

Metal dusting is a particularly aggressive form of high-temperature (mostly) CO corrosion, which has received particular attention in connection with the manufacture of CO-rich synthesis gases using steam reforming or catalytic partial oxidation.

Although certain details of the mechanism of metal dusting, and particularly the potential for countermeasures, remain the subject of some debate, the essentials have been understood for many years (Hochmann 1972; Grabke, Brach-Troconis, and Müller-Lorenz 1994). Under gasification conditions, the Boudouard reaction (2-4) oxidizes carbon with CO_2 to form carbon monoxide. However, during syngas cooling a point is reached where the equilibrium favors the reverse reaction. At sites on the surface of materials containing particularly iron or nickel, the reverse Boudouard reaction is catalyzed. The carbon thus formed destroys the matrix, and the surface metal is lost as a "dust" into the gas phase. The temperature range in which metal dusting can occur is from a lower limit of about 450°C, where the reaction kinetics cause the corrosion to slow down to a rate that may be considered acceptable, up to the Boudouard equilibrium temperature of the syngas (Hohmann 1996). A typical sign of metal dusting during process operation is the appearance of the products of corrosion as a "magnetic soot" in the process condensate.

Metal dusting is unusual in gasification plants, because sulfur is an excellent inhibitor for this type of corrosion (Gommans and Huurdeman 1994) and most gasifier feeds contain some sulfur. Where a desulfurized natural gas is used to manufacture syngas for Fischer-Tropsch products, however, the potential for metal dusting is very high, since the CO partial pressures will tend to be considerably higher than in a steam reformer plant. Furthermore, the corrosion product, which is very fine and has a large surface area, provides an ideal catalyst for the very exothermic methanation of the CO in the syngas.

There is still a consensus that although careful choice of metallurgy can slow down the effects of metal dusting, there is no real metallurgical solution. The best way to avoid this form of corrosion is to avoid gas-metal contact in the endangered temperature range. For a reformer tube outlet, Hohmann (1996) determined the area of the material at risk and introduced a purge gas at this location with success.

REFERENCES

Alstom Power Energy Recovery. Company brochure.

Anderson, B., Huynh, D., Johnson, T., and Pleasance, G. "Development of Integrated Drying and Gasification of Browncoal for Power Generation." Paper presented at IChemE Conference, "Gasification: The Gateway to a Cleaner Future," Dresden, 1998.

API. *API Publication 941.* 2nd ed. Washington, D.C.: American Petroleum Institute, April 1990.

Barin, I. *Thermochemical Data of Pure Substances.* Weinheim: VCH Verlagsgesellschaft, 1989.

BASF. Technical leaflet, BASF Catalyst K 8-11. Ludwigshafen: BASF AG, (undated).

BBP (Babcock Borsig Power) Power Plants. "Gasification Technology." Company brochure. 2001.

Becker, J. "Examples for the Design of Heat Exchangers in Chemical Plants." *verfahrenstechnik* 3(8) (1969):335–340.

Beeg, K., Schneider, W., and Sparing, M. "Einfache Bestimmung umweltrelevanter Metall-carbonyle in technischen Gasen mittels Flammen-AAS." *Chemische Technik* 45(3) (June 1993):158–161.

Bockelie, M. J., Denison, M. K., Chen, Z., Linjewile, T., Senior, C. L., Sarafim, A. F., and Holt, N. A. "CFD Modeling for Entrained-flow Gasifiers." Paper presented at Gasification Technologies Conference, San Francisco, October 2002.

Boerrigter, H., den Uil, H., and Calis, H.-P. "Green Diesel from Biomass via Fischer-Tropsch Synthesis: New Insights in Gas Cleaning and Process Design." Paper presented at Pyrolysis and Gasification of Biomass and Waste. Expert Meeting, Strasbourg, October, 2002.

Chambert, L. "Device for Feeding Bulk Material into a Pressurized Space." International Patent PCT WO 93/00282, January 1993.

Clayton, S. J., Stiegel, G. J., and Wimer, J. G. *Gasification Technologies: Gasification Markets and Technologies—Present and Future: An Industry Perspective.* U.S. Department of Energy, July 2002.

Davey, W. L. E., Taylor, E. L., Newton, M. D., Larsen, P. S., and Weitzel, P. S. "Atmospheric Pressure, Entrained-flow, Coal Gasification Revisited." Paper presented at Gasification Technologies Conference, San Francisco, October 1998.

de Graaf, J. D., and Magri, A. "The Shell Gasification Process at the Agip Refinery in Sannazzaro." Paper presented at IChemE Conference, "Gasification: The Clean Choice for Carbon Management," Noordwijk, April 2002.

Dogan, C. P., Kwong, K.-S., Bennett, J. P., Chinn, R. E., and Dahlin, C. L. "New Developments in Gasifier Refractories." Paper presented at Gasification Technologies Conference, San Francisco, October 2002.

Gommans, R. J., and Huurdeman, T. L. "DSM's Experience with Metal Dusting in Waste Heat Boilers." Paper presented at AIChE Ammonia Safety Symposium, Vancouver, 1994.

Gorges, I., Renzenbrink, W., and Wischnewski, R. "Improvements in Raw Gas Cooling for Lignite Gasification According to the HTW Process." Paper presented at IChemE Conference, "Gasification: The Gateway to a Cleaner Future," Dresden, September 1998.

Grabke, H. J., Brach-Troconis, C. B., and Müller-Lorenz, E. M. "Metal Dusting of Low Alloy Steels." *Werkstoffe und Korrosion* 45(1994):215–221.

Gumz, W. *Gas Producers and Blast Furnaces.* New York: John Wiley & Sons, 1950.

Higman, C. A. A. "New Developments in Soot Management." Paper presented at IChemE Conference, "Gasification: The Clean Choice for Carbon Management," Noordwijk, April 2002.

Higman, C. A. A. "Perspectives and Experience with Partial Oxidation of Heavy Residues." Paper presented at AFTP Conference, "L'Hydrogène, Maillon Essentiel du Raffinage de Demain," Paris, June 1994.

Hochman, R. F. "Fundamentals of the Metal Dusting Reaction." In *4th International Congress on Metal Corrosion*, National Association of Corrosion Engineers (NACE), 1972, pp. 258–263.

Hohmann, F. W. "Improve Steam Reformer Performance." *Hydrocarbon Processing* 75(3) (March 1996):71–74.

Holt, N. A. "Coal Gasification Research, Development, and Demonstration Needs and Opportunities." Paper presented at Gasification Technologies Conference, San Francisco, October 2001(a).

Holt, N. A. II "Integrated Gasification Combined Cycle Power Plants." In *Encyclopedia of Physical Science and Technology*, 3rd ed. London: Academic Press, 2001(b).

IPCS (International Programme on Chemical Safety). International Chemical Safety Card ICSC No. 0064 (Nickel tetracarbonyl), October 2001.

IPCS (International Programme on Chemical Safety). International Chemical Safety Card ICSC No. 0168 (Iron pentacarbonyl), March 1995.

Keintzel, G., and Gawlowski, L. "Planung und Auslegung der Dampferzeuger für Kohle-Kombi-Kraftwerke." Paper presented at VGB Conference "Kohle-Kombi-Kraftwerk Buggenum," Maastricht, November 1993.

Kerfoot, D. G. E. "Nickel." In *Ullmann's Encyclopedia of Industrial Chemistry*, 5th ed., vol. A17. Weinheim: VCH Verlagsgesellschaft, 1991, p. 191.

Klosek, J., Sorensen, J. C., and Wong, M. "Air versus Oxygen for Gasification Combined Cycle Power." Paper presented at VGB Conference, "Buggenum IGCC Demonstration Plant," Maastricht, November 1993.

Koss, U., and Meyer, M. "'Zero Emission IGCC' with Rectisol Technology." Paper presented at Gasification Technologies Conference, San Francisco, 2002.

Kriebel, M. "Gas Production." In *Ullmann's Encyclopedia of Industrial Chemistry*, 5th ed., vol. A12. Weinheim: VCH Verlagsgesellschaft, 1989, p. 265.

Lascelles, K., Morgan, L. G., and Nicholls, D. "Nickel Compounds." In *Ullmann's Encyclopedia of Industrial Chemistry*, 5th ed., vol. A17. Weinheim: VCH Verlagsgesellschaft, 1991, p. 239.

Leininger, T. "Design, Fabrication, and Testing of an Infrared Ratio Pyrometer System for the Measurement of Gasifier Reaction Chamber Temperature: A Status Report." Paper presented at the 19th Annual International Pittsburgh Coal Conference, Pittsburgh, September 2002.

Lohmann, C., and Langhoff, J. "The Development Project 'Ruhr 100'—an Advanced Lurgi Gasifier." Paper presented at 15th World Gas Conference, Lausanne, 1982.

Lorson, H., Schingnitz, M., and Leipnitz, Y. "The Thermal Treatment of Wastes and Sludges with the Noell Entrained-flow Gasifier." Paper presented at IChemE Conference, "Gasification: An Alternative to Natural Gas," London, November 1995.

Marsch, H. D. "Explosion of Synloop Ammonia Separator." Paper presented at AIChE Ammonia Plant Safety Symposium, San Diego, 1990.

McDaniel, J. E., and Hornik, M. J. "Polk Power Station IGCC: 4th Year of Commercial Operation." Paper presented at Gasification Technologies Conference, San Francisco, October 2000.

National Association of Corrosion Engineers "Protection of Austenitic Stainless Steels and Other Austenitic Alloys from Polythionic Acid Stress Corrosion Cracking during Shutdown of Refinery Equipment." NACE Standard RP0170-93. 1993.

O'Keefe, W. "New Feeder Helps Plant Burn More Anthracite Waste." *Power* (June 1994).

Partridge, L. J. "Production of Ammonia Synthesis Gas by Purification and Shift Conversion of Gas Produced from Coal." AMPO 78 (1978).

Pelofsky, A. H. *Heavy Oil Gasification.* New York: Marcel Dekker, 1977.

Perry, R. H., and Chilton, C. H. *Chemical Engineer's Handbook.* 5th ed. New York: McGraw-Hill, 1973.

Pickrell, G., Zhang, Y., and Wang, A. "Development of a Temperature Measurement System for Use in Coal Gasifiers." Paper presented at the 19th Annual International Pittsburgh Coal Conference, Pittsburgh, September 2002.

Reid, W. T., and Cohen, P. "Factors Affecting the Thickness of Coal-Ash Slag on Furnace-Wall Tubes." Transactions of the A.S.M.E., November 1944.

Reimert, R. "Schleusen für Druckreaktoren—Weiterentwicklungen," *Chemie-Ingenieur-Technik* Vol 58 No 11, pp. 890–892, 1986

Reimert, R. "Schleusen für Druckreaktoren—Konzepte und Ausführungen." *Chemie-Ingenieur-Technik* 53(5) (1981):335–344.

Reimert, R., and Schaub, G. "Gas Production." In Ullmann's Encyclopedia of Industrial Chemistry, 5th ed., vol. A12. Weinheim: VCH Verlagsgesellschaft, 1989.

Rutkowski, M. D., Klett, M. G., and Maxwell, R. C. "The Cost of Mercury Removal in an IGCC Plant." Paper presented at Gasification Technologies Conference, San Francisco, 2002.

Schingnitz, M., Gaudig, U., McVey, I., and Wood, K. "Gasifier to Convert Nitrogen Waste Organics at Seal Sands, UK." Paper presented at IChemE Conference "Gasification for the Future" Noordwijk, April 2000.

Simbeck, D. R., and Karp, A. D. "Air-Blown versus Oxygen-Blown Gasification." Paper presented at IChemE Conference "Gasification: An Alternative to Natural Gas," London, 1995.

Skrzypek, J., Sloczyński, J., and Ledakowicz, S. *Methanol Synthesis*. Warsaw: Polish Scientific Publishers, 1994.

Slack, A. V., and James, G. R. *Ammonia, Part II* New York: Marcel Dekker, 1974.

Smoot, L. D., ed. *Fundamentals of Coal Combustion for Clean and Efficient Use*. Amsterdam: Elsevier, 1993.

Soyez, W. "Slag-Related Risks in Partial Oxidation Plants." Paper presented at AIChE Ammonia Safety Symposium, Denver, 1988.

Staudinger, G., and van der Burgt, M. J. "Process for Quenching Product Gas of Slagging Coal Gasifier." U.S. Patent, 4,054,424, 1977.

Strelzoff, S. "Partial Oxidation for Syngas and Fuel." *Hydrocarbon Processing* 53(12) (December 1974): 79–87.

Stultz, S. C., and Kitto, J. B. *Steam: Its Generation and Use*. 40th ed. Barberton, OH: Babcock & Wilcox, 1992.

Supp, E. *How to Make Methanol from Coal*. Berlin: Springer, 1990.

Trapp, W. "Eastman Coal Gasification: 19 years of Reliable Operation." Paper presented at IChemE Conference, "Gasification: The Clean Choice for Carbon Management," Noordwijk, April 2002.

U.S. Department of Energy. Available at: www.netl.doe.gov/coalpower/environment/mercury/description.html, 2002.

U.S. Department of Energy. "The Tampa Electric Integrated Gasification Combined-Cycle Project: An Update." U.S. DoE Topical Report No. 19, July 2000.

U.S. Department of Energy. "The Wabash River Coal gasification Repowering Project, Final Technical Report", U.S. DoE, August 2000.

van der Burgt, M. J. "Techno-Historic Aspects of Coal Gasification in Relation to IGCC Plants." Paper presented at 11th EPRI Conference on Gas-Fired Power Plants, San Francisco, 1992

van der Burgt, M. J. "Process and Burner for the Gasification of Solid Fuel." European Patent Specification 0-129-921, 1990.

van der Burgt, M. J. "Pressurizing Solids via Moving Beds." Dutch Patent NETE 0118929, 1983.

van der Burgt, M. J. "Apparatus for the Supply of Fuel Powder to a Gas-Pressurized Vessel." U.S. Patent 4,120,410, 1978.

van der Burgt, M. J. "Method and Device for the Feeding of Finely Divided Solid Matter to a Gas-Containing Vessel." U.S. Patent 4,360,306, 1982.

van der Burgt, M. J., and van Liere, J. "The Optimal Gasification Combined Cycle (OGCC)." Paper presented at Electric Power Research Institute (EPRI) Conference on New Power Generation Technology, San Francisco, 1994.

van Liere, J., Bakker, W. T., and Bolt, N. "Supporting Research on Construction Materials and Gasification Slag." Paper presented at VGB (Technische Vereinigung der Großkraftwerkstbetreieber) Conference, "Buggenum IGCC Demonstration Plant," Maastricht, November 1993.

van Loon, W. "De vergassing van koolstof met zuurstof en stoom." (The gasification of carbon with oxygen and steam). Ph.D. diss., Delft University, 1952.

Visconty, G. J. "Method of Transporting Powder into Advanced Pressure Zone." U.S. Patent 2,761,575, 1956.

Weigner, P., Martens, F., Uhlenberg, J., and Wolff, J. "Increased Flexibility of Shell Gasification Plant." Paper presented at IChemE Confernce, "Gasification: The Clean Choice for Carbon Management," Noordwijk, April 2002.

Weiss, M. M. "Selection of the Acid Gas Removal Process for IGCC Applications." Paper presented at IChemE Conference, "Gasification Technology in Practice," Milan, February 1997.

Wildermuth, E., Stark, H., Ebenhöch, F. L., Kühborth, B., Silver, J., and Rituper, R. "Iron Compounds" In *Ullmann's Encyclopedia of Industrial Chemistry*, 5th ed., vol. A14, p. 596. Weinheim: VCH Verlagsgesellschaft, 1990.

Wilhelm, S. M. "Effect of Mercury on Engineering Materials Used in Ammonia Plants." Paper presented at AIChE Ammonia Plant Safety Symposium, San Diego, 1990.

Chapter 7
Applications

7.1 CHEMICALS

The two chief components of synthesis gas, hydrogen and carbon monoxide, are the building blocks of what is often known as C1 chemistry. The range of products immediately obtainable from synthesis gas extends from bulk chemicals like ammonia and methanol, through industrial gases, to utilities such as clean fuel gas and electricity. Furthermore, there are a number of interesting by-products, such as CO_2 and steam. As can be seen from Figure 7-1, many of these direct products are only intermediates toward other products closer to the consumer market, such as acetates and polyurethanes.

Synthesis gas is an intermediate that can be produced by gasification from a wide range of feedstocks and can be turned into an equally wide range of products. And although every combination of gasifier feed and end product is technically possible, this does not mean that every combination makes economic or even technical sense. In North Dakota, synthesis gas generated from coal is successfully processed to manufacture synthetic natural gas (SNG). In Malaysia, partial oxidation of natural gas is used to generate the synthesis gas feed for a synthetic liquid fuels operation. Yet it would clearly make no sense to generate synthesis gas by partial oxidation of natural gas to manufacture SNG.

Given that this broad range of products is available from the single intermediate of synthesis gas, there is no technical reason why one could not produce more than one product from the same gas source. In fact, many operators of gasification plants do precisely this. This is known, in an analogy with co-generation (electricity and heat), as polygeneration. Some even go a step further and install surplus downstream capacity compared with the available syngas generation capacity. In this manner, such operators are able to "swing" production from one product, say ammonia, to another, say methanol, or peaking power in accordance with market demand and are thus in a position to optimize revenue from the gasification plant. In a reverse manner, there are other operators using different feedstocks, and even where appropriate different technologies, to generate their syngas. In such a case, the opportunity is to work with the cheapest feedstocks, topping up with more expensive ones only as required.

This inherent flexibility associated with syngas production and use provides a multitude of choices that is increased by the variety of utility systems, in particular

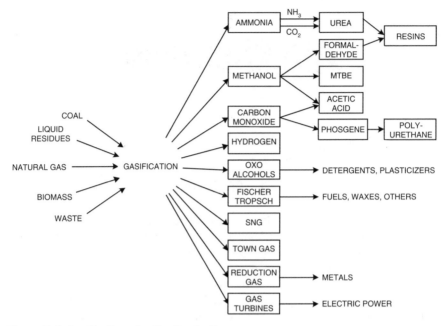

Figure 7-1. Applications for Synthesis Gas

the broad possibilities for steam system configuration. It is therefore useful to look at some typical gas processing designs for a number of the commoner applications and review the considerations behind them.

7.1.1 Ammonia

Market

Over 90% of the world's ammonia production capacity of 160 million t/y in 2001 is based on steam reforming of natural gas or (in India) naphtha. Almost all the rest, some 10 million t/y, is based on gasification of either coal or heavy oil.

The worldwide production of ammonia is by most measures the largest of any bulk chemical. The principle use of ammonia is as nitrogenous fertilizer for agriculture.

Typical plant sizes today are 1500–2000 t/d. Process licensors are currently revealing plans for plants up to 4000 or 5000 t/d size (Davey, Wurzel, and Filippi 2003; Parkinson 2001).

Synthesis Gas Specification

Ammonia synthesis takes place at high pressure over a catalyst that is usually iron, although one process uses ruthenium according to the reaction:

$$N_2 + 3\,H_2 \leftrightharpoons 2NH_3 \qquad\qquad -92\,MJ/kmol-\,N_2 \qquad\qquad (7\text{-}1)$$

A typical specification for ammonia synthesis gas is (Mundo and Weber 1982):

$N_2{:}H_2$	1:3	(For some modern processes nitrogen excess is required)
$CO + CO_2$	<30 ppmv	(As sum of total oxygen containing species)
H_2O		(in principle as for CO and CO_2, but it can be washed out with product ammonia in the synthesis unit itself)
Sulfur	<1 ppmv	
Phosphorus, Arsenic, Chlorine		(These are also poisons. Appl [1999] gives an upper limit of 0.1 ppm for chlorine)
Inerts (including methane)	<2% minimum	

Ammonia plants based on gasification technologies normally surpass these specifications when a liquid nitrogen wash is the last stage of purification.

Most ammonia plants are built in conjunction with urea plants, the CO_2 from the ammonia plant being used directly for urea production. According to the reaction

$$2\,NH_3 + CO_2 \leftrightharpoons NH_2CONH_2 + H_2O \qquad\qquad (7\text{-}2)$$

one mole of CO_2 is required for every two moles of ammonia. Typical requirements for the CO_2 are as follows:

CO_2	>98.5 mol%
$H_2S + COS$	<2 mg/Nm^3
H_2	<0.15 mol%
Methanol	<10 ppmv

Design Considerations

To process a raw synthesis gas to conform to this specification a number of different tasks must be completed:

- Tar and volatiles removal (if present in raw gas)
- Desulfurization
- CO shift ($CO + H_2O => H_2 + CO_2$)
- CO_2 removal
- Final removal of carbon oxides and water
- Adjustment of $N_2{:}H_2$ ratio

The order in which these tasks are performed may be modified to the extent that there is a choice between performing the desulfurization before or after the CO shift. Whereas older plants may have used a copper liquor wash for CO removal, most modern gasification-based ammonia plants combine the tasks of CO removal and nitrogen addition in a liquid nitrogen wash. In addition, the gas needs to be compressed to the operating pressure of the synthesis, which can vary between 90 and 180 bar.

Prior to developing a block flow diagram, two (or perhaps three) key decisions have to be taken.

First, a decision on the overall pressure profile of the plant needs to be determined. In general, there is an energy advantage to running the gasifier at the highest possible pressure, since the energy required to compress the synthesis gas is considerably more than that required for compression of the feed materials. However, there are three variables that one needs to consider:

- *The maximum pressure of the gasifier selected.* With oil feed, Texaco has numerous references with 80 bar gasification. Shell can also operate SGP at this pressure, but generally recommends operation at closer to 60 bar to reduce organic acid formation in the system. For coal gasifiers, current commercial experience is limited to 40 bar with one single unit at about 60 bar.
- *The oxygen compression system.* Here the choice is between compression of gaseous oxygen and liquid oxygen pumping. Many 900 t/d or larger ammonia plants operating with a gasifier at 60 bar have centrifugal compressors. Plants significantly smaller than this would require the use of a reciprocating machine with installed spare. Most of the 80 bar plants use liquid oxygen pumping. The overall compression energy demand for liquid pumping (i.e., air plus oxygen plus nitrogen) is some 3–7% higher than for the equivalent compression system. This cancels some of the syngas compression advantages of an 80-bar system. There are some safety arguments claimed in favor of liquid pumping, but the excellent record of oxygen turbo-compressors in this service does not argue against their use. All in all, there is not that much to choose between the two systems.
- *The synthesis pressure.* In the 1970s and 1980s, typical operating pressures in the ammonia synthesis were 220 bar. Today, most plants using conventional magnetite catalyst are designed to operate at 130–150 bar, occasionally up to 180 bar. Kellogg now uses a ruthenium catalyst operating at about 90 bar. The energy demand for the synthesis does not change much over the range 130–180 bar, since syngas compression gains made by operating at the lower pressure tend to be compensated for by an increased refrigeration demand. Furthermore, at higher pressures the volume of catalyst can be reduced, making the converter smaller and, despite the increased design pressure, also cheaper. Often, therefore, the optimization consists of selecting pressure levels for the gasifier and synthesis loop, which minimize the number of casings on the synthesis gas compressor but exploit that minimum number to the maximum extent.

- An additional factor to be aware of, even if it will not substantially influence the choice of operating pressure, is that with increased pressures physical solvents for desulfurization and CO_2 removal, gas losses through coabsorption of CO and H_2 will tend to increase. In the range up to 80 bar, however, this remains within acceptable limits.

The second important decision is the selection of the acid-gas removal system and its integration with the CO shift.

- When reviewing alternative gas treatment systems, the one immutable parameter is the specification of the syngas. For ammonia production from gasification it is not only a matter of eliminating carbon oxides, as is the case downstream of a steam reformer. One needs to remove other components, some of which are general to all gasification systems such as ammonia and HCN, others of which may be feedstock or gasifier specific, such as the hydrocarbons produced by low-temperature gasifiers. The system, which has proved itself capable of producing on-spec gas behind practically all gasification processes, is Rectisol, which uses cold methanol as a wash liquor. As a low-temperature process Rectisol is expensive. However, in the ammonia environment this is not as serious as in other applications, since a number of synergies can be achieved. All ammonia syntheses use some refrigeration to condense the ammonia from the loop gas, and the product ammonia is often stored in low-pressure tanks at a temperature of $-33°C$. The integration of the refrigeration systems for Rectisol and ammonia synthesis allows some savings compared with the stand-alone case.

  Similarly, it is possible to integrate the refrigeration demand of Rectisol and the liquid nitrogen wash, which operates at a temperature of $-196°C$. An additional advantage of a physical wash is that CO_2 required for urea production can in part be recovered under pressure, thus saving energy in CO_2 compression.

- One solution that is typical for use in conjunction with a syngas cooler is an immediate desulfurization of the raw gas. After desulfurization the gas is "clean" but also dry. In order to perform the CO shift reaction, it is necessary to saturate the gas with water vapor in a saturator tower using water preheated by the exit gas of the shift converter. CO_2 removal takes place in a separate step. The gas has thus to be cooled and reheated twice in the process of acid-gas removal and CO shift. The necessary heat-exchange equipment causes considerable expense and pressure drop, a fact that has to be counted as a disadvantage of this system.

  The raw gas from gasification of a typical refinery residue can contain about 1% H_2S and say 4% CO_2. A simple desulfurization of this raw gas will provide a sour gas of about 20% H_2S, sufficient for direct treatment in an oxygen-blown SRU. Concentration of the H_2S content up to 50% does not require much additional expense. Furthermore, the CO_2 formed in the shift is free from sulfur and can be used for urea production or emitted to the atmosphere without further treatment.

- When operating with a quench reactor, the gas emerges saturated with water vapor at about 240°C. This temperature is too low to be able to generate high-pressure steam, so it makes sense to utilize the water vapor immediately in a raw gas shift. In addition to saving the saturator tower, this has a number of minor advantages, for example, that COS in the raw gas is also converted to H_2S, and that HCN is hydrogenated to ammonia (BASF undated). The sour gase components, H_2S and CO_2, are then removed all in a single step. Attractive as this appears, it is not without its difficulties. In this case the "natural" sour gas has an $H_2S:CO_2$ ratio of about 1:50. This requires considerable expense to concentrate the H_2S to an acceptable level for a Claus furnace and to clean the CO_2 before emission to the atmosphere.

Ultimately, however, there is not much to choose between these two systems. A detailed study performed in 1971 (Becker et al.) compared a 60 bar scheme using syngas cooling and clean gas shift with a 90 bar scheme using quench and raw gas shift. It showed very little difference between the two routes. There was a slight energy advantage for the syngas-cooler route, and the difference in investment amounted to only 2%—lower than the estimating tolerance. This result continues to be valid to this day, as can be judged by the commercial and operating success of both schemes.

Example

In order to elucidate these matters in more detail, we provide a worked example of a 1000 t/d ammonia plant based on heavy oil feed and using a syngas cooler and clean gas shift. Based on the selected gasifier pressure of 70 bar and an overall pressure drop of 15 bar over the gas treatment train, the syngas compressor suction would be 55 bar. This would allow a synthesis loop at 135 bar. For the sake of the example, it is assumed that a liquid oxygen pump will be applied.

Before entering into a detailed description of the block diagram in Figure 7-2, there are a number of further design considerations that need reviewing:

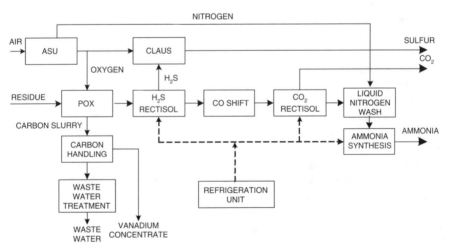

Figure 7-2. Residue-Based Ammonia Plant

- *Oxygen quality.* Considering the fact that the final ammonia synthesis gas contains 25 mol% nitrogen, it is worth reviewing the extent to which this could and should be brought into the system with the oxygen. The energy required for oxygen production shows a flat minimum between about 90% and 95% purity. On the other hand, an increase of nitrogen in the oxygen decreases the cold gas efficiency or yield of $H_2 + CO$ per kg residue. A detailed study will show an optimum at around 95% O_2. This conclusion is also valid for an IGCC application where nitrogen in the gas turbine burner is in any case required for NO_x suppression. For many other chemical applications (e.g., methanol), however, this would not be true, since N_2 is an unwanted inert in the synthesis loop.
- *Steam system.* The high-pressure steam must for safety reasons be capable of entering the gasification reactor under all conditions, and it is logical to generate steam in the syngas cooler at the same selected pressure. For our example plant, we will use 100 bar as the saturation pressure of the high-pressure steam and allow a pressure drop of 8 bar across the superheater and controls. Provision will be made for 25-bar medium-pressure steam and 10-bar low-pressure steam.
- *Compressor drivers.* With a syngas cooler installed after the gasifier, there is usually sufficient high pressure steam to satisfy the demand of the CO shift, the syngas compressor, and the nitrogen compressor. An external energy source is required for the air compressor and (in this case) the nitrogen circulator required for oxygen evaporation in the ASU. In the event of using an oxygen compressor, this would substitute for the latter. An external energy source is also required for a refrigeration compressor. Generally, two alternatives are available, electric power (provided the grid is stable enough to cope with starting a 12 MW electric motor) and steam, which is generated in an auxiliary boiler on site. This is an economic decision, but it should be borne in mind that start-up steam is required in any case. Furthermore, having a substantial boiler in operation all the time can help in stabilizing the overall steam system.
- *Refrigeration system.* The ammonia synthesis will require refrigeration capacity at typically about 0°C. For final product cooling to atmospheric storage and for the Rectisol acid-gas removal unit, additional refrigeration capacity at about −33°C is required. Ammonia compression and absorption systems are available for this duty. Absorption systems are generally more expensive in capital cost, but they can operate on low-level waste heat that might otherwise remain unrecovered. Studies performed in the context of ammonia plants like our example have shown that it is more economical to use waste heat in boiler feed-water preheat than in absorption refrigeration. On the other hand, if the waste heat really is "waste," then it pays to invest in an absorption system. Not only have both systems been used in various locations, but they have also been built in combination, where an absorption system was used to bring the refrigerant to 0°C and a booster compressor was used for the −33°C duty.

Looking at the results of this discussion in Figure 7-2 together with the material balance in Table 7-1 we see the following:

Table 7-1
Mass Balance for 1000 t/d Ammonia Plant

	Oxygen	Steam	Raw Gas	Desulf. Gas	Sour Gas	Shift Gas	Raw H_2	Pure CO_2	N_2 to LNW	NH_3 Syngas
CO_2, mol%			3.67	4.16	66.71	33.82	0.00	99.96		
CO, mol%			49.28	49.46	4.37	3.20	4.74	0.02		
H_2, mol%			45.65	45.92	0.88	62.66	94.81	0.02		75.00
CH_4, mol%			0.27	0.27	0.08	0.19	0.26			
N_2, mol%	0.10		0.09	0.09	0.02	0.06	0.09		100.00	25.00
A, mol%	0.40		0.10	0.10	0.00	0.07	0.10			
H_2S, mol%			0.90		26.75					
COS, mol%			0.04		1.19					
O_2, mol%	99.5									
H_2O, mol%		100.00								
Dry gas (kmol/h)	1014	801	4250	4223	143	6116	3940	496	1321	4942
Pressure (bar)	65	72	55.1	52.5	1.5	49.5	47.6	1.5	50.0	44.5
Temp. (°C)	120	380	45	35	10	45	-60	20	45	35

Oxygen and nitrogen are manufactured in the ASU, where the compressors are all driven by condensing steam turbines. The oxygen is pumped in the liquid phase to a pressure of 80 bar and evaporated with gaseous nitrogen, which returns the cryogenic cold to the cold box. The vacuum residue is gasified in the partial oxidation reactor with oxygen and steam at 70 bar and about 1300°C. The raw gas from the reactor contains soot and ash, which is removed in a water wash.

The raw gas, freed of solid matter is cooled down to about −30°C in the Rectisol unit, where it is washed with cold methanol to a residual total sulfur content of less than 100 ppbv. The sulfur-free gas is then heated up and saturated with water at about 220°C in a saturator tower in the CO shift. Additional steam is added that reacts over the catalyst with carbon monoxide to form hydrogen and CO_2. The gas at the outlet of the CO shift has a CO slip of about 3.2% and a CO_2 content of about 34%. This gas reenters the Rectisol unit and is washed again with cold methanol, this time at about −60°C. The CO_2 content is reduced to about 10 ppmv. The resulting gas is a raw hydrogen with about 92% H_2 and about 5% CO, the rest being nitrogen, argon, and methane. This gas is cooled down to about −196°C and washed with liquid nitrogen.

Simultaneously, the amount of nitrogen required for the ammonia synthesis is added. The gas is then compressed to the pressure required for the synthesis loop.

The mass balance in Table 7-1, based on gasifying 32 t/h visbreaker residue in the partial oxidation unit to produce 1000 t/d ammonia, illustrates this gas treatment scheme. The feed quality is as described in Table 4-10.

7.1.2 Methanol

Market

Approximately 3.3 million metric tons per year, or about 9% of the estimated world methanol production, is based on the gasification of coal or heavy residues.

Methanol is an important intermediate and, as can be seen from Figure 7-3, approximately two-thirds of the production goes into the manufacture of formaldehyde and MTBE (methyl tertiary-butyl ether). The demand for methanol has varied substantially from year to year, creating some dramatic price swings when supply has failed to keep up with demand. At the time of this writing, the future of MTBE as a component in reformulated gasoline is still uncertain, and this has its effects on the market.

During the 1990s the typical size of world-scale natural gas–based plants increased from 2000 to about 3000 mt/d. Current designs go up to 5000 t/d, and three plants of this size are in the design stage (Göhna 1997). Most plants based on gasification technologies are somewhat smaller. The largest, in the Leuna refinery in Germany, has a nameplate capacity of 2060 mt/d.

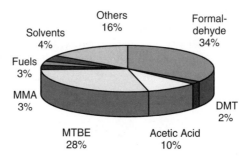

Figure 7-3. Relative Consumption of Methanol by Usage (*Source: Methanol Institute website*)

Synthesis Gas Specification

Methanol synthesis takes place by the reaction of hydrogen with carbon oxides according to the following reactions:

$$CO + 2\,H_2 => CH_3OH \qquad -91\,MJ/kmol \qquad (7\text{-}3)$$

and

$$CO_2 + 3\,H_2 => CH_3OH + H_2O \qquad -50\,MJ/kmol \qquad (7\text{-}4)$$

Modern low-pressure methanol synthesis takes place today at a pressure between 50 and 100 bar over a copper catalyst. An ideal synthesis gas specification is

CO_2	3% mol	
$\dfrac{H_2 - CO_2}{CO + CO_2}$	2.03	Stoichiometric Ratio = SR
H_2S	<0.1 ppmv	
Inerts (including methane)	minimum	

For the methanol synthesis it is important to recognize that the above specification for the stoichiometric ratio $(SR = (H_2 - CO_2)/(CO + CO_2))$ and carbon dioxide content represent an optimized synthesis gas. This is not the quality produced by the majority of plants using steam reforming of natural gas, as shown in Table 7-2.

The data in Table 7-2 show that the conventional steam reforming process operates with a considerable hydrogen surplus $(SR = 2.7)$ and high CO_2 content. Combined reforming using a steam reformer with an oxygen-blown secondary reformer is able to supply an optimized stoichiometric ratio, but it still has a high CO_2 content. Since the conversion rate of CO_2 is considerably less than that of CO,

Table 7-2
Comparison of Methanol Synthesis Gas Analyses

Process	Conventional Reforming	Combined Reforming	Gasification
Feedstock	Natural Gas	Natural Gas	Heavy Residue
CO_2, mol%	7.30	7.68	3.52
CO, mol%	16.80	21.62	27.86
H_2, mol%	72.10	67.78	67.97
CH_4, mol%	3.70	2.84	0.21
Inerts, mol%	0.10	0.08	0.44
$\dfrac{H_2 - CO_2}{CO + CO_2}$	2.7	2.05	2.05

it is preferable to keep the CO_2 content low if reasonably possible; however, a small amount of CO_2 is required to ensure a high CO conversion. The optimum CO_2 content lies between 2.5 and 3.5 mol%.

Supp (1990) provides a detailed explanation for these optima in his book *How to Produce Methanol from Coal*, so these aspects of methanol synthesis will only be touched upon here. Furthermore, he has described the manufacture of methanol from coal in considerable depth, and those interested in the topic are referred to his work. As a practical example, we will therefore review the process of making methanol from petroleum residues.

Design Considerations

The main considerations to be applied in developing a synthesis gas production scheme for methanol manufacture are the same as for ammonia, namely, selection of gasification pressure, syngas cooling arrangement, and acid-gas removal system. Contrary to the ammonia case, the optimization of oxidant quality is not a consideration, since any inerts in the syngas lower the conversion in the synthesis. The oxygen should simply be as pure as reasonably possible, which in effect means 99.5% purity.

- The selection of the exact pressure to run the methanol synthesis loop will depend on an OPEX/CAPEX optimization. For medium-size units of, for instance, 600 t/d the loop would operate at about 50 bar, that is, without any intermediate syngas compression. For a large unit of say 2,000 t/d, the pressure would be somewhat higher. The principles are shown in Figure 7-4. For the smaller unit, the clean gas from the CO_2 removal unit is fed to the suction side of the loop gas circulator,

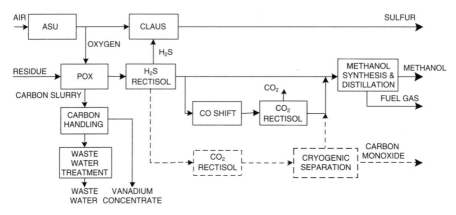

Figure 7-4. Residue-Based Methanol Plant

while the CO shift bypass gas has sufficient pressure to enter the loop on the discharge side. For the larger plant, the gases are mixed together at the suction of the booster compressor.

- For the production of CO-rich gases, the use of a syngas cooler is usually the better selection. It provides an efficient use of the sensible heat in the raw gas leaving the gasifier, where the production of steam required for CO shift in a quench is of no advantage. The methanol case lies halfway between the extremes of all hydrogen and all CO production, in that about 50% of the gas must be shifted. In most actual plants, the syngas cooler option is used for the whole raw-gas stream, and a portion of the desulfurized gas is passed over a shift catalyst. It is, however, possible to divert some of the gas from a quench reactor into a syngas cooler, employing a so-called combi-reactor, and to use a raw-gas shift (Jungfer 1985). For our example, we will cool all the gas as described above.

- As with ammonia, Rectisol is the most advantageous acid-gas removal system. It is the only wash that will achieve the desired degree of desulfurization. Alternative systems would require an additional stage of COS hydrogenation and a subsequent zinc-oxide bed for final clean up. In the case of a methanol plant, Rectisol has the added advantage that the wash liquor is the plant product itself, thus enabling some saving of infrastructure (although it should not be overlooked that an inventory tank for H_2S or CO_2 contaminated methanol is still required).

- It is worth pointing out that the use of an optimized synthesis gas quality also influences the choice of synthesis technology. Methanol formation from CO has a significantly higher heat of reaction than that from CO_2. The higher proportion of methanol produced from CO when using syngas from coal or oil gasification means that in such a plant more attention must be paid to the issue of heat removal than, for instance, in a natural gas-fed steam reformer-based plant. It is necessary not only to remove the larger quantity of heat. It is also necessary to perform this in a manner that prevents the slightest local overheating in order to avoid by-product formation, since the production of impurities from side reactions increases with increasing temperature. The intense and intimate cooling provided by the boiling water in an isothermal reactor has therefore made it the preferred reactor system for gasifier-based methanol plants. Over 90% of gasification-based methanol production operates with isothermal reactors. This includes all the large capacity ones.

The above applies for gas-phase syntheses. A demonstration liquid phase synthesis has been built at the Eastman plant in Tennessee, which is also reported to be giving good service (Benedict, Lilly, Kornosky 2001).

Example

The block flow diagram for a 1000 t/d methanol plant based on gasification of 29.5 t/h visbreaker residue in Figure 7-4 shows the result of these deliberations. The mass balance is given in Table 7-3. The residue with quality, as in Table 4-10, is gasified with 99.5% purity oxygen at a pressure of 60 bar. The raw gas is then desulfurized in a Rectisol wash to a residual total sulfur level of less than 0.1 ppmv. Approximately 50% of the desulfurized gas is subjected to a CO shift and CO_2 removal in a second Rectisol stage, providing a raw hydrogen with sufficient CO_2 slip to meet the overall synthesis gas specification. This gas is fed to the circulator of the methanol synthesis loop at a pressure of 48.8 bar. The remainder of the desulfurized gas contains about 48% CO and has a pressure of 52.5 bar. This gas is then added to the loop on the discharge side of the circulator. The exact ratio of the two streams of desulfurized gas is controlled to maintain the correct stoichiometric ratio.

The flowsheet shows with dotted lines an alternative means of adjusting the H_2, CO, and CO_2 flows to maintain the synthesis gas specification, which is particularly appropriate in an environment where the methanol is utilized for acetyls production. Instead of converting CO to hydrogen in the shift unit, CO is removed from part of the desulfurized gas in a cryogenic unit and recovered as pure CO for acetic acid production. In this case, all the residual CO_2 in the desulfurized gas draw-off must be removed in an additional stage of the Rectisol unit prior to cryogenic treatment.

7.1.3   Hydrogen

Market

The market for hydrogen is extremely diversified. The type of industry served ranges from petroleum refiners with plants ranging in size from 20,000–100,000 Nm^3/h to the

Table 7-3
Mass Balance for 1000 t/d Methanol Plant

	Oxygen	Steam	Raw Gas	Desulf. Gas	Sour Gas	Shift Bypass	Shift Feed	Shift Gas	Raw H₂	MeOH Syngas
CO_2, mol%			3.67	2.73	66.71	2.73	2.73	33.17	4.29	3.50
CO, mol%			49.28	50.20	4.37	50.20	50.20	3.20	4.50	27.59
H_2, mol%			45.65	46.60	0.88	46.60	46.60	63.31	88.31	67.24
CH_4, mol%			0.27	0.28	0.08	0.28	0.28	0.19	1.46	0.86
N_2, mol%	0.1		0.09	0.09	0.02	0.09	0.09	0.06	0.57	0.33
A, mol%	0.4		0.10	0.10	0.00	0.10	0.10	0.07	0.87	0.48
H_2S, mol%			0.90	0.00	26.75	0.00	0.00			
COS, mol%			0.04	0.00	1.19	0.00	0.00			
O_2, mol%	99.5									
H_2O, mol%		100.00								
Dry gas (kmol/h)	965	738	3905	3823	131	2021	1802	2623	1978	3999
Pressure (bar)	65	72	55.1	52.5	1.5	52.5	52.5	49.5	48.8	48.8
Temp. (°C)	120	380	45	25	10	25	25	45	25	35

food industry with requirements of about 1000 Nm³/h or less. Similarly, feedstocks and technologies differ widely, the largest plants being based on steam reforming of natural gas or resid gasification. At the smaller end of the scale, steam reformers can still hold their own, but methanol or ammonia cracking and electrolysis of water are also commercially available. An additional source is as a by-product of chlorine production. Inside refineries, much of the hydrogen demand is met from the naphtha reformer.

The estimated total world hydrogen production (excluding ammonia and methanol plants as well as by-product hydrogen) is about 16 million Nm³/h. Of this, over 500,000 Nm³/h is produced by gasification. Practically all the gasification-based hydrogen production falls into the category of "large plants," having capacities of 20,000 Nm³/h upwards. The largest plant is Shell's 112,000 Nm³/h facility in its Pernis, Netherlands refinery. This reflects current economics, and in particular the opportunities for resid-based hydrogen production in refineries. However, this may not always remain true.

Hydrogen Specification

Just as the range of hydrogen consumers is wide, so are the demands on quality. Typical specifications are:

For silicon wafer production	99.9999%
For optic-fiber cable production	99.99%

In the refinery environment it is necessary to define the quality of the hydrogen product on a case-by-case basis. Many hydrocracking processes will accept a 98% H_2 purity, which can be produced by the traditional shift, CO_2 removal, and methanation route. A number of processes do, however, require higher purities of 99% H_2 or higher, in which case the final purification step will have to be pressure swing adsorption (PSA). The higher purity is achieved, however, at the cost of a lower hydrogen yield (about 85–90% instead of 98%) and the production of a relatively large quantity of low-pressure, low-Btu fuel gas that one may or may not be able to accommodate in the refinery–fuel gas balance. It is important in this context to review the hydrogen purity specification carefully with the hydrocracker requirements, since all too often a purity of >99.5% is specified on the basis of the economics of the conversion unit alone, or on the assumption that hydrogen will be generated from a steam reformer that can accommodate the PSA tail gas internally, without reference to the economics of the overall configuration. A careful review of site-specific parameters will produce different solutions. Thus, for instance, the Pernis plant uses the methanation route; others use PSA units (de Graaf et al. 1998; Kubek et al. 2002).

Design Considerations

Many of the considerations required for the design of a hydrogen plant are similar to those discussed earlier in the ammonia and methanol cases (Sections 7.1.1 and 7.1.2, respectively), so those will not be repeated here. Specific to hydrogen manufacture are the following:

- Generally in a refinery, particularly one where the product slate is favorable to the use of gasification, the hydrogen consumers, hydrocracker or hydrotreater, operate at high pressures. PSA units can operate over a wide pressure range of about 20–70 bar. However, at the higher end of this pressure range, co-adsorption of the hydrogen increases with a consequent drop in yield. On the other hand, operation at a high gasifier pressure can provide significant savings in compression energy. Thus the pressure selection is a matter for optimization studies on a site-specific basis.
- As for ammonia, the decision on the process for the syngas cooling and CO shift (quench plus raw-gas shift, versus syngas cooler plus clean-gas shift) is finely balanced. Issues such as on-site energy-integrated or offsite stand-alone oxygen production can tip the scales in one direction or the other. For our example, we will use raw-gas shift, so as to contrast with the clean-gas shift used above for ammonia—not a real-life criterion!

  If the clean-gas shift route is chosen, then the question often arises; need one include a CO_2 wash upstream of a PSA unit? This question must be answered in the affirmative, at least as a bulk removal, even though in a steam reformer plant this practice is seldom, if ever, applied. The reason lies in the quantity of CO_2 to be removed. Examination of the shift gas quality in such a plant (similar up to this point to the ammonia plant) shows a CO_2 content of over 30 mol%. Feeding such a gas to the PSA unit has two deleterious effects. First, the hydrogen yield of the PSA drops from say 85% to about 70%. Second, the tail gas contains some 60% CO_2 and has a heating value of about 4500 kJ/Nm^3 (LHV), requiring the use of a support fuel to combust it satisfactorily.
- For the desulfurization, a Selexol wash has been chosen. As a physical wash it has many of the same characteristics as the Rectisol process used in the previous examples. It has the advantage of operating at ambient or near ambient temperatures, thus eliminating the refrigeration load of the Rectisol unit. The disadvantage that Selexol would have in handling unshifted gas, namely its poorer COS solubility, is of no import in the shift gas case, since the COS is already largely converted to H_2S on the shift catalyst. Selexol is able to concentrate the sour gas stream to an acceptable level for processing in a Claus sulfur recovery unit (Kubek et al. 2002).
- The use of membrane technology is a worthwhile additional consideration in multiproduct plants, where hydrogen is only a small part of the overall product slate. One example is shown in Section 7.1.4. Another could be a side

stream of hydrogen coming from an IGCC plant, such as those discussed in Section 7.3.

- A further aspect to consider if refrigeration is required for acid-gas removal is the choice of refrigerant. In the ammonia plant, it is natural to use ammonia as the refrigerant. In the refinery, it may be an unfamiliar medium. An alternative in such a situation would be propane or propylene, fluids with which the refinery operator will be accustomed and for which no unit-specific additional safety measures are required.

Example

The resultant flow diagram and mass balance for a 50,000 Nm^3/h hydrogen plant are shown in Figure 7-5 and Table 7-4, respectively. 20.42 t/h visbreaker residue, as described in Table 4-10, is gasified in a quench reactor. The gas leaves the reactor at a temperature of 240°C saturated with water vapor. In this condition it is passed over the raw-gas shift catalyst. In addition to the CO shift reaction

$$CO + H_2O \leftrightarrows CO_2 + H_2 \qquad (2\text{-}7)$$

COS is shifted to H$_2$S and HCN is hydrogenated to ammonia. The shift gas is cooled and is then desulfurized in the Selexol wash. Also in the Selexol wash the bulk of the CO$_2$ is removed so that the raw hydrogen entering the PSA unit already has a hydrogen content of 97 mol%. Final purification of the hydrogen takes place in the PSA unit.

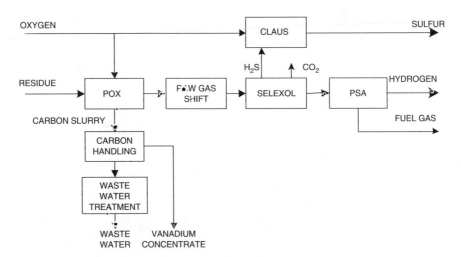

Figure 7-5. Residue-Based Hydrogen Plant

Table 7-4
Mass Balance for 50,000 Nm³/h Hydrogen Plant

	Oxygen	Steam	Raw Gas	Shift gas	Desulf. Gas	Sour Gas	CO₂ Vent	Pure H₂	PSA Tail Gas
CO_2, mol%			3.67	35.16	5.28	29.12	94.92		33.43
CO, mol%			49.28	0.50	0.72		0.07		4.57
H_2, mol%			45.65	63.40	93.57		4.96	99.99	59.27
CH_4, mol%			0.27	0.18	0.26		0.03		1.63
N_2, mol%	0.1		0.10	0.06	0.09	17.69			0.55
A, mol%	0.4		0.09	0.06	0.09				0.54
H_2S, mol%			0.90	0.63	0.00	52.59			
COS, mol%			0.04	0.00	0.00	0.35	4 ppm		
O_2, mol%	99.5					NH_3 0.25			
H_2O, mol%		100.0							
Dry gas (kmol/h)	646	510	2708	4022	2649	48	977	2232	309
Pressure (bar)	73	72	60	55	53	2.1	1.5	52	1.3
Temp. (°C)	120	380	240	35	35	45	24	25	25

7.1.4   Carbon Monoxide and Oxo-Alcohols

Carbon Monoxide Market

Pure carbon monoxide is a raw material for a number of organic chemicals such as acetic acid, phosgene, which is an intermediate for polyurethane manufacture, and formic acid (see Figure 7-6). The toxic nature of CO makes it difficult to store or transport. For safety, inventories are usually kept to a minimum. For these reasons pure carbon monoxide plants tend to be located close to the point of use of the product and are accordingly fairly small. Approximately 500 kt/y of CO is used for producing acetic acid.

Oxo-Alcohols Market

The term oxo-alcohols covers a range of higher alcohols, from C_3-alcohol to C_{18} +. Lower alcohols (propanol, butanol, and pentanol) find usage as solvents. In the range C_8 to C_{10} they are used as plasticizers. Higher alcohols (C_{12}+) are required for the manufacture of synthetic detergents. Oxo-alcohols are manufactured by reacting olefins with syngas (50:50 H_2:CO) to form aldehydes. The aldehyde is then hydrogenated with high purity hydrogen to produce the oxo-alcohol end product. World production of oxo alcohols is about 8100 kt/y, over half of which is produced in two countries alone, the United States and Germany.

CO and Oxo Syngas Specifications

Typical specifications for carbon monoxide and oxo-synthesis gas are:

	CO	**Oxo Syngas**
H_2/CO	–	about 1
CO_2	–	<0.5 mol%
H_2	<0.1 mol%	about 49 mol%
CO	>98.5 mol%	about 49 mol%
Inerts		<0.5 mol%
Exact numbers depend on use and process.		

Design Considerations

The example will show combined production of 5000 Nm³/h pure carbon monoxide and 5000 Nm³/h oxo-synthesis gas, based on natural gas feed. For a plant of this size it is assumed that oxygen is available from a pipeline or a multicustomer gas supply facility.

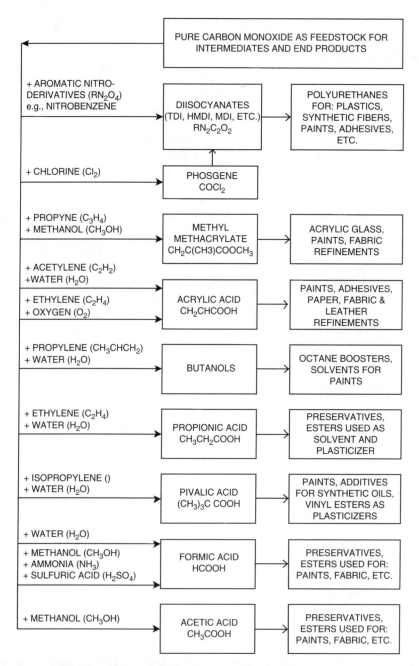

Figure 7-6. Major Applications of High Purity Carbon Monoxide
(*Source: Lath and Herbert 1986*)

- There are three principle processes for the manufacture of synthesis gas from natural gas, steam reforming, catalytic autothermal reforming, and partial oxidation. The hydrogen:carbon monoxide ratio of the syngas is an important characteristic distinguishing between these three processes. Unless there is the possibility of importing CO_2, the typical range for the three processes with and without CO_2 recycle is:

Process	H_2/CO ratio	
	with CO_2 recycle	without CO_2 recycle
Steam reforming	2.9	6.5
Catalytic autothermal reforming	1.7	3.7
Partial oxidation	1.55	1.81

Thus the desired ratio of hydrogen and carbon monoxide in the product streams is an important factor in process selection. Note, however, that with partial oxidation, the CO_2 produced is small and so also the effect of CO_2 recycle. For this reason, CO_2 recycle is seldom applied with partial oxidation units.

The other determining issue is primarily economic, namely, the availability of oxygen for both autothermal reforming and partial oxidation. For small plants it is seldom economic to build a dedicated air-separation plant, so where no pipeline oxygen is available or synergies with a gas supplier cannot be realized, steam reforming would be applied, despite the potential hydrogen surplus that can only be used as fuel.

For both oxo-synthesis and pure CO production, all processes supply excess hydrogen so partial oxidation, which produces the lowest H_2/CO ratio, is often selected if oxygen is available.

- Looking at the flowsheet in Figure 7-7, one observes that no desulfurization step has been expressly included. The decision regarding what to do about desulfurization will

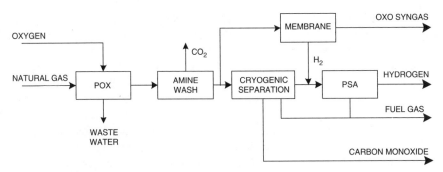

Figure 7-7. CO and Oxo-Syngas Plant

depend heavily on the amount of sulfur in the gas. One frequently applied possibility is to include a zinc-oxide bed in the natural gas preheat train. However, one needs to pay careful attention to the matter of metal dusting corrosion, particularly at the hot gas inlet of a syngas cooler. The alternative is to place the zinc-oxide bed in the syngas line, upstream of the amine wash, allowing the sulfur to act as a corrosion inhibitor (see Section 6.11). An additional advantage of this choice is that sulfur also inhibits the methanation reaction. Spontaneous methanation at the temperatures prevailing is a rare occurrence, but in the presence of catalytic impurities in the gas it can take place. A third possibility, if the sulfur level in the natural gas is sufficiently low, is to remove it with the CO_2 in the amine wash. The latter must, however, be designed to remove COS as well as H_2S, so as to meet the oxo-gas specification, and the sulfur content in the CO_2 must be within environmentally permitted levels. For the purposes of our example this matter is not included in the mass balance, Table 7-5.

Example

Natural gas is fed to the partial oxidation reactor where it is reacted with oxygen and without any steam moderator to produce a raw synthesis gas. The raw gas is desulfurized with zinc oxide at the outlet of the partial oxidation unit after cooling in a syngas cooler. The raw gas is washed with MDEA to achieve a residual CO_2 content of 10 ppmv, which meets the oxo-syngas specification and the requirements of the cold box for the cryogenic separation. In the membrane unit, sufficient hydrogen is extracted from the clean gas as permeate to leave the H_2/CO ratio of the non-permeate at 1:1 as required by the oxo-synthesis process.

In the cold box of the cryogenic separation, a 98.5 mol% CO product is obtained, which must be compressed to feed the downstream CO consuming units. Both membrane and cryogenic separation produce a raw hydrogen, which is purified in a PSA unit so that it can be used in the hydrogenation stage of the oxo process or for other purposes. Tail gases from the cryogenic unit and the PSA are available for use as a low-pressure fuel gas.

7.2   SYNFUELS

For transportation and upgrading reasons there is often a need for converting one fuel into another fuel. On the one hand this may concern the conversion of coal or remote natural gas into a liquid fuel, and on the other hand the conversion of coal into substitute or synthetic natural gas (SNG).

Virtually all modern coal gasification processes were originally developed for the production of synthesis gas for the subsequent production of chemical feedstocks or hydrocarbon liquids via Fischer-Tropsch synthesis. The only place in the world where the process sequence of coal gasification to Fischer-Tropsch is currently practiced is at the Sasol complex in South Africa. For the production of SNG from coal, only one plant is in operation in Beulah, North Dakota. For the conversion of remote natural gas via partial oxidation and Fischer-Tropsch synthesis into hydrocarbon liquids, plants are currently in operation in Malaysia and in South Africa.

Table 7-5
Mass Balance for CO and Oxo-Syngas Plant

	Natural gas	Oxygen	Raw Gas	Clean gas	CO₂	Membrane Feed	Oxo Syngas	Permeate	Coldbox Feed	CO	PSA Feed	Hydrogen	Fuel Gas
CO_2, mol%	1.2		3.22	0.00	97.14	0.00	0.00	0.00	0.00	98.51			33.65
CO, mol%			37.10	38.33	1.12	38.33	48.56	11.40	38.33		9.02		60.25
H_2, mol%			57.65	59.58	1.74	59.58	48.56	88.57	59.58		90.97	99.99	1.16
CH_4, mol%	87.7		0.30	0.31		0.31	0.43		0.31				
C_2H_6, mol%	5.1												
C_3H_8, mol%	1.0												
C_4H_{10}, mol%	0.4												
C_5+, mol%	0.3												
N_2, mol%	4.3	0.1	1.55	1.60		1.60	2.20	0.03	1.60	1.49	0.01	0.01	4.26
A, mol%		0.4	0.18	0.18		0.18	0.25	0.00	0.18				0.68
O_2, mol%		99.5											
H_2O, mol%													
Dry gas (kmol/h)	397	274	1071	1035	35	308	223	85	727	226	478	380	194
Pressure (bar)	36	33	28	27	1.3	27	26	17	27	25	17	16	1.3
Temp. (°C)	25	120	35	35	45	35	35	35	25	35	35	35	35

This last option is especially attractive when low-cost natural gas is available that cannot be economically transported to markets by pipeline or as liquefied natural gas (LNG). In principle, there are two liquid products that can be produced: methanol and Fischer-Tropsch (FT) liquids. For the production of methanol, the reader is referred to Section 7.1.2, where the production of methanol has been discussed.

Classically, two different FT synthesis process types are available: the ARGE and the Synthol synthesis. In the ARGE process, synthesis gas is converted into straight chain olefins and paraffins over a cobalt containing catalyst at temperatures of about 200°C and pressures of 30–40 bar. The reaction takes place in a large number of parallel fixed-bed reactors that are placed in a pressure vessel containing boiling water for cooling and ensuring an essentially isothermal process.

The product is subsequently hydrogenated in cases where straight paraffins are the desired product. Such products are immanently suitable for the production of solvents and waxes, as the product is completely free from sulfur and nitrogen compounds as well as from aromatics. By adding an acidic function to the hydrogenation catalyst, some iso-paraffins are produced as well that improve the low-temperature characteristics of the premium fuels that can be produced by the ARGE process. Moreover, the boiling range of the products can be controlled within a wide range as the acidic function of the catalyst can be used for hydrocracking the heavier fractions.

Due to the absence of aromatics, the kerosene fraction has a very high smoke point and is a excellent blending component for aviation turbine fuels. For the same reason, the gasoil fraction has a very high cetane number (>70) and is a valuable blending component for automotive diesel fuels. However, a warning is appropriate since the lack of sulphur in unblended FT products can create problems in standard fuel pumps.

In the Synthol process, synthesis gas is converted into an aromatic-rich product over an iron-containing catalyst at temperatures of about 250°C and pressures of 30–40 bar. The reaction takes place in large fluid-bed reactors. The product is rich in aromatics and is used for the production of motor gasoline and as a diesel blending component. This process is being used at the Sasol plant in Secunda and in Mossel Bay, both in South Africa.

In recent years further developments have been made. The Shell Middle Distillate Synthesis (SMDS) process uses a fixed-bed reactor similar to that of ARGE. Sasol has developed its advanced slurry-bed reactor. Exxon, BP, Statoil, and others have demonstration plants in operation or under construction. The effects of various synthesis characteristics on the gas production facility are discussed in what follows.

7.2.1 Gas to Liquids

There are considerable attractions to producing liquid hydrocarbon fuels from remote sources of natural gas. On the one hand, it provides a means of bringing energy resources from remote locations to the market, in a form that is not limited by the small number of receiving terminals, as is the case for LNG, but that allows the utilization of the existing large and flexible infrastructure in place for the transport of liquid hydrocarbons.

On the other hand, the quality of Fischer-Tropsch products enable them to be sold at a premium price. All FT products are sulfur-free (typically <1 ppm), but particularly the high-quality diesel cut with no aromatic content and a cetane index of over 70 can contribute significantly to achieving the U.S. Environmental Protection Agency (EPA) standards valid from 2006 (Mulder 1998; Agee 2002).

The basic Fischer-Tropsch process produces a mixture of straight-chain hydrocarbons from hydrogen and carbon monoxide according to the reaction

$$CO + 2\,H_2 = -[CH_2]- +\ H_2O \qquad -152\,MJ/kmol \qquad (7\text{-}5)$$

where $-[CH_2]-$ is the basic building block of the hydrocarbon molecules. The product mixture depends on the catalyst, the process conditions (pressure and temperature), and the synthesis gas composition. The product slate follows the Schulz-Flory distributions. The selectivity of two typical processes is shown in Table 7-6. See also Tables 7-7 and 7-8.

Different FT syntheses require different H_2:CO ratios in the syngas. Furthermore, additional hydrogen is often required for product work-up. These differences demand an individual approach to syngas generation for each project, depending on

Table 7-6
Selectivity (Carbon Basis) of Fischer-Tropsch Processes at Sasol

Product	ARGE	Synthol
CH_4	4	7
C_2 to C_4 Olefins	4	24
C_2 to C_4 Paraffins	4	6
Gasoline	18	36
Middle Distillate	19	12
Heavy Oils and Waxes	48	9
Water Soluble Oxygenates	3	6

Source: Mulder 1998

Table 7-7
Operating Characteristics of ARGE and Synthol

	ARGE	Synthol
Temperature (°C)	220–225	320–340
Pressure (bar)	25 bar	23
H_2:CO ratio	1.7	2.54

Source: Derbyshire and Gray 1986

Table 7-8
Synthesis Gas Specifications

Synthesis	H$_2$/CO	Remarks
ARGE	1.3–2.3	
SMDS	1.5–2	
Synthol	2.6	
Methanol	2.4–3	$(H_2-CO_2)/(CO+CO_2) = 2.05$

Source: Higman 1990

the synthesis process selected, the desired product slate, and the product work-up scheme (Higman 1990).

Proven and operating syngas production routes from natural gas include partial oxidation and combined reforming (steam reformer followed by an oxygen-blown secondary reformer as used for methanol production). Both routes are described in Higman (1990). In principle, an autothermal reforming or gas-heated reformer-based scheme can also be applied, although no plant of this nature has yet been proven at the sizes likely to be required for a world-scale GTL facility. In conformity with the scope of this book, however, our example is partial-oxidation–based.

A typical specification for Fischer-Tropsch syngas is shown in Table 7-9. From this we can see that the gas must be sulfur-free, since sulfur is a catalyst poison. Whereas with catalytic reforming processes the desulfurization must be performed upstream of gas generation to protect the reforming catalyst, with partial oxidation one also has the option of desulfurizing in the syngas. In fact, syngas desulfurization has a number of advantages over natural gas desulfurization. First, organic sulfur

Table 7-9
Specification for Fischer-Tropsch Synthesis Gas

Gas Component	Max. Allowable Concentration
H$_2$S + COS + CS$_2$	<1 ppmv
NH$_3$ + HCN	<1 ppmv
HCl + HBr + HF	<10 ppbv
Alkaline metals	<10 ppbv
Solids (soot, dust, ash)	essentially nil
Tars including BTX	below dewpoint
Phenols and similar	<1 ppmv

Source: Boerrigter, den Uil, and Calis 2002

species in the natural gas are converted to H_2S (and traces of COS) in the partial oxidation reactor, thus obviating the need for an upstream hydrogenation stage complete with hydrogen recycle. Second, the syngas from a typical partial oxidation unit has a very high CO partial pressure and high Boudouard equilibrium temperature (about 1050°C), which makes it an extremely aggressive metal dusting agent. Syngas desulfurization leaves the sulfur in the syngas while it is in the dangerous temperature range. The sulfur is therefore able to act as an effective corrosion inhibitor. And third, sulfur-free synthesis gas can be subject to spontaneous methanation at temperatures above about 400°C, given the right conditions. Against these benefits is the fact that the volume of the synthesis gas to be treated is about three times that of the natural gas feed. The level of desulfurization is therefore less, and equipment will tend to be larger. Nonetheless, in applications with waste-heat recovery, which would be typical for a Fischer-Tropsch environment, the authors would recommend syngas desulfurization.

The Fischer-Tropsch synthesis produces methane and other light fractions that, depending on the tail gas recovery arrangement, may contain olefins. Some of this stream may be recycled to the partial oxidation unit for reprocessing to synthesis gas. A 100% recycle is, however, not possible, since the inerts (argon and nitrogen) will build up excessively in the recycle loop. In a reformer-based system it is possible to create an inerts purge by using FT tail gas as reformer fuel. In the partial oxidation scenario, hydrogen production provides the opportunity for a purge, as shown in Figure 7-8. Care should be taken, however, with FT tail gas as reformer feed. The olefin content will tend to coke on the reformer catalyst. CO in the tail gas may methanate on the catalyst for olefin hydrogenation. Nonetheless, with suitable pretreatment, tail gas can be used as reformer feed.

The overall flowsheet can be seen in Figure 7-8. Natural gas is fed directly to the partial oxidation unit undesulfurized. (In this context, it is assumed that any bulk sulfur removal, LPG recovery, or the like has been conducted at the wellhead. Clearly this has to be taken into account in the economics of a project for handling remote gas, for which no wellhead treatment would otherwise be available. See Chapter 8.) Desulfurization takes place on a zinc-oxide/copper-oxide adsorber bed in the syngas stream before the latter is fed to the synthesis unit. In the tail gas

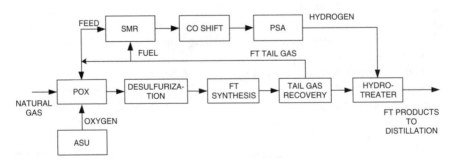

Figure 7-8. Block Flow Diagram of Liquids Production Using Partial Oxidation and FT Synthesis

recovery unit, light, gaseous products are recovered and partially recycled. The rest is used for hydrogen manufacture. Heavy oils and waxes are then hydrotreated as part of the product work-up.

7.2.2 SNG from Coal

The energy crisis of the 1970s and the accompanying concerns about a shortage of natural gas gave rise not only to intensive research into hydrogenating gasification systems but also a large number of projects for the manufacture of synthetic natural gas (SNG). Of these only one was ever built, in Beulah, North Dakota. The plant still operates today and in 2000 has broken new ground by making the CO_2 from the acid-gas removal unit available for enhanced oil recovery (Dittus and Johnson 2001). Given the current availability of natural gas it is unlikely that another SNG facility will be built in the near- or even medium-term. Nonetheless it is instructive to look at a number of issues connected with its manufacture from coal.

SNG consists primarily of methane, which is synthesized by the reaction of carbon oxides with hydrogen over a nickel catalyst according to the equations:

$$CO + 3\,H_2 \rightleftarrows CH_4 + H_2O \qquad -206\,MJ/kmol \qquad (7\text{-}6)$$

$$CO_2 + 4\,H_2 \rightleftarrows CH_4 + 2\,H_2O \qquad -165\,MJ/kmol \qquad (7\text{-}7)$$

Specifications for SNG require a maximum hydrogen content of 10% and, depending on the heating-value requirement, a limitation on CO_2. Typically, this results in a requirement on the stoichiometry of the synthesis gas such that the stoichiometric number ($SN = H_2/(3\,CO + 4\,CO_2)$) has a value between 0.98 and 1.03.

The selection of the coal gasifier cannot be made in isolation from the quality of the coal itself (see Chapters 4 and 5). Looked at from the point of view of the application alone, there is a distinct advantage to processes that produce a high methane content ex gasifier, since this reduces the volumes of gas to be treated and, if necessary, compressed between gasification and synthesis. For this reason, a Lurgi dry-ash gasifier has been selected. (See Figure 7-9.)

With the selection of a Lurgi dry-ash gasifier a decision has to be taken on the matter of tar handling. The tar can be used as a raw material for the manufacture of tars and phenols. Alternatively, it could also be gasified to generate additional synthesis gas, thus reducing the coal throughput requirement. The former solution has been chosen in the example. Phenosolvan and CLL units have been incorporated to recover ammonia from the waste water.

The raw gas contains more CO than that required by the methane synthesis, so a partial stream is shifted on a raw gas shift catalyst. In the downstream acid-gas removal, the residual hydrocarbons, sulfur and nitrogen compounds (NH_3 and HCN), must be removed. To achieve the correct stoichiometry, a partial CO_2 removal is also required. The sulfur specification for the nickel catalyst is maximum 100 ppbv or even lower. The only system capable of all these tasks is Rectisol, and

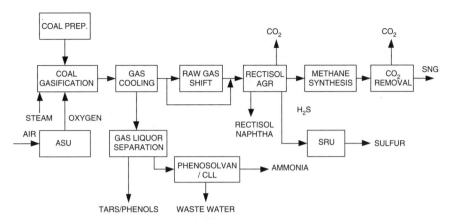

Figure 7-9. Block Flow Diagram of SNG Manufacture from Coal

this is also included in the example flowsheet. In order to achieve the low CO_2 specification of the SNG, CO_2 may be removed in a final Rectisol stage, which also provides the necessary drying function.

7.3 POWER

Much of the world's 3.1 million installed MW power generation capacity is over thirty years old, particularly in the industrialized countries. Typically, the average efficiency of coal-fired power stations is 30–35%. For gas-fired power stations the efficiency varies from 35–43% for open cycles and 50–58% for combined cycles. Although the efficiency of coal-fired units with modern combustion technology is above 40%, this state of affairs has (rightly) caused considerable concern, both in terms of conservation of resources and of CO_2 emissions (van der Burgt, Cantle, and Boutkan 1992). Over the last 20 years a vast amount of work has gone into improving existing combustion technologies as well as investigating alternatives. Of the potential alternatives the use of gasification to produce a suitable fuel for highly efficient gas turbines, the IGCC has continually proved to be a leading contender.

The power industry is worldwide the largest man-made emitter of CO_2 with 33% of the total. (The transport sector is the next largest with 25%.) The increasing use of natural gas as a feedstock for gas turbines (the "dash-to-gas") has allowed the power sector to reduce emissions on new capacity substantially, even though much of the motivation lies in the reduced investment and shorter construction times associated with gas turbine technology. However, the medium-term picture has to take account of the limits in natural gas supply. By 2030 Europe will be reliant on gas imports for some 70% of its supply (European Commission 2001). So alternatives to this simple solution are required.

Typical efficiencies of the current generation of steam cycle power plant, including flue gas desulfurization and NO_x abatement are 40–42%. Ultra-supercritical cycles in operation have 43%, and IGCCs 38–43%. Serious assessments project between 46% and 50% as being possible with these technologies in the next 10 years. Looking at these figures, one should bear in mind that a four-point increase in efficiency represents about 10% less CO_2 production per unit of power produced.

Already there is a capacity of some 3800 MW in installed IGCCs plus 3500 MW in projects currently under development or construction. Much of this is built in association with oil refineries, where synergies in the residue or petroleum coke disposal as well as in hydrogen production have encouraged their development. Coal-based projects are following.

7.3.1 Comparison with Combustion

The maximum theoretical efficiency of a machine for the conversion of chemical combustion energy into power is given by the formula for a reversible Carnot process:

$$\eta = \frac{w}{q} = 1 - \frac{T_L}{T_H}$$

η is the fraction of the power w produced by the cycle over the heat of combustion q added to the cycle. T_L and T_H are the lowest and the highest absolute temperatures of the cycle. The lowest temperature is almost always the ambient (or cooling water) temperature, and hence the formula immediately shows how important the maximum temperature is for the cycle efficiency. A graphical representation of the above formula is given in Figure 7-10.

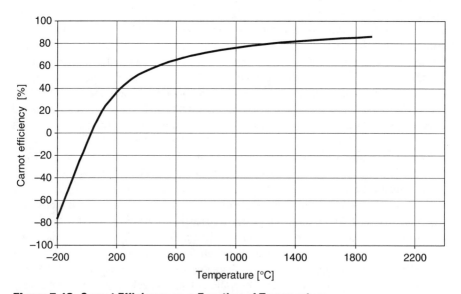

Figure 7-10. Carnot Efficiency as a Function of Temperature

Table 7-10
Theoretical and Practical Efficiencies of Various Power Plant Cycles

Cycle	Fuel	T_{Low}	T_{High}	Carnot	Actual	Actual as % Carnot
Conventional steam power plant	Coal	27	540	63	40	63
Ditto Ultra supercritical	Coal	27	650	67	45	67
IGCC	Coal	27	1350	82	46	56
Open-gas turbine cycle	Gas	27	1210	80	43	54
Combined cycle	Gas	27	1350	82	58	71
Tophat cycle	Gas	27	1350	82	60	73
Low-speed marine diesel	Heavy fuel oil	27	2000	87	48	55
Low-speed marine diesel with supercharger	Heavy fuel oil	27	2000	87	53	61

The column headers are grouped: **Temperatures, °C** spans T_{Low} and T_{High}; **Efficiencies, % LHV** spans Carnot, Actual, and Actual as % Carnot.

The Carnot efficiency for various cycles are given in Table 7-10, together with some real values.

As can be seen, the potential for any particular cycle as represented by its Carnot efficiency is by no means the only consideration when looking at the merits and limitations of different cycles. The efficiencies offered by gas turbines, which operate with an upper temperature of between 1200 and 1400°C (compared with 500–650°C for steam turbines), are restricted by the fact that the gas turbine itself requires a clean gaseous or liquid fuel, whereas the conventional combustion processes can handle dirty fuels including solids. The diesel engine cycles benefit from the high temperatures (2000°C), but precisely this property contributes to the extremely high NO_x emissions connected with this technology.

In summary, the conclusion can be drawn that in the actual processes only 54–73% of the Carnot efficiency is realized. The best results can be obtained in a so-called Tophat cycle that is a single cycle based on a gas turbine, as discussed in Section 7.3.3. The Carnot efficiencies as given in Figure 7-10 for cryogenic cycles have been calculated for a T_H of 27°C and a T_L of −100 and −200°C. These data are solely given here as they illustrate how energy-intensive cryogenic cycles, as applied in ASUs, are. The negative efficiency of −80% of a cryogenic cycle in an ASU that has a minimum temperature of about −200°C is in absolute terms about equal to a cycle with a maximum temperature of 1200°C. A negative efficiency for cryogenic cycles is caused by the fact that the energy for such cycles constitutes a loss. Before dealing with the pro's and con's of various more complex cycles, the basic principles of the steam and gas turbine cycles will be discussed.

Steam Cycles

The simplest cycle for the conversion of heat, which may be derived from full or partial combustion or other sources such as nuclear energy, is the open steam cycle depicted as a flow scheme and T-s diagram in Figures 7-11 and 7-12, respectively. The corresponding points in the cycle are labelled A-F in both figures. The top of the curve in the T-s diagram is the critical point. Points on the curve to the right of the critical point represent saturated steam, and points to the left water.

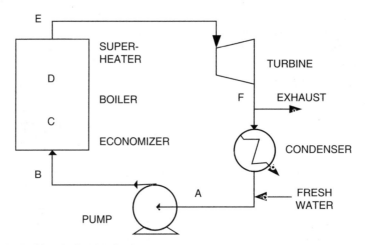

Figure 7-11. Simple Steam Cycle

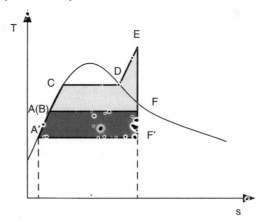

Figure 7-12. T-s Diagram: Simple Steam Cycle

In the open cycle, corresponding to an atmospheric back-pressure turbine, the pump raises the pressure of the water from atmospheric pressure at the point A to the working pressure of the boiler at point B. The water is then heated to boiling temperature (C) and evaporated at constant temperature (D). Superheating is represented in the T-s diagram by the line D-E. Expansion to exhaust at atmospheric pressure (the vertical line E-F) takes place in a steam engine such as a turbine. The work extracted from the process is represented by the upper shaded area of the T-s diagram. The heat input is represented by the shaded area plus that of the rectangular area below the line A-F.

In the closed cycle with a condenser, known as the Rankine cycle, the expansion continues to the point F′ at a subatmospheric pressure, and the steam is condensed, returning it to the state A′. As can be seen from the T-s diagram, the work extracted is increased by the area of the lower shaded area.

In the last century many improvements have been made to the Rankine cycle. As regards equipment the most important improvement has been the use of steam turbines instead of piston expansion machines. Another feature that has increased the efficiency is the use of low level heat from the turbine for preheating the boiler feed water, thus using less high level heat in the boiler. High-level heat should preferably be used only where topping-heat is required. Doing this is making good use of the exergy present in the fuel. For a good understanding of exergy, the reader is again referred to treatises on thermodynamics (Shvets et al. 1975). Knowledge about exergy is of utmost importance for energy efficient designs (Dolinskovo and Brodianski, 1991). For the purpose of understanding the cycles discussed in this book it is sufficient to know that high-level heat should preferably not be used for low-level heat requirements, and furthermore, that the pressure energy in media should not be throttled away unnecessarily. This last point is especially important for gases.

Apart from adding a condenser and boiler feed water preheat, there are three major other improvements to the steam cycle. These are the application of higher steam pressures, higher superheat temperatures, and reheat cycles. For details the reader is referred to the literature (e.g., Shvets et al. 1975).

Supercritical Cycles

The biggest competition for gasification based power plants are advanced so-called supercritical steam cycles. These are called supercritical as both the maximum temperature and the maximum pressure of the steam are above the critical values of 374°C and 221 bar. By combining the benefits of the above features for improving the efficiency, the state-of-the-art steam cycles can now reach about 45% LHV basis for plants that are equipped with DeSOx and DeNOx facilities. The costs of these plants are very competitive, and gasification-based power stations must have even higher efficiencies and lower cost in order to be successful.

Examples of modern supercritical units are the pulverized coal (PC) coal-fired double-reheat unit at Aalborg in Denmark (285 bar/580°C/580°C/580°C) and the single-reheat unit at Matsuura in Japan (241 bar/593°C/593°C). In Germany and in Japan units are now under design or construction for single-reheat plants of

260 bar/580°C/600°C and 250 bar/600°C/610°C respectively. The next step will feature plants at steam conditions of 300 bar/600°C/620°C. A simplified flow scheme for such a plant is given in Figure 7-13. The new steels that are required are a major issue for all these developments (Viswanathan, Purgert, and Rao 2002).

As a result of all these improvements, it has been possible to increase the efficiency of steam cycles to 45% LHV basis, that is 67% of the maximum Carnot efficiency, as shown in Table 7-10.

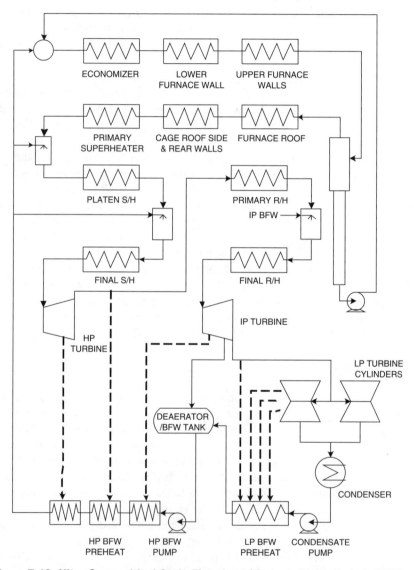

Figure 7-13. Ultra-Supercritical Cycle Flowsheet (*Source: Welford et al. 2002*)

This result is obtained with water as a working fluid, which is in some ways the worst choice for a Rankine cycle because of the very high heat of evaporation and the high specific heat of liquid water. Yantovski (1996) has studied extensively the merits of using cycles in which CO_2 is used as the main working fluid, which has the advantage that it has a low heat of evaporation. Such a cycle has advantages when CO_2 has to be captured and sequestered. A major disadvantage of CO_2 is that it is more difficult to handle, as the vapor pressure of liquid CO_2 is already 40–50 bar at ambient temperatures.

The reason for discussing in some detail the steam cycle is twofold. On the one hand the steam cycle plays an important role in combined cycles (CC), and on the other hand much can be learned from its development that is also of importance for the Brayton or Joule cycle as used in gas turbines.

Gas Cycles and Combined Cycles (CC)

The most widespread gas turbine cycle is the aircraft engine, which operates an open cycle using essentially air as a working fluid. In power applications the open cycle is mainly used for peak-shaving installations. The flow scheme and the T-s diagram of the Joule cycle are given in Figures 7-14 and 7-15 respectively below. Air (A) is compressed in a compressor, and the pressurized air (B) is used for the combustion of the fuel. The resulting very hot gases (C) then enter the turbine, and the still hot gases (D) are sent to the stack. Although this simple cycle has at best about the same efficiency (43%) as a fairly advanced steam cycle, the efficiency as a percentage of the Carnot potential is about 10% lower (see Table 7-10). This problem can be remedied by using the hot gases leaving the gas turbine as a heat source for a steam cycle in a heat recovery steam generator (HRSG). In this case the efficiency increases to almost 60%, which corresponds to over 70% of the Carnot potential.

The flow scheme and the T-s diagram for this combined cycle (CC) is given in Figures 7-16 and 7-17, respectively. In the T-s diagram it is clearly shown that part of the space below the area of the Joule cycle is occupied by a relatively small steam cycle. The ratio of the shaded areas to the area below the upper line of the cycle, which is a measure for the overall efficiency, is thus considerably increased.

The open cycle is used in airplanes and for peak shaving in power stations. Larger and more efficient gas turbines are now entering the marketplace with firing temperatures of about 1500°C. These advanced turbines claim additional economies of scale, reduced capital costs, and higher overall net efficiencies of 45–50% (LHV basis). This illustrates how gas turbine manufacturers have concentrated on higher turbine inlet temperatures. But the efficiency gain achievable by this means has its limits. Taking the Carnot criterion (Figure 7-10) illustrates that increasing the inlet temperature from 1200 to 1500°C will increase the efficiency from 80 to 83%. Assuming a high 60% thereof can be realized, in actual practice this implies that the cycle efficiency will increase by a mere 2%. Further, one should keep in mind that blade cooling and low NO_x requirements, which are problems that increase

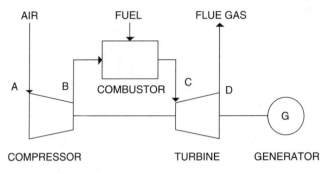

Figure 7-14. Gas Turbine with Open Cycle

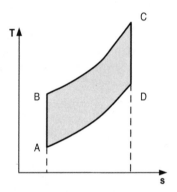

Figure 7-15. T-s Diagram for Open-Cycle Gas Turbine

with temperature, will imply a higher parasitic power consumption, all of which casts some doubt on too great expectations of further increases in gas turbine inlet temperatures.

In base load power stations combined cycles (CC) are generally used. Although the Joule cycle is not very efficient, it has the advantage of a very low capital cost. From the point of view of the Joule cycle it is advantageous to achieve as low a gas outlet temperatures as possible. Typically, the gas outlet temperatures of the gas turbine are about 550°C. This is, however, a disadvantage for the steam part of the combined cycle, since high superheat temperatures cannot be achieved. The steam cycle is therefore relatively capital intensive (expressed in dollars per installed kW). The main disadvantage of the use of gas turbines is that only gas and light petroleum distillates can be used as a fuel. To take full advantage of the high efficiency of the CC in combination with dirty feedstocks as coal and heavy oils, these fuels must first be gasified.

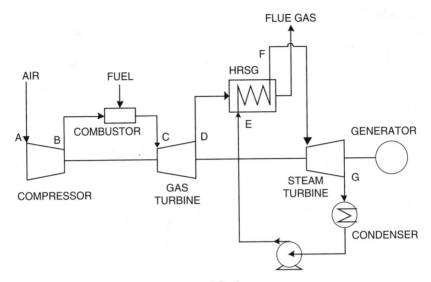

Figure 7-16. Gas Turbine with Combined Cycle

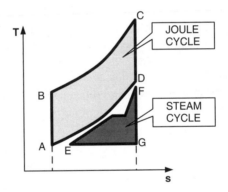

Figure 7-17. T-s Diagram for Combined Cycle

7.3.2 State-of-the-Art IGCC

Flow Schemes

The integrated gasification combined cycle (IGCC) is in its present form basically a combination of gasification and a CC plant. The advantage of incorporating gasification into these plants is to convert solid and residual liquid fuels into a form that gas turbines can accept. Furthermore, gasification provides a means of desulfurization of the fuel before combusting the gas in the CC plant to levels not achievable at reasonable cost with current flue gas desulfurization (FGD) technologies. Thus, with a gasifier and the proper gas treating train behind it, one obtains a fuel that can be combusted as simply and as environmentally friendly as natural gas. IGCC plants with coal as feedstock can be found in power stations, whereas IGCC plants with heavy residual fractions as feedstock are found in refineries.

Although the two component parts of an IGCC (gasification and gas turbines) are both well developed technologies, the combination is nonetheless relatively new. As is typical for a technology in such a stage of development, there is considerable variation in the optimization of flowsheets for IGCC. This is particularly true in the matter of effective integration of the two core technologies involved (Holt 2002).

The most important integration that is applied in almost all IGCC plants is the integration of the steam system. The steam cycle of a standard CC has an efficiency of about 38%. The pinch problem caused by the evaporation of the water is the main reason why it is so low. On the other hand, a steam system that derives its heat only from the syngas cooler of a gasifier has also a low efficiency of perhaps 38% because superheating is difficult in a syngas cooler. Combining both steam sources alleviates the restrictions, as the pinch in the HRSG is eased by the large evaporating duty in the syngas cooler and more saturated steam can be superheated in the HRSG, albeit at a lower superheat temperature. It is therefore not surprising that when combining the heat sources from HRSG and syngas cooler the efficiency becomes about 40%. This is represented by the "conventional integration" case shown in Figure 7-18.

Two other possibilities for integration that are applied in some IGCCs both involve the air separation unit. One possibility is air integration, which involves

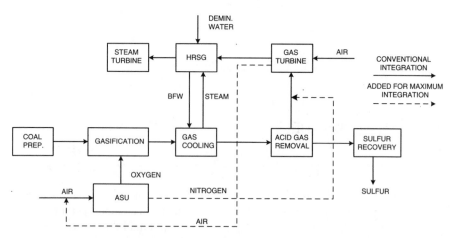

Figure 7-18. Block Flow Diagram of an Integrated IGCC Power Plant

using extraction air from the combustion turbine compressor as feed air for the ASU. This results in an overall somewhat lower compression duty. The ASU main column then operates at a higher than the normal atmospheric pressure (about 10 bar instead of 5 bar) that results in the production of oxygen and nitrogen at an elevated pressure of 5 bar instead of 1 bar. Hence the oxygen for the gasification requires less compression. Nitrogen integration uses the nitrogen stream from the ASU as diluent for the clean fuel gas to reduce the flame temperatures and hence NO_x emissions. Also, here it is advantageous that the nitrogen already has an elevated pressure.

Although the efficiency of the IGCC increases with a high degree of integration, there is a risk of the availability suffering. A compromise that is applied in new plants is one where part of the air for the ASU is supplied as extraction air from the compressor of the gas turbine and part from a dedicated compressor in the ASU. This adds an additional piece of rotating equipment to the plant, but it facilitates the start-up and improves the overall availability of the plant.

The extraction air leaving the adiabatic gas turbine compressor has first to be cooled before it can be used in the ASU. This is a further cycling of the temperature of a main gas stream, which is one of the reasons why the efficiencies of IGCC plants are still relatively low. In the next section some ideas are discussed to improve this situation.

Fuel Gas Expansion

Texaco and others have proposed the use of a quench IGCC design in which the gasification is conducted at a higher pressure of about 70 bar (e.g., Allen and Griffiths 1990). Additional power can then be generated by expansion of the clean fuel gas from 70 to about 20 bar followed by gas reheat and firing in the gas turbine. The gas cooling by an expander can also be used to provide much of the refrigeration demand of a physical absorption desulfurization system (Zwiefelhofer and Holtmann 1996). Fuel gas expansion is being used in two of the heavy oil IGCC plants at Falconara and Priolo in Italy, which also increases the power generated in these plants.

Efficiencies

The efficiency of gasification is at best about 80%, which, assuming 60% for the CC, implies that the overall efficiency of an IGCC will not be much higher than $80 \times 60/100 = 48\%$. However, for virtually all gasifiers an oxygen plant (ASU) is required and the gas treating also requires energy, and hence the efficiencies of the best first-generation plants are all below 45%. And even the lower figures that are reached require a fair amount of integration.

For a further understanding of the cycle, a Sankey diagram of a refinery residue-based IGCC is shown in Figure 7-19.

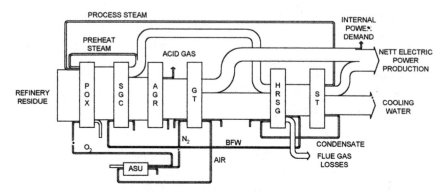

Figure 7-19. Sankey Diagram of a Residue-Based IGCC

7.3.3   Advanced Cycles

Options for Improving the Joule Cycle

Although the CC cycle already has a high efficiency, it has the disadvantage that two cycles are involved and much of the elegance of the simple Joule cycle is lost. In relation to this it is interesting to explore whether the Joule cycle cannot be improved in a similar way as the Rankine cycle has been improved over time. Although both the Rankine cycle and the Joule cycle have made use of the benefits of higher pressure ratios and higher temperatures, little has been done in the Joule cycle to make use of features such as reheat and heat recuperation to improve the cycle efficiency.

Cycles can be made more efficient by preheating the medium of the cycle (water in case of the Rankine cycle, and air in case of the Joule cycle) with low level heat that comes available during the cycle. In the Rankine cycle this is, for example, accomplished by using extraction steam from the expansion turbine to preheat the boiler feed water. In the Joule cycle such preheat is only possible in gas turbines with a low-pressure ratio of 7-10 where the exhaust gas of the turbine has a higher temperature than the exhaust air from the compressor. In that case the sensible heat in the exhaust gas from the turbine may be used to preheat the air leaving the compressor in a so-called recuperator. In more advanced industrial turbines and aero-derivative turbines with pressure ratios of 15 and higher, the turbine and the compressor have about the same outlet temperature, which makes recuperation impossible. This is a serious disadvantage of more advanced gas turbines and is due to the fact that virtually all gas turbine compressors feature adiabatic rather than isothermal compression. The same reason limits the use of reheat in gas turbine cycles. The very high pressures of 70–200 bar, which would be necessary for a realistic reheat cycle, prohibit the use of adiabatic compression.

More Isothermal Compression

The advantage of isothermal compression is not only that it makes reheat and recuperation possible but also that it requires less energy than adiabatic compression, as is illustrated by the following formulae for the isothermal and adiabatic compression of an ideal gas:

Isothermal

$$E = RT_{in} \cdot \ln\left(\frac{p_{high}}{p_{low}}\right)$$

Adiabatic

$$E = RT_{in} \cdot \frac{k}{k-1} \cdot \left[\left(\frac{p_{high}}{p_{low}}\right)^{\frac{k-1}{k}} - 1\right]$$

where E is the energy in J/mol, R is the universal gas constant of 8.314 J/mol.K, k is the isentropic exponent C_p/C_v of the isobaric heat capacity divided by the isochoric heat capacity, T_{in} is the inlet temperature of the compressor in K, and p_{high}/p_{low} is the pressure ratio of the compressor.

Taking air of 300 K, for which k is about 1.4 and a pressure ratio of 10, it is easily shown that for this case adiabatic compression requires 1.41 times the energy of isothermal compression. One should not be surprised by this large difference, as it must be kept in mind that the energy for heating of the gas during compression is coming from shaft power that otherwise could have generated power or electricity. This heating is thus equivalent to electric resistance heating! In case of isothermal compression the air will be heated with additional fuel in the combustor, but there the heating takes place with virtually 100% efficiency, whereas via the shaft the heating takes place with an efficiency of about 40% based on fuel.

Many attempts have been made over the years to accomplish a more isothermal compression. The most obvious solution is to split the compressor in various parts and to apply indirect intercooling between the various stages. Although this will lower the energy required for compression, it has the disadvantage that the compressor is split up in various parts and that the heat exchangers result in additional pressure drop for the air flow.

An example of a more positive approach to a more isothermal compression is the Sprint gas turbine that features a water spray injection between the two compressor stages of a General Electric LM6000 aero-derivative gas turbine (McNeely 1998). This so-called wet compression as a means to accomplish a more isothermal compression has often been proposed in the past but was never applied (Milo AB 1936; Brown, Boveri & Cie 1968; Beyrard 1966; Société Rateau 1952). The main purpose of the water spray in the Sprint gas turbine is to increase the capacity of the turbine for power generation, but at ambient temperatures above 5°C it also increases the efficiency of the power plant.

Combinations of More Isothermal Compression and Recuperation

All major improvements in gas turbine–based cycles concern the use of a more isothermal compression and recuperation. The highest efficiencies are achieved with so-called humid air turbines (HAT) in combination with heat recuperation from the turbine exhaust gases. The cycles involved are called HAT cycles. Two HAT cycles will be discussed: the HAT cycle and the Tophat cycle.

The HAT Cycle

In the HAT cycle a flue gas heat recuperator replaces the heat-recovery steam generator (HRSG). In the recuperator the sensible heat in the hot exhaust gases leaving the turbine are used to preheat humidified combustion air and water (Schipper 1993). The combustion turbine air compressor is also intercooled and cooled after final compression (aftercooling). The heat recovered in these cooling steps preheats additional water, and the hot water humidifies the pressurized combustion air in a multistage, countercurrent saturator. The major disadvantage of this scheme is that instead of a compressor and a turbine as in a normal gas turbine, many more pieces of rotating equipment are required in the form of an additional compressor and pumps. Moreover, large spray columns are required for humidifying the water to be injected into the air, and finally, the heat in the hot gases leaving the turbine is used for the low-temperature service of preheating and evaporating water. Various modifications of the HAT cycle have been proposed, such as the cascaded HAT (CHAT) cycle (Nabhamkin 1995), but all suffer from one or more of the disadvantages mentioned above. The only exception is the Tophat cycle discussed in the next section.

The Tophat Cycle

The reason why wet compression, that is humidifying the air during compression inside the compressor, has found so little application is most likely due to the fact that most atomizing devices available today can only produce water droplets with a diameter of $30\,\mu$ or larger. Smaller droplets can be made, but this requires generally complex or bulky equipment. An alternative for small droplets are spray towers as used in HAT cycles, which result in additional pressure drops. This is a pity, as in wet compression good use is made of the unique high heat of evaporation of water, whereas this same quality is a disadvantage for the Rankine cycle.

Only recently an elegant and compact method has been proposed for making small droplets. This has made it possible to inject an extremely large amount of water into the atmospheric air entering the air compressor or in the compressor itself.

Flow Scheme

The flow scheme of the Tophat cycle is shown in Figure 7-20 (van der Burgt and van Liere 1996). The water is injected in the air A entering the compressor in such

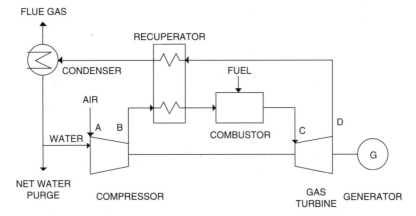

FLUE GAS

RECUPERATOR

CONDENSER

FUEL

AIR

A B

WATER

COMBUSTOR

C

D

G

NET WATER
PURGE

COMPRESSOR

GAS
TURBINE

GENERATOR

Figure 7-20. Tophat Cycle

a way that the compressor does not suffer from a parasitic pressure drop. It is injected in the form of very fine droplets of a mean diameter of about $1-3\,\mu$. These droplets, that can be made by combining flash evaporation with efficient atomizers as in the "swirl flash technology" (van Paassen and van Liere 1980), are so small that the droplets will (a) evaporate in the milliseconds available in the compressor, (b) will not cause erosion problems, and (c) follow the path of the gas stream without being centrifuged out. The humidified air B leaving the compressor at the required pressure is essentially saturated with water. In a recuperator, the humidified air is heated with the hot exhaust gases leaving the turbine to a temperature of, say, 50–100°C below the turbine outlet temperature before being routed to the combustor, where the hot air is used for the combustion of the fuel. The hot pressurized flue gas C then enters the turbine. The exhaust gas D leaving the turbine preheats the humidified air as well as the water used for humidifying the air and, if required, the fuel. After leaving the recuperator the water in the exhaust gas is routed to a condenser, after which the dry exhaust gas leaves via the stack. The condensate is partly recycled, and the surplus is purged from the system.

Quasi-Isothermal Compression

The compression as proposed for the Tophat cycle is not completely isothermal but quasi-isothermal. In practice it results in a compressor outlet temperature for the humidified air, which varies from about 100 to 175°C for discharge pressures of 8 to 32 bar, respectively, when starting with ISO air (15°C and a moisture content of 1.19 mol%) and using injection water of 200 °C. This is clearly illustrated in Figure 7-21A (data for an isentropic efficiency of the compressor of 87%).

Quasi-isothermal compression hence requires less energy per unit (kmol/s)/ISO-air than adiabatic compression. This advantage increases with the pressure ratio, as illustrated in Figure 7-21B.

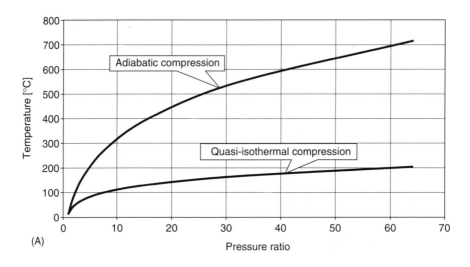

(A)

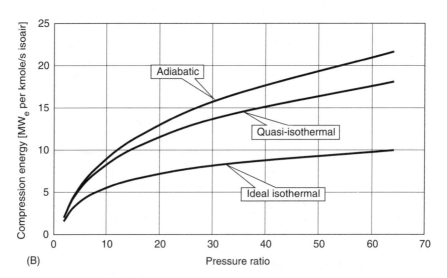

(B)

Figure 7-21. (A) Compressor Discharge Temperatures as Function of Compression Ratio (B) Compression Energy for Adiabatic and Isothermal Compression

The greatest advantage of a more isothermal compression is that now it becomes advantageous to have a recuperator in which the sensible heat in the gases leaving the turbine is used to preheat the air leaving the compressor. This heat, which in a combined cycle is used to drive an additional steam cycle, is now used in the more efficient and less costly Joule cycle itself.

The Recuperator

Recuperators—or flue gas heated air pre-heaters—play an important part in some synthesis gas technologies such as steam reforming but have not found favour in connection with gas turbines, whether in IGCC or standard applications. This is a logical outcome of the concentration on adiabatic compression and the fact that the outlet temperatures of air compressor and gas turbine are too close for a recuperator to have any important effect. Also the use of a recuperator with conventional quasi-isothermal compression with the use intercoolers as used in process gas compressors does not have any beneficial effect, since even if the heat removed via the intercoolers is used for say boiler feedwater preheat, it transfers heat from the gas turbine cycle to the less efficient steam cycle. This is different in the Tophat cycle since the heat is used to increase the mass entering the gas turbine by evaporating water into the combustion air.

Arguments are sometimes raised against recuperators because of the poor heat transfer and large surface area involved. These arguments are however generally superficial. The steam superheater in the HRSG is also a gas-gas exchanger with similar heat transfer coefficients as is the air preheater in a steam reformer and both are successful components in their respective environments.

Furthermore, the construction of the headers and so on is much lighter than the equivalent HRSG steam superheater, which reduces problems related to thermal shock. Assuming that the Tophat stations will be started up and shut down as frequently as the alternative of combined cycles, the point of thermal shock is not very relevant. The reason is that the metal temperatures and temperature cycles are about the same when the preheat temperature of the humidified air is restricted to the superheat temperature of the steam in a CC.

Moreover, the recuperator has a very smooth temperature profile in the steady state. Because the heat exchange is restricted to the exchange of sensible heat (gas-gas and gas-water) the enthalpy supply and demand lines are almost parallel, as is illustrated in the typical example in Figure 7-22. In this example the humidified air and the fuel gas are both preheated to 500°C, and the water used for evaporation during compression is preheated to 200°C. As can be seen, there is hardly any pinch. For the case in question, the hot exhaust gas, after having preheated the humidified air, the natural gas, and the water, has a temperature of about 145°C.

The Water Cycle

The distillate quality water required for injection can be obtained by condensing the water in the exhaust gas. This gas has then to be further cooled after it leaves the recuperator. This can advantageously be accomplished in a two-stage direct contact condenser. The first condensate, comprising 5–10% of the water present in the exhaust gas, contains virtually all the solids contained in the combustion air and the fuel that have acted as condensation nuclei. This water can be used as a purge in order to avoid build-up of solid contaminants in the system. The pure condensate from the second stage can then be used for humidifying the air.

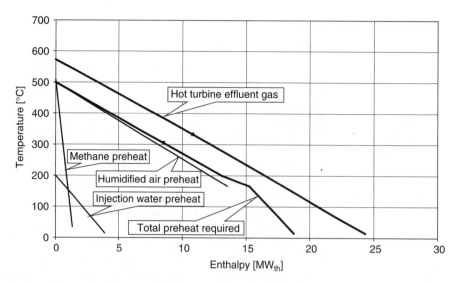

Figure 7-22. Typical Recuperator Enthalpy-Temperature Diagram

The use of indirectly cooled condensers does not look attractive because of the large amount of inert gases in exhaust gas, as this results in very large heat exchange surfaces and hence in costly equipment. Sometimes indirect cooling may be economical though, for example, when the heat utilized in a combined heat and power system involving, for example district heating or seawater distillation.

An important point is, of course, whether cooling water is available. On ships and for offshore applications, this will never present a problem. In arid areas, air-cooling or a cooling tower must be used. As all fuels contain hydrogen, there is always a net production of water. In arid areas this is advantageously used for irrigation. In the case of natural gas the mass of net water produced is about equal to the mass of the fuel.

The Tophat Cycle Efficiency

The efficiency of the Tophat cycle is very dependent on the temperature difference between the hot turbine exhaust gases entering the recuperator and the humidified air leaving the recuperator. Typically for a 30°C decrease in the recuperator temperature difference there will be an increase of about one percentage point on the overall cycle efficiency.

Also, in the Tophat cycle the turbine inlet temperature is a factor in relation to efficiency, although it should be realized that raising the turbine inlet temperature of the turbine is not so important for the Tophat cycle as for a CC. The reason is that because of higher inlet temperatures, both a high pressure ratio is required and the temperature of the gases leaving the turbine is generally increased. Hotter exhaust gases would lead to higher maximum temperatures in the recuperator and hence imply the use of more expensive steels for this service. For this reason the maximum preheat temperatures of the humidified air was limited to 500°C so as to keep the maximum metal temperatures in virtually all cases to below 550°C. With these restrictions there is not much effect in raising the inlet temperatures above 1300°C.

Station Efficiency and NO_x Control

The biggest advantage regarding NO_x control of the Tophat cycle is the fact that the stoichiometric adiabatic flame temperatures (SAFTs) are so low. As is well known, lower SAFTs result in lower NO_x emissions. In the standard Joule cycle, higher station efficiencies are obtained by increasing both the pressure ratio and the turbine inlet temperatures, resulting in higher SAFTs. Using quasi-isothermal compression as applied in the Tophat cycle generally leads to lower SAFTs for stations with a higher efficiency. This is clearly illustrated in Figure 7-23, where SAFTs are plotted against station efficiencies for various cases: a Joule cycle, a Tophat cycle, and a case where only quasi-isothermal compression (without recuperator) is used. The reason for the low SAFTs of the quasi-isothermal compression cases is the low oxygen content and the higher moisture content of the air (see Figure 7-24).

Applications

The high efficiency of the Tophat cycle of 60% or more makes it attractive for many applications apart from as a replacement for combined cycle stations. The fast

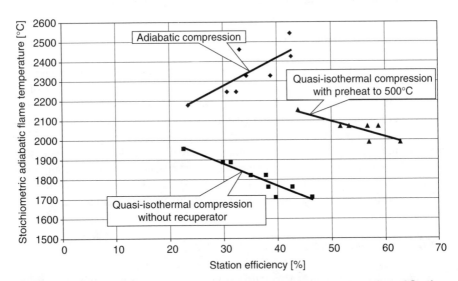

Figure 7-23. Stoichiometric Adiabatic Flame Temperature as a Function of Station Efficiency

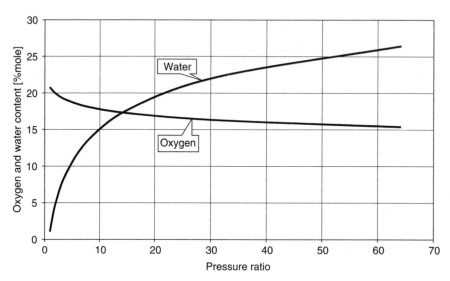

Figure 7-24. Oxygen and Water Content in Humidified Air

start-up and the absence of a steam cycle make it attractive for many applications where now open cycles are used. Examples include peak shaving, the use of gas turbines in ships, offshore applications and liquid natural gas (LNG) plants, combined heat and power schemes, and so on. The fact that Tophat cycles can be applied for duties from, say, 500 kW onwards, means that they can even be considered for trucks, locomotives, off-the-road vehicles, and mining equipment.

7.3.4   Flue Gas Treatment

The loss of efficiency involved in cycling of gas temperatures between hot and cold parts of an IGCC have been pointed out on many occasions, and has been one of the principle driving forces behind the so far unsuccessful attempts at a hot, or at least warm, fuel gas clean-up. All current development efforts are based on performing all the gas clean up on the fuel gas side of the gas turbine, not least because many of these techniques are available from chemical applications. In this section we wish to review the possibilities for reducing the fuel gas treatment to the absolute minimum required by the gas turbine—which after all can operate on (sulfur-containing) fuel oil—and perform the rest of the gas clean-up as flue gas treatment as in the rest of the power industry.

IGCC Temperature Profiles

The ideal process for power generation from fossil fuels using combustion would feature a steady rise followed by a steady drop in temperature and pressure in such a way that little heat and pressure energy is wasted. An open Joule cycle with heat recuperation comes close to this ideal. A conventional IGCC is, however, very far from this situation as, on the one hand, part of the gas stream has to be cooled to $-190°C$ in the ASU, and on the other hand, the gas leaving the high temperature gasifier has to be cooled for desulfurization before being heated for the second time in the gas turbine combustor.

Figure 7-25 shows the temperature profile as encountered in a conventional IGCC with fuel gas treatment (front profile) and in an IGCC with flue gas treatment for an entrained-flow bed gasifier (middle profile). For comparison, an air-blown fluid-bed gasifier with flue gas treatment is also shown (back profile). The temperature scale is logarithmic, because it reflects somewhat better the thermodynamic repercussions of the temperature cycling, since the ratios between the various temperatures are more important than the absolute temperature differences.

Each kink in the diagram refers to the process unit mentioned along the abscissa. Straight lines between kinks imply that the units in between are not present. It is clear that flue gas treating gives a temperature pattern with fewer temperature swings. It should be noted, however, that the cryogenic temperature applies only to the air going to the ASU, whereas the high gas inlet temperature of the turbine applies to a gas mass flow that is about a factor 5 higher.

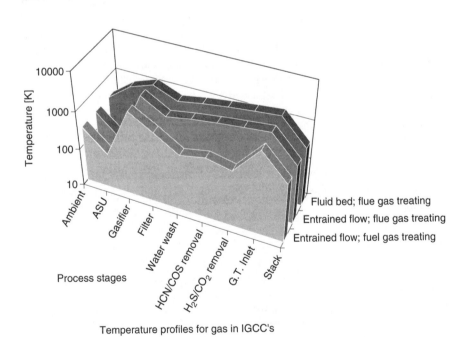

Temperature profiles for gas in IGCC's

Figure 7-25. Temperature Profiles of a Various Types of IGCC Power Stations

With flue gas treating instead of fuel gas treating, the sensible heat in the fuel gas, corresponding to 5–10% of the LHV of the fuel, and the heat of combustion of H_2S and COS, corresponding to 0.5–1% of the LHV of the fuel gas, can be utilized in the CC and no COS removal unit is required.

Comparisons of Gas Treating Schemes

We have made some simple comparisons for various configurations so as to investigate the potential represented by flue gas treatment without, at this stage, the constraint that all technology must be existent and proven today. All calculations were made on a consistent basis as described in Appendix E. Before presenting and discussing the results of these studies, two important aspects need to be discussed, namely the gas cleaning concept and the effect of flue gas recycle, which are an integral component of the concept.

Gas Treating

Developing a minimum gas treatment concept for the gas turbine involves removal of any solid particulate material in the gas and removing any gaseous components that could become solid under conditions that might be experienced in the turbine or HRSG. For this reason, particulate removal must take place at a temperature that is sufficiently low that alkali compounds can be removed as solids and so present no risk of forming corrosive alkali (hydro-) sulfates. This temperature is about 500°C, which represents an upper bound for the particulate removal. The lower bound is governed by the ammonium chloride sublimation temperature and lies at about 280–300°C. After filtering out the fly ash on which the volatile alkali compounds and other metal compounds have been deposited (the fly ash acting as a substrate), volatile sulfur, nitrogen, and arsenic compounds as well as mercury may pass through the turbine without causing problems. Within this allowable range of 280–500°C it is preferable to remain close to the upper limit of 500°C for efficiency reasons. Filtering at a temperature of 500°C is possible with candle filters. Looking purely at the gas turbine requirements, this represents then the minimum fuel gas cleanup concept. All other pollutants will require post-combustion removal as in current conventional PC technologies.

Flue Gas Recycle

The potential of flue gas recycle is a neglected area of cycle design. It has significant advantages for natural gas-fired gas-turbine–based power generation (including IGCC) where gaseous components, for example, CO_2, have to be removed from the flue gas as well as in terms of extremely low thermal NO_x emissions.

Before discussing its use in syngas applications we will review its effect on a natural gas-fired combined-cycle plant. The basic concept is to replace most of the excess air to the burner by recycled fuel gas, which is compressed in the air compressor. A flowsheet is given in Figure 7-26. The mass and energy balances, the gas compositions, and temperatures and pressures for plants without and with flue gas recycle are given in Tables 7-11 and 7-12, respectively.

The calculations have been made on basis the 1 kmole of methane fuel, and LHVs are used throughout. The gas turbine used has a pressure ratio of 32, an inlet temperature of 1350°C, and an isentropic efficiency of 90%. Neither temperature losses nor pressure losses have been taken into account. The overall efficiencies are calculated as follows: it is assumed that 30% of the HRSG duty is converted into power via a Rankine (steam) cycle, and that there are 2% mechanical losses in the Joule (gas) cycle,

$\{(794-427)\times0.98+436\times0.30\}\times100/803=61.1\%$ for the classical combined cycle base case and:

$\{(764-401)\times0.98+440\times0.30\}\times100/803=60.7\%$ for the cycle with flue gas recycle.

These figures should, of course, only be used on a comparative basis. What may be concluded, though, is that the efficiencies are about equal, but that the quantity of gas to be treated in case of the recycle is only $10.7/25.9\times100\%=41\%$ of that of the

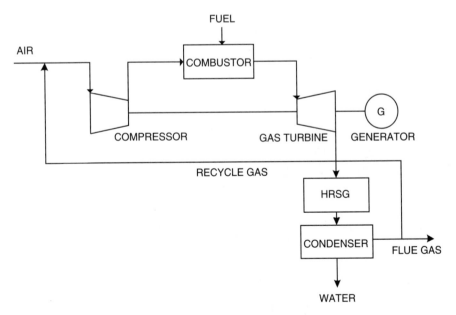

Figure 7-26. Flow Scheme Combined Cycle Power Station with Flue Gas Recycle

Table 7-11
Flow Data and Gas Compositions for Methane-Fired Combined Cycle

Flow	Air to Compr.	Comp. Duty	Air ex Compr.	Fuel	Gas to Turbine	Gas ex Turbine	Turbine Duty	HRSG Duty	Water ex Scrubber	Stack Gas
kg	762		762	16	778	778			27	751 = 25.9 kmol
MJ	0	427	427	803	1230	436	794	436	0	0
Temp. °C	25		556	25	1350	534			25	25
Press. bar	1		32	32	32	1			1	1
mol%										
CO_2	0.03		0.03		3.67	3.67				3.88
H_2O	1.19		1.19		8.43	8.43			100	3.15
O_2	20.70		20.70		12.66	12.66				13.39
N_2+A	78.08		78.08		75.24	75.24				79.58
CH_4				100						

Table 7-12
Flow Data and Gas Compositions for Methane-Fired Combined Cycle with Recycle of Flue Gas

Flow	Gas to Comp. Air	Gas to Comp. Rec. gas	Comp. Duty	Gas ex Comp.	Fuel	Gas to Turbine	Gas ex Turbine	Turbine Duty	HRSG Duty	Water ex Scrubber	Stack Gas
kg	331	403		734	16	750	750			33	$314 = 10.7\,\text{kmol}$
MJ	0	0	401	401	803	1204	440	764	440	0	0
Temp. °C	25	25		539	25	1350	551			25	25
Press. bar	1	1		32	32	32	1			1	1
mol%											
CO_2	0.03	9.41		5.13		8.76	8.76				9.41
H_2O	1.19	3.15		2.26		9.82	9.82			100	3.15
O_2	20.70	3.51		11.35		3.27	3.27				3.51
$N_2 + A$	78.08	83.93		81.26		78.15	78.15				83.93
CH_4					100						

base case. Furthermore, the CO_2 concentration is 9.4 mol% versus 3.9 mol% for the base case. In other words, it brings the concentration of a natural gas-fired unit up to that of a conventional boiler or furnace. The cold stack gas after a wash is a problem that has to be solved in all cases having scrubbing/wash facilities in the stack gas. The gas may also have to be heated for boyancy reasons.

An important additional advantage is that the SAFT, which is a good indicator for thermal NO_x emissions, is for the recycle case only 1640°C and for the base case 2380°C. This will result in an extremely low thermal NO_x formation. As this is in the case of natural gas firing, the only source of NO_x, this means that then the only gas that may have to be removed is CO_2, assuming no sulfur is present in the gas.

Low NO_x burners are not required, but standard gas turbine burners probably have to be modified because of the lower oxygen content in the gas. Furthermore, good mixing of the air and the recycle gas is required.

Neither the compressor nor the turbine itself need modifications, as the gases they have to cope with are both in quality and quantity not very different from those in the base case.

Comparison Results

The results of various calculations are shown in Table 7-13, in which efficiencies are given for IGCC plants with fuel gas treating and with flue gas treating. In both cases efficiencies with and without CO_2 removal are given.

Some comparisons were already made in Section 5.3. The following additional conclusions can be drawn from these data:

- The penalty in efficiency when including 90% CO_2 removal as part of fuel gas treatment is 4–5 efficiency points.
- In all cases, flue gas treating could well be a more attractive means for CO_2 removal than fuel gas treating. In cases where all acid gases can be removed together from the flue gas and can then be sequestered together, this would mean a major advantage in terms of efficiency and in terms of process simplification.
- Apart from flue gas recycle over the gas turbine (to the air inlet), attention should be paid to the fact that the fuel gas to the gas turbine will have a large volume due to the presence of inerts and because of the preheat. As a result of the inerts, the gas has a relatively low heating value.
- Due to the flue gas recycle to the air inlet, the air is diluted and has a low oxygen content. The dilution of both the fuel gas and the air will require special measures in the combustion chamber. On the other hand, the fact that both gases will be pre-heated will make the combustion less difficult. It may be expected that not only the thermal NO_x formation but also the SO_3 formation will be low.
- Many power station operators are already used to flue gas treating.
- Using nitrogen as quench gas or dilution gas in combination with flue gas treating may not be attractive, as it lowers the concentration of acid gases in the flue gas.

Table 7-13
Efficiencies of Various IGCC Power Stations with and without
CO$_2$ Removal Facilities

	Process		Fuel Gas Treating		Flue Gas Treating	
Feed	Gasifier conditions (bar/°C)	Syngas cooling	Without CO$_2$ removal	With CO$_2$ removal	Without CO$_2$ removal	With CO$_2$ removal
Slurry	64/1500	Water quench	37.8	35.5	43.0	39.7
Slurry	64/1500	Gas quench	43.6	39.4	43.1	39.8
Extreme preheat slurry	64/1500	Gas quench	48.8	43.7	49.6	46.3
Dry	32/1500	Gas quench	50.0	44.5	50.6	47.3
Dry	32/1500/1100	Coal quench	50.9	45.5	51.5	48.2
Dry	32/1100	Water quench	–	–	51.5	48.2
Supercritical steam power plant					45	41.7

Note: Efficiencies based on standardized, idealized conditions of Appendix E.

- Flue gas treating with recycle opens the possibility of applying high-efficiency air-blown fluid-bed gasifiers in power generation schemes because of the relatively low temperature difference between the gasifier and the high-temperature filtering step. The combination of fluid-bed gasification and flue gas treating will be only attractive when the capital costs are low and the efficiency is not lower than that of most entrained-flow slagging oxygen-blown gasifiers.
- Gasification-based power stations have the potential of about 5 percentage points better efficiencies than a conventional coal-fired power station featuring a super-critical steam cycle.

When CO$_2$ must be removed from the flue gas, an additional complication is introduced. One solution is to shift all the CO to hydrogen and remove the CO$_2$ from the fuel gas. This removes 7–10% of the LHV of the fuel gas from the CC. Altogether, fuel gas treating loses between 10 and 25% of the LHV in the raw gas and transfers it to the steam cycle, which has an efficiency of 40% as opposed to the 60% of the

CC. This corresponds to a penalty of 2–5% in station efficiency when compared with flue gas treating. However, where CO_2 has to be removed, more than half of the energy gain for flue gas treating will be lost due to the additional CO_2 compression required.

Additional Observations

Although flue gas treating has advantages, the following should be observed:

- Flue gas treating is only a possible option in power generation schemes. In the case of syngas production, fuel gas treating is always the way to go.
- Flue gas recycle is mandatory for flue gas treating as otherwise the concentrations of the components to be removed are too low and the amount of gas to be treated is too high.
- Thermal NO_x production can be made extremely low by flue gas recycle or quasi-isothermal compression. The NO_x production originating from organic nitrogen in the feed that has been converted into HCN, or NH_3, which yield NO_x upon combustion is low and may be acceptable in some cases. Research, for example, using CFD may show how the formation of HCN and NH_3 during gasification can be reduced. Moreover, it is worth exploring whether the conversion of these compounds into NO_x can be reduced by lower SAFTs in the gas-turbine combustors. CFD may help to solve this problem as well.
- Flue gas treating will become easier if there is no NO_x to be removed.
- Flue gas treating for sulfur removal is as yet generally not better than 95%, but this may be acceptable in many cases. It should be kept in mind that the SO_x emissions per kWh with 95% removal by stack gas scrubbing of an internationally traded coal having a sulfur content of 1% is about the same as for fuel gas treating with a 4% sulfur coal of which 99% of the sulfur is removed by fuel gas treating.
- In case of CO_2 sequestering, it is important to explore the possibility to remove and sequester all contaminants together.
- Mercury, arsenic, and antimony are present in coals having a high pyrite/cinnabar content. To date these compounds can only be removed by fuel gas treating. In feedstocks with a low pyrite/cinnabar, these compounds may not result in unacceptably high concentrations in the flue gas. This is certainly the case for heavy petroleum-based residues.

Gas Turbine Improvements

In this section on advanced cycles, we have sketched out a number of possibilities through which IGCC efficiencies could be improved. It must, however, be stated that their realization is dependant on changes being made to the design of existing gas turbines, whether it be in the use of quasi-isothermal compression or accommodating a flue gas recycle. This will require considerable research and development effort and investment by the manufacturing organizations.

7.3.5 Miscellaneous

Fuel Cells

Another approach that has received a lot of attention is the combination with fuel cells that can convert the chemical energy in pure hydrogen directly into electricity. The gas produced by gasification is especially suitable for this purpose, as the main additional treatment required is a CO shift that will convert all CO in the gas to additional hydrogen. Much development work has still to be done on fuel cells. Several types of fuel cell are being developed, including solid oxide, molten carbonate, phosphoric acid, and proton exchange membranes. At this stage of development it appears that the higher-temperature solid oxide and molten carbonate systems would best integrate with gasification-based systems for central station power production. The problem is that the gas for the fuel cells must be extremely pure. The target for natural gas is 70% efficiency, but our expectation is that at the time this can be realized the same efficiency can be reached with more advanced IGCC or similar cycles.

7.3.6 Energy Storage

Following the fluctuating grid demand for electricity is a problem as old as power stations. In principle, there are two solutions to cope with this problem: frequent shutting down and starting up, or storing energy in such a way that the power station can run continuously while following the demand pattern. In both cases, additional equipment is required that stands idle for part of the time. Even in the extreme case that electricity could be stored as such, additional equipment would be required.

When low capital cost power stations with a high efficiency can be built that can be started up and shut down within minutes, then in most cases no energy storage would be required. Topping units using only gas turbines and diesel generators are relatively low cost but have a low efficiency of about 40%. The Tophat cycle discussed in Section 7.3.3, with an efficiency of about 60% comprising only a gas turbine, could be a possible candidate in case fast start-up gas turbines such as aero-derivatives are used. But even in this case it is more advantageous to run the gas turbine proper continuously and opt for energy storage.

Of the various options for energy storage, such as flywheels, magneto-hydrodynamic rings, reversible chemical reactions, pressurized air in underground strata, and hydro, only the latter has become commercially successful. All the other options appear to have severe drawbacks that make them at least in the short- and medium-term unlikely candidates to solve the problem of efficient large-scale energy storage. On a smaller scale as required for IGCC power stations there are a number of options that are discussed below. These can be divided into the production of intermediate peak shaving fuels, methanol, and storage of liquid oxygen.

Peak Shaving Fuels

In the past it has been suggested to produce methanol, Fischer-Tropsch liquids, or dimethyl-ether from synthesis gas during the off-peak hours, and then to use these clean fuels as additional gas turbine fuels during the peak hours. It should be realized, though, that all these options are capital intensive, and that the "battery" efficiency of all these options is low and at best 70–80%. Moreover, these options require very pure synthesis gas and hence additional gas-treating facilities and CO shifting. All these conversion processes mean that the overall efficiency of the fuel conversion train will be negatively affected.

Methanol

Of these options methanol is still the most attractive, as this product can be made with an almost 100% conversion of the synthesis gas in a single-stage reactor and has the lowest heat of reaction and hence the highest conversion efficiency. Fuel-grade methanol can be used that will reduce the capital cost and increase slightly the process efficiency. Further, methanol may be reformed into synthesis gas with low-level heat of 300–350°C that will increase the efficiency of the overall fuel conversion train. The battery efficiency of methanol is over 80%.

Liquid Oxygen Storage

Some storage of liquid oxygen is always required in IGCC power stations in order to cope with sudden changes in demand. In order to use the oxygen for peak shaving, generally a much larger storage capacity is needed. This does increase the capital cost of the plant, but it has the advantage that the ASU does not have to be designed for the peak demand but for the average daily demand. So it is a matter of balancing these two cost items against each other. It should be mentioned that apart from the additional storage capacity required for liquid oxygen, means also have to be installed to recuperate the cold generated by evaporation of the oxygen.

REFERENCES

Agee, M. "The Natural Gas Refinery." *Petroleum Economist* (January 2002).

Allen, R. W., and Griffiths, J. "The Application of Hot Gas Expanders in Integrated Gasification Combined Cycles." Paper presented at IMechE Conference, "Power Generation and the Environment," London, November 1990.

Appl, M. *Ammonia: Principles and Industrial Practice.* Weinheim: Wiley-VCH, 1999.

BASF. Technical leaflet, BASF Catalyst K 8–11. Ludwigshafen: BASF AG. (Undated).

Becker, P. D., Hiller, H., Hochgesand, G., and Sinclair, A. M. "Heavy Fuel Oil as Ammonia Plant Feedstock." *Chemical and Process Engineering* (November 1971).

Benedict, D. E., Lilly, R. D., and Kornosky, R. M. "Liquid Phase Methanol Process Development Demonstration Plant Availability." Paper presented at Gasification Technologies Conference, San Francisco, October 2001.

Beyrard, M. N. R. "Perfectionnements aux installations fixes de turbine à gas." French patent 1467142, December 1966.

Boerrigter, H., den Uil, H., and Calis, H.-P. "Green Diesel from Biomass via Fischer-Tropsch Synthesis: New Insights in Gas Cleaning and Process Design." Paper presented at Pyrolysis and Gasification of Biomass and Waste Expert Meeting, Strasbourg, October 2002.

BP. *BP Statistical Review of World Energy*. BP PLC, June 2002.

Brown, Boveri, & Cie. "Gasturbinenanlage mit Wassereinspritzung." Swiss Patent 457039, 1968.

Davey, W. L. E., Wurzel, T., Filippi, E. "'Megammonia'—the Mega-Ammonia Process for the New Century." Paper presented at Nitrogen 2003, Warsaw, February 2003.

de Graaf, J. D., Zuideveld, P. L., Posthuma, S. A., and van Dongen, F. G. "Initial Operation of the Shell Pernis Residue Gasification Project." Paper presented at IChemE Conference, "Gasification: The Gateway to a Cleaner Future," Dresden, September 1998.

Derbyshire, F., and Gray, D. "Coal Liquifaction." In *Ullmann's Encyclopedia of Industrial Chemistry*, 5th ed., vol. A7. Weinheim: VCH Verlagsgesellschaft, 1986, pp. 197–243.

Dial, R. E. "Refractories and Insulation." *Industrial Heating* (November 1974).

Dittus, M., and Johnson, D. "The Hidden Value of Lignite Coal." Paper presented at Gasification Technologies Conference, San Francisco, October 2001.

Dolinskovo, A. A., and Brodianski, B. M. *Ексергетические Расчеты Технических Систем*. Kiev: Naukova Dumka, 1991.

European Commission. "Green Paper: Towards a European Strategy for the Security of Energy Supply," 2001.

Göhna, H. "Concepts for Modern Methanol Plants." Paper presented at 1997 World Methanol Conference, Tampa, December 1997.

Haupt, G. "IGCC Stations as Best Option for CO_2 Removal in Fossil-Fueled Power Generation." Paper presented at IChemE Conference, "Gasification: The Clean Choice for Carbon Management," Noordwijk, April 2002.

Hebden, D. "Further Experiments with a Slagging Pressure Gasifier." *Gas Council Research Communication* GC112 (November 1964).

Hebden, D. "Experiments with a Slagging Pressure Gasifier." *Gas Council Research Communication* GC50 (November 1958).

Higman, C. A. A. "Methanol Production by Gasification of Heavy Residues." Paper presented at IChemE Conference "Gasification: An Alternative to Natural Gas" London, November 1995.

Higman, C. A. A. "Synthesis Gas Processes for Synfuels Production." Paper presented at Eurogas '90, Trondheim, May 1990.

Higman, C. A. A., and Grünfelder, G. "Clean Power Generation from Heavy Residues." Paper presented at IMechE Conference, "Power Generation and the Environment," London, November 1990.

Hochgesand, G. "Gaserzeugung." In *Ullmann's Encyclopedia of Industrial Chemistry*, 4th ed., vol. A17. Weinheim: VCH Verlagsgesellschaft, 1977, p. 191.

Holt, N. A. "Integrated Gasification Combined Cycle Power Plants." In *Encyclopedia of Physical Science and Technology*, 3rd ed. London: Academic Press, September 2001.

Hoy, H. R. "Behavior of Mineral Matter in Slagging Gasification Processes." *IGE Journal* (June 1965):444–469.

Jungfer, H. "Synthesis Gas from Refinery Residues." *Linde Reports on Science and Technology*, No 40 May 1985, pp. 14–20.

Kubek, D. J., Polla, E., and Wilcher, F. P. "Purification and Recovery Options for Gasification." Paper presented at 1996 Gasification Technologies Conference, San Francisco, October 1996.

Kubek, D. J., Sharp, C. R., Kuper, D. E., Clark, M. E., DiDio, M., and Whysall M. "Recent Selexol, PolySep, and PolyBed Operating Experience with Gasification for Power and Hydrogen." Paper presented at Gasification Technologies Conference, San Francisco, October 2002.

Lacey, J. A. "The Gasification of Coal in a Slagging Pressure Gasifier." The Gas Council, The Midlands Research Station. Solihull, England: Preprints Div. Fuel Chem. Am. Chem. Soc. 10 (1966) 4:151–167.

Lath, E., and Herbert, P. "Make CO from Coke, CO_2, and O_2." *Hydrocarbon Processing* 65 (August 1986):55–56.

Linzer, V., and Haug, W. "Temperaturverlauf und Wärmefluss in Stampfmasse und Stift von abgestampften Feuerraumwänden bei Strahlungsbeheizung." *Brennstoff-Wärme-Kraft* vol. 27 no. 5 (May 1975):206–210.

Lurgi Oel. Gas. Chemie GmbH "Company Profile," October 2001, p. 15.

Mansfield, K. "Methanol Production: Experience with Alternative Feedstocks." Paper presented at 1994 World Methanol Conference, Geneva, November 1994.

McNeely, M. "Intercooling for LM6000 Gas Turbines." *Diesel and Gas Turbine Worldwide* (July/August 1998).

Methanol Institute. "World Methanol Supply and Demand." Available at: www.methanol.org.

Milo AB. "Gasturbine-Installatie met een of meer gasturbines en een of meer daardoor gedreven compressoren." Dutch patent No. 39361, 1936.

Mulder, H. "From Syngas to Clean Fuels and Chemicals via Fischer-Tropsch Processes." Paper presented at IChemE Conference, "Gasification: The Gateway to a Cleaner Future," Dresden, September 1998.

Müller, W.-D. "Gas Production." In *Ullmann's Encyclopedia of Industrial Chemistry*, 5th ed., vol. A17. Weinheim: VCH Verlagsgesellschaft, 1991.

Mundo, K., and Weber, W. "Anorganische Stickstoffverbindungen." In *Chemische Technologie*. Winnacker and Küchler, vol. 2. Carl Hanser Verlag, 1982.

Nakhamkin, M. "The Cascaded Humidified Air Turbine (Chat) and Its Integration with Coal Gasification (IGCHAT)." Paper presented at EPRI Conference on New Power Generation Technology, San Francisco, October, 1995.

Parkinson, G. "Krupp Uhde Raises the Limit on Ammonia-Plant Capacity." *Chemical Engineering* (November 2001):17.

Reid, W. T., and Cohen, P. "Factors Affecting the Thickness of Coal-Ash Slag on Furnace-Wall Tubes." *Transactions of the ASME* (November 1944).

Schipper, D. "HAT Cycle Power Plant Study For High Efficiency." Novem publication DV2-SO2.088 94.2, 1993.

Shvets, I., Tolubinski, V. I., Kirakovski, N. F., Neduzhy, I. A., and Sheludko, I. M. *Heat Engineering*. Moscow: Mir Publishers, 1975.

Société Rateau. "Perfectionnements aux turbo-moteurs." French patent 1007140, 1952.

Supp, E. *How to Produce Methanol from Coal*. Berlin: Springer, 1990.

van der Burgt, M. J. "Shell's Middle Distillate Synthesis Process." Paper presented at AIChE Spring meeting, New Orleans, 1988.

van der Burgt, M., Cantle, J., and Boutkan, V. K. "Carbon Dioxide Disposal from Coal-based IGCCs in Depleted Gas Fields." *Energy Conversion Management* 33 (no. 5–8, 1992):603–10.

van der Burgt, M. J., and Sie, S. T. "Liquid Hydrocarbons from Natural Gas." Paper presented at Petro-Pacific Conference, Melbourne, September 1984.

van der Burgt, M., and van Liere, C. "Applications of the Tophat Cycle in Power Generation." EPRI, October 1996.

van Paassen, C. A. A., van Liere, J. J. C. "Überblick über die Forschungsarbeit 'Einspritzkühlung' an der Technischen Universität Delft," *VGB Kraftwerkstechnik* (December 1980):958–969.

VIK. *Statistik der Energiewirtschaft 1999/2000*. Essen: Verlag Energieberatung, 2001.

Viswanathan, R., Purgert, R., and Rao, U. "Materials for Ultra Supercritical Coal-Fired Power Plants." Paper presented at 19th International Pittsburgh Coal Conference, September 2002.

Welford, G. B., Panesar, R. S., Reed, A., Chaimberlain, R., Effert, M., Ghiribelli, L., Toste de Azevedo, J. L., and Toledo, R. "Innovative Supercritical Boilers for Near Term Global Markets." Paper presented at Powergen Europe 2002 Conference, Milan, June 2002.

Yantovski, E. T. "Stack Downward Zero-Emission Fuel-Fired Power Plants Concept." *Energy Conversion and Management* 37 (Issues 6–8, 1996):867–877.

Zwiefelhofer, U., and Holtmann, H.-D. "Gas Expanders in Process Gas Purification." Paper presented at VDI-GVC Conference, "Energy Recovery by Gas Expansion in Industrial Use," Magdeburg, March 1996.

Chapter 8

Auxiliary Technologies

8.1   OXYGEN SUPPLY

The oxygen supply to a gasifier is one of the most expensive single parts of any gasification project. Cost estimates for various different IGCC projects put the ASU with its associated compressors at between 10 and 15% of the total plant cost. It also makes a significant contribution to overall operating costs, the power requirement for compression being of the order of magnitude of 5–7% of gross generator output. The arrangement for oxygen supply is one of the most important early decisions in any gasification project.

This decision on oxygen supply is not only technical but to a high degree also commercial. This is because the industrial gas market is dominated by a small number of highly competitive companies who are able not only to build oxygen plants for third parties, but are also willing to build and operate their own plants close to or even on a client's site and supply oxygen and other gases "over the fence." In some parts of the world, these companies have their own extended pipe-line networks, which can also be integrated into the oxygen supply strategy for a gasification project.

8.1.1   Technologies

Cryogenic Processes

Since the commercialization of the Linde-Fränkle process in the 1920s, oxygen supply has been dominated by cryogenic technology.

The principle features of cryogenic air separation are shown in Figure 8-1. Air is compressed, dried in a prepurification unit, and then cooled to its liquefaction temperature. The liquid air is then distilled into its two main constituents, oxygen and nitrogen. These separated products are then heated and vaporized. This basic flow scheme has formed the basis for all processes, or cycles as they are known, to this day, although many detail improvements have been made over the years to decrease costs and improve efficiency. In addition to these improvements, the size of air separation units has risen dramatically over the last 40 years. The 1400 t/d (40,000 Nm^3/h) ASU supporting a gasification plant with 65 t/h visbreaker residue

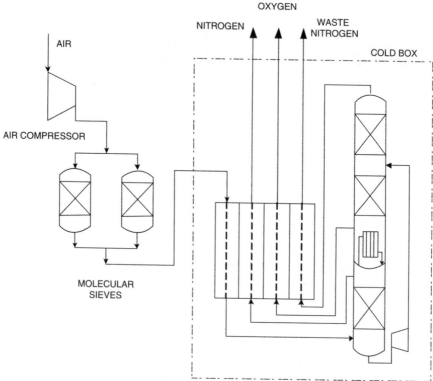

Figure 8-1. Air Separation Unit

feed that came on stream in 1972 was at that time the largest in the world (Butzert 1976), but there are several plants now running with a capacity of about 90,000 Nm^3/h (3200 t/d), including for instance that at Rozenburg, The Netherlands, which among others provides the oxygen for the Shell Pernis gasifiers.

The operating pressure of modern plants varies with the application. Low-pressure (LP) cycles, which supply by-product gaseous nitrogen at only atmospheric pressure, operate at about 5–7 bar, depending on oxygen pressure and required energy efficiency. Where much of the nitrogen is required at higher pressures, it can be advantageous to operate the ASU at elevated pressure (EP) above this level. This is particularly appropriate in an IGCC with air and N$_2$ integration, since the air is in any case available from the gas turbine compressor at the higher pressure. Operating at the higher pressure also has the advantage of tending to reduce equipment size and cost.

The prepurification unit (PPU) has seen substantial changes since the original Fränkle regenerators. After a period when most plants were equipped with reversing exchangers, in which water and CO$_2$ were frozen onto the surfaces of the main heat exchangers and evaporated off in the regeneration part of the cycle with waste nitrogen, the 1970s saw the introduction of molecular sieves for prepurification. Most modern PPUs have twin beds of silica gel and molecular sieve. They hold back not only water and CO_2 but also potentially dangerous hydrocarbons that may be in the atmosphere.

The partially liquefied air enters the lower, high-pressure column where pure nitrogen is drawn off as overhead product. The bottoms, an oxygen-rich liquid, is expanded through a turbine into the upper, low-pressure column, where in the simplest cycles, pure oxygen is drawn off as bottoms product. The overhead is an impure nitrogen stream, which like the main products is used for chilling the incoming air before being used as regeneration gas in the PPU and then being discharged to the atmosphere. Where large quantities of pure nitrogen (<10 vppm O_2) are required as in ammonia plants, a reflux can be added to the top of the LP column and some of the impure nitrogen also recovered as pure product. Typically, the maximum amount of pure product obtainable is about 70% of the incoming air.

Production of argon is possible by tapping the middle of the LP column, where the argon concentration is highest, and adding a further distilling stage. Final purification of the argon takes place by the catalytic reduction of final traces of oxygen with hydrogen. This additional processing stage requires a higher operation pressure of the ASU at the upper range of LP cycles.

For the production of pressurized oxygen, two cycles are used: the compression cycle, and the pumped liquid cycle. In the former gaseous oxygen leaves the cold box at slightly above atmospheric pressure and is compressed in a compressor. Alternatively, the liquid oxygen can be pumped up to the required pressure and the vaporized under pressure. This latter cycle, also known as internal compression, required when introduced about 5–7% more energy, but cycle development has reached the point where there is little difference between compression and pumped liquid cycles.

Air separation has a widely recognized reputation for reliability. This is important for the gasification process, since oxygen production is at the beginning of the flowsheet, and loss of oxygen brings the whole downstream facility to a standstill. Traces of water or CO_2 can slip past the PPU and over a period of time freeze out on the heat exchangers. If this goes beyond a tolerable limit for the operation, then the coldbox must be reheated to ambient temperature and derimed. Typically, under normal operation this may be necessary every two years. The deriming itself may take about one or two days.

The vulnerability of a gasification plant to interruptions of oxygen supply makes the consideration of building some liquid storage capacity an important issue for the planning of any gasification project. The only economic method of oxygen storage is as a liquid at low temperature. Typically, under such conditions a boil-off of some 0.2–0.5%/d, depending on the size of tank, must be expected. The principle aspects to be considered are the response time in which oxygen from the storage is available at the gasifier (seconds), and the size of the storage. For the latter consideration, some hours worth of storage to cover a compressor trip and restart would be a minimum. A storage volume to cover a coldbox deriming period is unrealistic.

Alternative Processes for Small Quantities

For small units, other processes are available. They cannot, however, reach the purity obtainable with a cryogenic unit. Pressure swing adsorption units are available up to a capacity of about 140 t/d (4000 Nm^3/h), but they can only reach a purity of about 95% O_2. The product purity obtainable with polymer membrane technology is much less—about 40% O_2—and such units are available for capacities of 20 t/d (600 Nm^3/h). The by-product capability of both these technologies is poor, but they both have the advantage of quick start-up compared with cryogenic units (Smith and Klosek 2001). Given that most small gasification facilities are for chemical applications where even 5% nitrogen in the oxygen is unacceptable, their use in connection with gasification is likely to be limited. The most probable gasification application could be with biomass power applications where sizes are also at the lower end of the scale.

Future Developments

Oxygen production by means of ion transport membranes is the subject of intense research and development (Allam, Foster, and Stein 2002).

The principle of these devices is based on the use of nonporous ceramic membranes that have both electronic and oxygen ionic conductivity when operated at high temperatures, typically 800 to 900°C (Figure 8-2).

Oxygen from the feed side of the membrane adsorbs onto the surface of the membrane, where it dissociates and ionizes by electron transfer from the membrane. The oxygen diffuses through the lattice of the membrane, which is stoichiometrically deficient in oxygen. The driving force is the differential partial pressure of oxygen across the membrane. Oxygen ions arriving at the product side of the membrane release their electrons, recombining and desorbing from the surface as molecular oxygen. Electrons flow in counter current from product to the feed side of the membrane.

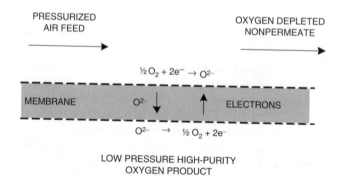

Figure 8-2. Principles of Ion Transport Membrane (*Source: Allam et al. 2000*)

This ion-transport mechanism is specific to oxygen so that, discounting leakage through any cracks or seals and so on, the product is 100% oxygen.

Considerable work has been performed in scale-up and demonstration of production facilities (Armstrong et al. 2002). Integration with both the gasification and/or the combustor of the gas turbine is mandatory for its success. Commercialization is currently expected "in the 2006 to 2008 time frame."

8.1.2 Pipeline and Other Supply Possibilities

Pipelines

In areas where there is already an established pipeline network, a gas supply contract can be especially attractive. For small plants close to a pipeline, this may be a very simple affair, simply drawing on existing capacity. For large plants, purpose-built capacity will have to be added, but synergies are often possible with a need for increased capacity to feed the pipeline system.

The following is a list of major pipeline operators and the location of their networks.

Air Products	Houston—Port Arthur, Texas
	Decatur, Alabama
	Rotterdam, The Netherlands
	Humberside, England
	Mab Ta Phut, Thailand
	Onsan/Ulsan, South Korea
Air Liquide	Dunkirk – Antwerp – Rotterdam – Liege
	Corpus Christi – Lake Charles
	Weswego – Geismar/Baton Rouge (Louisiana – Mississippi River Corridor)
Praxair	Northwest Indiana
	Houston area
	Wilmington, California
	Detroit, Michigan (in Ecorse) and in other enclaves throughout the United States.
BOC	Various locations
Linde	Ruhr district, Germany
	Leuna district, Germany
Messer Griesheim	Dortmund-Cologne, Germany
	Saarland, Germany
	Tarragona, Spain

Over the Fence

Even in locations where no pipelines are available, it is possible to buy oxygen "over the fence." In this business model, which has a long tradition in oxygen supply to the steel industry, a gas supply company will own and operate a dedicated air separation unit within the gasification complex. By-products such as argon, which may be of little interest to the gasifier operator, can make such an arrangement very attractive. The specialist operating know-how of the gas companies enable them to achieve excellent plant availability. In some locations where this model is practiced, it has been extended to cover utility supply.

8.2   SYNTHESIS GAS TREATING

Although the number of gasification processes for any particular feedstock is limited, the range of choice for acid gas removal (AGR) processes and other gas treating tasks appears at first sight somewhat bewildering. Their selection requires a basic knowledge of the different classes of process and a systematic approach to defining the task in hand. Although the approach developed in this chapter cannot replace the expertise of the gas purification specialist, it can help to ensure that he or she has all the facts available to make an optimum selection.

The term *acid gas removal* is a general term that is often used as a synonym for desulfurization, but strictly speaking in the synthesis gas environment it includes also CO_2 removal, and it is in this inclusive sense that the term is used here.

A large number of different processes have been and are used for acid gas removal, but they are all based on one of the following principles:

- Absorption (physical or chemical) in a liquid solvent with a subsequent desorption step.
- Adsorption (again physical or chemical) onto a mass of solid particles.
- Diffusion through a permeable or semipermeable membrane.
- Chemical conversion, generally on a catalyst, often as a preparatory step to one of the above three methods.

Practically all commercially available processes operate with the syngas "cold," that is, at ambient temperatures or lower (chemical adsorption of trace H_2S on zinc oxide is the prominent exception). The loss of efficiency in IGCC applications, which is associated with gas cooling between gasifier and gas turbine, has generated considerable interest in the possibilities of hot desulfurization processes, and these are treated separately as a topic in its own right.

8.2.1   Selection Criteria

As part of the selection process it is necessary to consider the following criteria:

- *Gas purity*. The demands on syngas purity can vary extremely with application. A chemical application such as ammonia, methanol, or SNG can require desulfurization to 100 ppbv or lower. For an IGCC power application with a limit of 5 ppmv SO_2 in the flue gas, about 40 ppmv at the outlet of the AGR is satisfactory. But desulfurization is not the only purity criterion. Ammonia syngas requires 10 ppmv max CO_2. Other limitations may arise indirectly; in a hydrogen application for instance, any sulfur entering the PSA unit becomes concentrated in the tail gas, where emissions regulations may also create a sulfur slip limitation.
- *Raw gas composition*. The washing solution must be able to cope with impurities in the raw gas. Most gasification processes leave about 5% of the sulfur as COS, which in many AGR processes is not absorbed as well as H_2S, if at all, so where deep sulfur removal is required, it may be necessary to convert it to H_2S prior to the main sulfur removal stage. But the raw gas composition issue is not just a matter of the previous "gas purity criterion"; the action of minor impurities on the solution itself also have to be considered: for example, HCN in the raw gas reacts, particularly with primary amines causing solution degradation.
- *Selectivity*. The selectivity of a gas separation process is the ability, for example, to remove H_2S while leaving CO_2 in the synthesis gas. There can be a number of motivations leading to a desire for high selectivity.
  - o The capital and operating expense of most gas-washing systems correlates well with the amount of solvent in circulation. A nonselective wash that also washes out large quantities of CO_2, may (in the case of a chemical wash will) inflate the solution rate and therefore costs considerably. A good selectivity can therefore make a contribution to good economics.
  - o The acid gas removed from the syngas is usually processed in a Claus sulfur recovery unit (SRU). It is important for the Claus process that the H_2S stream to the SRU not be too dilute. For a typical air-blown SRU, the lower limit for practical operation is about 30 mol%. In the context of gasification, it is usually possible to operate the SRU in an oxygen-blown mode, since the SRU oxygen requirement is generally small compared to that necessary for the gasifier. An oxygen-blown SRU can operate with as little as 10 mol% H_2S in the sour gas. Either way, excess CO_2 in the sour gas will inflate the cost of the SRU. When looking at the selectivity of an AGR system, it is therefore necessary to consider the effect on the SRU.
  - o For an IGCC application CO_2 in the synthesis gas contributes to the total mass flow through the gas turbine and so to the power output. It is therefore in principle desirable to leave any CO_2 in the synthesis gas rather than washing it out. For many gasification processes with a low content of CO_2 (<5%) in the syngas this is not usually decisive, but with a higher CO_2 content it can be important.
- Other issues that need to be reviewed as part of the selection process are corrosion (often the practical limit on higher solvent loading), co-absorption of the useful gas components (in syngas application hydrogen and carbon monoxide, but in general this would include hydrocarbons), solvent losses through degradation or vapor

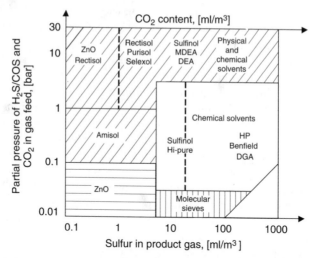

Figure 8-3. Initial Selection of AGR processes (*With permission: Lurgi*)

pressure, opportunity for waste heat integration, particularly for the solvent regeneration in chemical washes, and availability of the solvent and toxicity.
- *Economic boundary conditions.* In particular the depreciation rate or pay-out time specified for a project may influence the process selection. Typically, chemical washes will tend to require less capital investment than a physical wash, but at the expense of a higher utility demand for solvent regeneration.

Examination of the above criteria in any particular case will probably narrow the field down to three or four serious contenders, sometimes even less. The chart in Figure 8-3 provides assistance in this. Selection from this short list is then generally a matter of pure economics.

8.2.2 Absorption Systems

Absorption processes are characterized by washing the synthesis gas with a liquid solvent, which selectively removes the acid components (mainly H_2S and CO_2) from the gas. The laden solvent is regenerated, releasing the acid components and recirculated to the absorber. The washing or absorption process takes place in a column, which is usually fitted with (dumped or structured) packing or trays.

The absorption characteristics of a solvent depend either on simple physical absorption or on a chemical bond with the solvent itself. This provides the basis for the classification of AGR systems into physical or chemical washes, which have distinctly different loading characteristics.

The loading capacity of a physical wash depends primarily on Henry's law and is therefore practically proportional to the partial pressure of the component to be

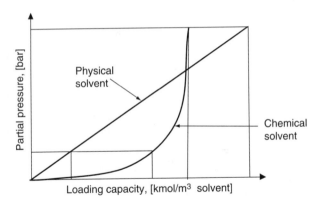

Figure 8-4. Equilibrium of Physical and Chemical Absorption

removed (Figure 8-4). This leads to the fact that the solution rate for any particular operating pressure is approximately proportional to the volume of raw gas to be processed.

In contrast, the loading capacity of a chemical wash is limited by the quantity of the active component of the solution. Once a saturation level is reached only a minor additional loading can be achieved by physical absorption in the solution. The solution rate is approximately proportional to the volume of acid gas removed.

Some mixed solvents have been developed using both effects. These are known as physical-chemical washes.

Generally, solvent regeneration is achieved by one of or a combination of flashing, stripping, and reboiling. Both flashing and stripping reduce the partial pressure of the acid component. In physical washes, reboiling raises the temperature and reduces the acid gas solubility. In chemical washes the increased temperature serves to break the chemical bond. In such systems the acid components are released in the same chemical form in which they were absorbed (Figure 8-5).

An additional class of washing systems, oxidative washes, regenerate the chemically absorbed sulfur by oxidizing the active component in the solvent and recovering the sulfur in elemental form.

Chemical Washes

Amines. Solutions of amines in water have been used for acid gas removal for over 50 years. The principle amines used for synthesis gas treatment are mono- and diethanolamine (MEA and DEA), methyldiethanolamine (MDEA), and di-isopropanolamine (DIPA), the latter particularly as a component of the Sulfinol solvent. Others amines used in natural gas applications, such as diglycolamine (DGA) or triethanolamine (TEA), have not been able to make any significant impact in syngas applications.

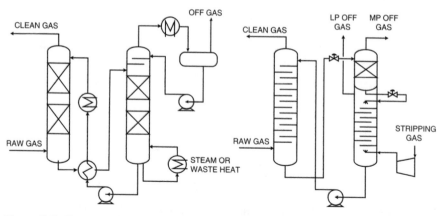

Figure 8-5. Regeneration by Reboiling (left) and Flashing or Stripping (right)

MDEA is the most widely used amine today. It is more selective than primary (e.g. MEA) or secondary (e.g. DEA) amines, due to the fact that CO_2 is absorbed more slowly than H_2S.

A number of proprietary formulations have been developed to address specific issues. For example, Ucarsol was developed to reduce corrosion with high CO_2 loading. BASF's aMDEA includes an activator to accelerate CO_2 absorption, where selectivity is not a requirement. Variation in the degree of promotion can influence the energy requirement for regeneration. Exxon developed the Flexsorb family of hindered amines specifically for high selectivity.

Typical performance data of different amine washes may be seen in Table 8-1. The flowsheet of a typical MDEA wash is shown in Figure 8-6.

Table 8-1
Properties of Amine Solvents

	Standard MEA	Inhibited MEA	DEA	MDEA
Molecular weight	61		105	119
CO_2 partial pressure, bar	<100	<100		
Gas purity CO_2, ppmv	20–50	20–50		
Solution strength, wt%	10–20	30	25–35	30–50
Solution loading, mol/mol	0.25–0.45		0.4–0.8	0.8
Energy demand, MJ/kmol CO_2	210	140		
Notes:				selective

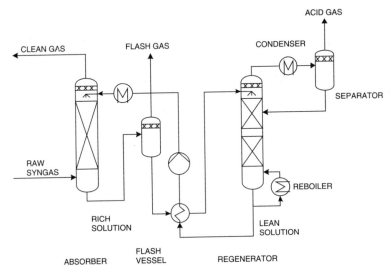

Figure 8-6. Typical MDEA Flowchart with Single Flash Stage

Physical Washes

The important characteristics for any successful physical solvent are:

- Good solubility for CO_2, H_2S, and COS in the operating temperature range, preferably with significantly better absorption for H_2S and COS compared with CO_2 if selectivity is an important issue for the application in hand.
- Low viscosity at the lower end of the operating temperature range. Although lowering the operating temperature increases the solubility, the viscosity governs in effect the practical limit to lowering the operating temperature.
- A high boiling point reduces vapor losses when operating at ambient or near ambient temperatures.

Rectisol. The Rectisol process, which uses cold methanol as solvent, was originally developed to provide a treatment for gas from the Lurgi moving-bed gasifier, which in addition to H_2S and CO_2 contains hydrocarbons, ammonia, hydrogen cyanide, and other impurities.

In the typical operating range of −30 to −60°C, the Henry's law absorption coefficients of methanol are extremely high, and the process can achieve gas purities unmatched by other processes. This has made it a standard solution in chemical applications such as ammonia, methanol, or methanation, where the synthesis catalysts require sulfur removal to less than 0.1 ppmv. This performance has a price, however, in that the refrigeration duty required for operation at these temperatures involves considerable capital and operating expense.

Methanol as a solvent exhibits considerable selectivity, as can be seen in Table 8-2. This allows substantial flexibility in the flowcharting of the Rectisol process and both standard (nonselective) and selective variants of the process are regularly applied according to circumstances.

As a physical wash, which uses at least in part flash regeneration, part of the CO_2 can be recovered under an intermediate pressure. Typically, with a raw gas pressure of 50 bar, about 60–75% of the CO_2 would be recoverable at 4–5 bar. Where CO_2 recovery is desired, whether for urea production in an ammonia application or for sequestration, this can provide significant compression savings.

Figure 8-7 shows the selective Rectisol variant as applied to methanol production. The incoming raw gas is cooled down to about −30°C, the operating temperature of the H_2S absorber. Both H_2S and COS are washed out with the cold methanol to a residual total sulfur content of less than 100 ppbv. The desulfurized gas is then shifted outside the Rectisol unit, the degree of shift being dependent on the final product. Carbon dioxide is then removed from the shifted gas in the CO_2 absorber to produce a raw hydrogen product. This column is divided into two sections: a bulk CO_2 removal section using flash regenerated methanol, and a fine CO_2 removal section in

Table 8-2
Properties of Physical Solvents

		Methanol	NMP	DMPEG
Chemical Formula		CH_3OH	$CH_3N-(H_2C)_3$ $C=O$	$CH_3O(C_2H_4O)_x$ CH_3
Mol. Weight	kg/kmol	32	99	178 to 442
Boiling point at 760 Torr	°C	64	202	213 to 467
Melting point	°C	−94	−24.4	−20 to −29
Viscosity	cP	0.85 at −15°C	1.65 at 30°C	4.7 at 30°C
		1.4 at −30°C	1.75 at 25°C	5.8 at 25°C
		2.4 at −50°C	2.0 at 15°C	8.3 at 15°C
Specific mass	kg/m³	790	1.027	1.031
Heat of evaporation	kJ/m³	1090	533	
Specific heat at 25°C	kJ/kg.K	0.6	0.52	0.49
Selectivity at working temperature	($H_2S:CO_2$)	1:9.5	1:13	1:9

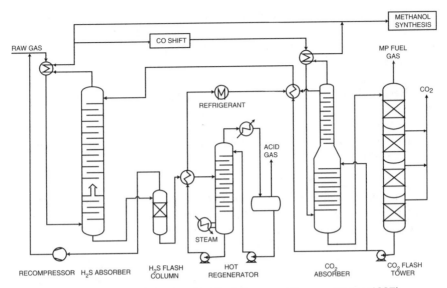

Figure 8-7. Flowsheet of Selective Rectisol Process (*Source: Weiss 1997*)

which hot-regenerated methanol is used. The CO_2 removal section operates at lower temperatures, typically about $-60°C$. The permissible CO_2 slip is dependent on the application. For methanol synthesis gas 1 mol% residual CO_2 in the raw hydrogen is quite adequate. For hydrogen production based on methanation, typically 100 ppmv would be appropriate. For ammonia where the gas is subsequently treated in a cryogenic nitrogen wash, 10 ppmv would be typical.

Following the solvent circuit, we see first an intermediate H_2S flash from which co-absorbed hydrogen and carbon monoxide are recovered and recompressed back into the raw gas. The flashed methanol is then reheated before entering the hot regenerator. Here the acid gas is driven out of the methanol by reboiling, and a Claus gas with an H_2S content of 25–30% (depending on the sulfur content of the feedstock) is recovered. Minor adaptations are possible to increase the H_2S content if desired.

The hot-regenerated methanol, which is the purest methanol in the circuit, is used for the fine CO_2 removal. The methanol from the CO_2 removal is subjected to flash regeneration in a multistage flash tower. The configuration shown is typical for the methanol applications with only atmospheric flash regeneration. For hydrogen or ammonia applications where better absorption is required, the final flash stage may be under vacuum, or it may use stripping nitrogen from the air separation plant. Finally, the loop is closed with the flash regenerated methanol returning to the H_2S absorber.

Water entering the Rectisol unit with the syngas must be removed, and an additional small water-methanol distillation column is included in the process to cope with this.

Typically, the refrigerant is supplied at between −30 and −40°C. Depending on application, different refrigerants can be used. In an ammonia plant, naturally, ammonia is used, and the refrigeration system is integrated with that of the synthesis. In a refinery environment, propane or propylene may be the refrigerant of choice.

The Rectisol technology is capable of removing not only conventional acid gas components but also, for example, HCN and hydrocarbons. Supp (1990, p. 83) describes a typical hydrocarbon prewash system. Mercury capture using Rectisol as a cold trap to condense out metallic mercury is also documented (Koss, Meyer, and Schlichting 2002).

Selexol. The Selexol process was originally developed by Allied Chemical Corporation and is now owned by UOP. It uses dimethyl ethers of polyethylene glycol (DMPEG). The typical operating temperature range is 0–40°C. The ability to operate in this temperature range offers substantially reduced costs by eliminating or minimizing refrigeration duty. On the other hand, for a chemical application such as ammonia, the residual sulfur in the treated gas may be 1 ppmv H_2S and COS each (Kubek et al. 2002) which is still more than the synthesis catalysts can tolerate. This is not an issue, however, in power applications where the sulfur slip is less critical. Selexol has a number of references for such plants including the original Cool Water demonstration unit and most recently the 550 MW Sarlux IGCC facility in Italy.

The ratio of absorption coefficients for H_2S, COS, and CO_2 is about 1:4:9 in descending order of solubility (Kubek, Polla, and Wilcher 1997). A plant designed for, say, 1 ppm COS in the clean gas would require about four times the circulation rate of a plant for 1 ppm H_2S, together with all the associated capital and operating costs. In a gasification environment it is therefore preferable to convert as much COS to H_2S upstream of a Selexol wash. In a plant using raw gas shift for hydrogen or ammonia, this will take place simultaneously on the catalyst with the carbon monoxide shift. Where no CO shift is desired, then COS hydrolysis upstream of the Selexol unit provides a cost-effective solution to the COS issue.

Other characteristics favorable for gasification applications include high solubilities for HCN and NH_3 as well as for nickel and iron carbonyls.

The Selexol flowsheet in Figure 8-8 exhibits the typical characteristics of most physical absorption systems. The intermediate flash allows co-absorbed syngas components (H_2 and CO) to be recovered and recompressed back into the main stream. For other applications, including H_2S concentration in the acid gas or separate CO_2 recovery, staged flashing techniques not shown here may be applied.

Purisol. NMP or n-methyl-pyrrolidone is the solvent used in Lurgi's Purisol process. The operating range is 15°C to 40°C. The selectivity for H_2S/CO_2 is extremely high and largely independent of the operating temperature (Grünewald 1989). Solvent properties are included in Table 8-2. The characteristics are in many ways comparable with Selexol.

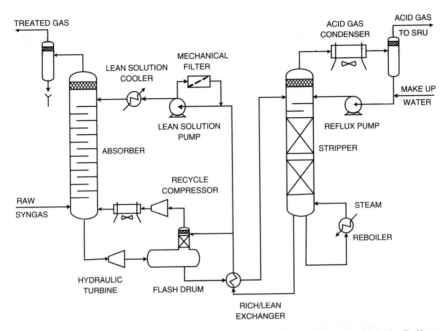

Figure 8-8. Selexol Flowchart for Selective H₂S Removal (*Source: Kubek, Polla, and Wilcher 1997*)

Physical-Chemical Washes

Some gas-washing systems exploit the principles of both physical and chemical absorption and are known as physical-chemical washes. They generally use an amine together with organic physical solvent. They can usually accept a higher loading than an aqueous amine solution, thus reducing solvent rates. Furthermore, the organic solvents applied in such systems accelerate the hydrolysis of COS to H_2S in the lower sections of the column, thus permitting an improved total sulfur removal performance than a pure amine system. Other aspects, which still need review when considering a physical-chemical system, are the potential for amine degradation, which is generally unchanged compared with the equivalent aqueous amine system. Their effectiveness at absorbing metal carbonyls is not documented and so must be considered as unproven.

Sulfinol. Shell's Sulfinol solvent in its original form was a mixture of DIPA and Sulfolane (tetrahydrothiophene dioxide). The former provides a chemical solvent and the latter a physical solvent. Meanwhile a modified solvent, known as m-Sulfinol has been developed that uses MDEA as the chemical component. The original Sulfinol formulation has been used successfully downstream of a large number of small oil gasifiers for the production of oxo-synthesis gas. The AGR at the Buggenum IGCC is an example of a larger m-Sulfinol unit.

Amisol. The Amisol process was developed by Lurgi using a mixture of MEA or DEA with methanol. It has been applied downstream of a number of oil gasification units, but it has not established a wide market. Details can be found in Supp (1990) and Kriebel (1989).

Oxidative Washes

Oxidative washes or liquid redox systems differ from other types of absorption system in that the H_2S in the acid gas is oxidized directly to elemental sulfur in the absorption stage. The active agent in the solution is regenerated in a separate oxidizing vessel, which also serves to separate the solid elemental sulfur from the solution. The solvents of oxidative washes absorb essentially only H_2S, but not CO_2 nor COS. This makes them suitable for applications where H_2S must be removed from a stream containing large quantities of CO_2, even if the H_2S partial pressure is low.

There is no known existing application in a gasification environment, but such washes exhibit potential as a substitute for a Claus plant, where the gasifier feed has very low sulfur content and the sour gas is unsuitable for treatment in a Claus plant.

Earlier plants, notably the Stretford and Takahax processes, used vanadium-based agents, which undergo a valence change from the pentavalent to the tetravalent state during the absorption stage. Modern processes, of which Lo-Cat and Sulferox are the best known, use chelated iron formulations.

The Lo-Cat process can be arranged in a number of different application-dependant configurations of which that shown in Figure 8-9 is typical. Acid gas enters the absorber, where the H_2S is absorbed into the aqueous chelated iron solution. The

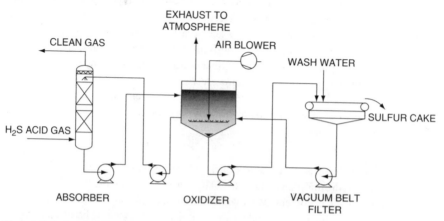

Figure 8-9. Lo-Cat Flowsheet (*Source: Adapted from Nagl 2001*)

ferric iron oxidizes the HS^- ion to elemental sulfur according to reaction 8-1. The iron is reduced to the ferrous state.

$$HS^- + 2Fe^{+3} \rightarrow S^0 + 2Fe^{+2} + H^+ \tag{8-1}$$

In the oxidizer, the sulfur settles out and is transferred to a vacuum filter where it is separated from the solution as a cake. Air is blown into the oxidizer, where oxygen is absorbed into the solution and oxidizes the ferrous iron back to the ferric state (reaction 8-2) for recirculation back to the absorber.

$$2Fe^{+2} + \tfrac{1}{2}O_2(l) + H_2O. \rightarrow 2Fe^{+3} + 2\,OH^- \tag{8-2}$$

The raw sulfur from the vacuum filter is typically 65% to 85% sulfur, the remainder being water and dissolved salts including iron. This product requires further treatment to meet generally accepted market quality, melting (to remove the water) and filtering being important process steps. Nonetheless, the usual "bright yellow" color specification for commodity sulfur is not met, and specialized applications need to be located in the marketplace.

As mentioned above, liquid redox systems are generally applied for small plants, in particular where H_2S concentrations are lower than can be handled by the Claus process (see Section 8.4). The Stretford process was regularly applied for Claus tail gas processing as part of the Beavon tail gas treating process. A similar application using an iron chelate process was put into service in 2001 (Nagl 2001).

8.2.3 Adsorption Systems

A second important group of gas treatment processes are based on the adsorption of impurities onto a solid carrier bed. Some of these processes, such as molecular sieve driers or pressure swing, allow in situ regeneration of the bed. Others, such as H_2S chemisorption onto zinc oxide, cannot be regenerated economically in situ, and the beds require regular exchange.

The quantity of a gaseous component, which can be carried by any particular adsorbent, depends not only on the characteristics of component and sorbent but also on the temperature and pressure under which it takes place. This increase in loading capacity with higher pressures and lower temperatures is illustrated in Figure 8-10 and is utilized for the in situ regeneration of such sorbents as activated carbon, activated alumina, silica gel, and molecular sieves.

The classic adsorption-desorption cycle uses both the temperature and pressure effect "swinging" between high pressure and low temperature for adsorption (point 1 in Figure 8-10) and low pressure and high temperature (point 2) for desorption. The differential loading $(L_1 - L_2)$ is extremely high. The pressure swing cycle operates at constant temperature T_0 between points 3 and 4. A temperature swing process operating at constant pressure between points 1 and 5 is possible but unusual in practice.

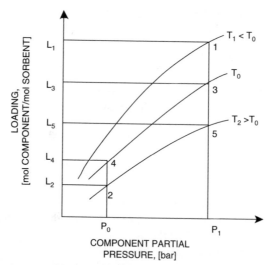

Figure 8-10. Adsorption Loading at Different Temperatures and Pressures

Molecular Sieves

The most common application of molecular sieves in connection with gasification plants is the removal of water and CO_2 upstream of cryogenic units. Processes working at cryogenic temperatures, such as air separation or cryogenic gas separation, require a feed gas completely free of these components, which would otherwise freeze and deposit on the inlet heat exchangers and finally block them.

The classic cycle described above is usually employed. In air separation duty, water and CO_2 are not the only considerations. The prepurification unit also prevents the ingress of hydrocarbons into the cold box as a safety measure. Recently the ingress of NO_x into the cold box has also become an issue of concern. For air separation. a combination of molecular sieve and silica gel is often used.

Pressure Swing Adsorption

Pressure swing adsorption (PSA) operates on an isothermal cycle, adsorbing at high pressure and desorbing at low pressure. The principle application of PSA is for hydrogen purification, although there are a number of others including air separation (see Section 8-1).

The optimum pressure for hydrogen purification lies in the range 15–30 bar. At higher pressures the hydrogen yield falls off, a point to be considered when integrating a hydrogen off-take from a gasification plant optimized for a different application.

The hydrogen yield of a modern PSA unit usually lies between 80% and 92%. Apart from the matter of pressure already mentioned, other influences are the quality

Table 8-3
Relative Strength of Adsorption of Typical Impurities

Non-adsorbed	Light	Intermediate	Heavy
H_2	O_2	CO	C_3H_6
He	N_2	CH_4	C_4H_{10}
	Ar	C_2H_6	C_5+
		CO_2	H_2S
		C_3H_8	NH_3
			H_2O

Source: Miller and Stoecker 1989

of the feed gas (the higher the quantity of impurities to be removed, the more hydrogen is lost with them) and the tail gas pressure. Where the tail gas is burned in dedicated burners, as for instance in a steam reformer hydrogen plant, the typical tail gas pressure is 0.3–0.4 bar gauge. Where the tail gas pressure is higher (e.g., 3–5 bar gauge), the drop in hydrogen yield can become very significant.

Additionally, the hydrogen purity can affect the yield, though only to a small degree. Typical purities range from 99 to 99.999 mol%. An additional common hydrogen specification is a limit on the amounts of carbon oxides (CO and CO_2). Levels of 0.1 to 10 ppmv are easily achieved. In the design of an overall gasification-to-hydrogen system, it is useful to have an idea about the performance of likely impurities in the PSA unit. A comparison of a number of components is shown in Table 8-3. In this connection it is important to note that water is strongly adsorbed and so will not contaminate the product. It is disadvantageous however to have large quantities in the feed gas since this requires excessively large beds. Usually, cooling to below 40°C with subsequent condensate separation is sufficient to provide an economic design.

A further design consideration is the number of adsorber vessels. Early plants used four beds, as is still the practice in smaller plants. Larger modern plants use as many as twelve adsorbers. Sophisticated cycles have been developed to minimize the loss of hydrogen on depressurization from the adsorption step to the desorption step, by using this hydrogen to repressurize a bed that has just completed its desorption step. Thus there can be a trade-off between a higher investment for an increased number of vessels (and valves) and operating savings from an increased hydrogen yield.

Zinc Oxide/Copper Oxide

Adsorption of H_2S onto zinc oxide is an effective method for removing trace quantities of sulfur from gas to achieve a purity of less than 0.1 ppmv, as is required by copper or nickel catalysts. It is therefore the standard method of desulfurization upstream of natural gas steam reformers. The adsorption takes place via the reaction of hydrogen sulfide with zinc oxide to form zinc sulfide. In situ regeneration is not possible, and this places a limitation on the amount of sulfur that the process can accept in the inlet gas.

There are two generally accepted designs for zinc oxide desulfurization units. In a guard bed function or where the sulfur load is low, a single bed is provided, sized to adsorb the total quantity of sulfur to be expected between planned turnarounds, say one or two years. Where the sulfur load is higher and a single bed would become unmanageably large, a two-vessel series arrangement is provided and provision is made for exchanging the adsorbent online. With this arrangement, the individual bed can be sized smaller, such as for a six-month interval between bed replacement.

Zinc oxide can adsorb sulfur present as H_2S almost completely. Performance with other sulfur compounds (COS, mercaptans) is not as good. In cases where sulfur is present other than as H_2S, it is necessary to hydrogenate these components to H_2S upstream of the zinc-oxide bed. This is normally done over a cobalt-molybdenum (CoMox) or nickel-molybdenum (NiMox) catalyst.

Zinc oxide adsorption is essentially a process for polishing or guard bed duty. This becomes clear when considering a zinc-oxide bed for the carbon monoxide plant described in Section 7.1.4. Operating in its optimum temperature range of 350 to 400°C, zinc oxide has a pick-up capacity of around 20% by weight. Assuming a sulfur content of 100 ppmv in the natural gas, the total sulfur intake is about 10 tons/year, requiring replacement of about 50 tons/year zinc oxide. Compare this with the nearly 30 t/d sulfur intake of the 1000 t/d methanol plant of Section 7.1.2, and the limitations become very apparent.

Given these numbers, zinc oxide in the gasification environment is limited either to guard bed duty, for example, upstream of a low temperature shift or methanator catalyst or to natural gas feeds. As discussed in Section 7.1.4, there are arguments for desulfurizing either upstream or downstream of the partial oxidation reactor.

Where extreme sulfur cleanliness is required, copper oxide can be used for final desulfurization down to 10 ppbv. Commercial adsorbents are available for this purpose, either in a mixed ZnO/CuO formulation or as a separate polishing bed.

8.2.4 Membrane Systems

Permeable gas separation membranes in syngas service utilize differences in solubility and diffusion of different gases in polymer membranes. The rate of transport of a component through the membrane is approximately proportional to the difference in partial pressure of the component on the two sides of the membrane. Polymer membranes have found increasing use in a number of applications, including natural gas processing (CO_2 removal) and in the synthesis gas environment for hydrogen separation out of the main syngas stream.

The design of a polymer membrane system exploits the different permeability rates of the components in the feed gas. An idea of the relative rates through a typical hydrogen separation polymer can be gained from Table 8-4. Thus a good separation can be achieved between, for example, hydrogen and CO or N_2. Separation from CO_2 will be only moderately satisfactory, however.

Membrane units are usually supplied packed, typically as a bundle of hollow tube fibers. The feed is supplied to the shell side of the bundle and the permeate (hydrogen rich stream), which passes through the fiber-tube walls, is collected on the tube side. Design variables are the pressure difference selected and the total surface area of the polymer.

For the system designer, the integration of a membrane unit has two important characteristics. First, permeable membranes provide the only system leaving the carbon monoxide at essentially the same pressure level as at the gas inlet (less hydraulic losses only) and the hydrogen on the low-pressure side. This is exactly the reverse of the pressure swing adsorber.

Second, as mentioned above, it must be recognized that since all permeable membranes work on the basis of different rates of diffusion, they can only have a limited selectivity. This can be disadvantageous, since in a hydrogen extraction application, the product hydrogen is not very pure, and the diffusion of CO through the membrane can be considered as a loss of high pressure gas.

Nonetheless, skilled integration of membrane and PSA technologies can together provide some extremely attractive solutions. Consider the following situation where 20,000 Nm^3/h pure hydrogen is required from a main stream of syngas in an IGCC (Figure 8-11 and Table 8-5). The membrane is used to produce a raw hydrogen at reduced pressure (but still adequate for PSA feed) with only a small loss of other syngas components for the gas turbine. The raw hydrogen has a purity of about 70–90 mol%, depending on syngas composition and pressure, which allows the PSA to have a significantly higher efficiency than would be the case with syngas feed. Furthermore, the much smaller quantity of tail gas to be adsorbed allows the PSA unit to be smaller too.

Care should be exercised with liquid carry over from an upstream AGR system. In some cases these can damage the membrane. Proper separation at the AGR outlet should however be sufficient to prevent problems (Collodi 2001).

Table 8-4
Relative Permeability Rates of Typical Syngas Components

Quick	Intermediate	Slow
H_2	CO_2	CO
He		CH_4
H_2S		N_2
Source: Kubek, Polla, and Wilcher 1997		

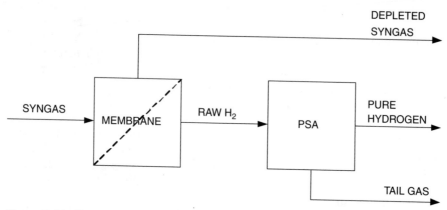

Figure 8-11. Membrane and PSA Combination

Table 8-5
Mass Balance for Membrane/PSA Combination

	Syngas In mol%	Syngas Out mol%	Raw H_2 mol%	Pure H_2 mol%	Tail Gas mol%
CO_2	8.6	8.4	9.4		31.4
CO	43.8	52.4	7.9		26.3
H_2	45.3	36.5	82.4	100.0	41.2
CH_4	2.3	2.7	0.3		1.1
Total (kmol/h)	6635	5360	1275	893	382
Pressure (bar)	50	49	25	24	1.3

Hot Gas Cleanup

For power applications the energy loss involved in cooling synthesis gas down to ambient or lower temperatures as required by current acid gas removal systems is reason enough for the interest in so-called "hot gas cleanup." Actually, hot gas cleanup is a misnomer, and these technological developments should rightly be called "warm gas cleanup," since the target operating temperature range is between 250 and 500°C.

Impurities that need to be considered in a warm gas cleanup system include particulates (fly ash and char) as well as gaseous compounds such as H_2S, COS, NH_3, HCN, HCl, and alkali species. At temperatures above about 500°C, alkaline species will pass through a particulate filter, and this together with materials issues is the principle reason why no attempts at hotter cleanup have been made.

Technologies for warm gas cleanup using zinc-based sorbents have been built at demonstration scale in Polk County and Piñon Pine without great success (Simbeck, 2002; U.S. Department of Energy 2002). In fact, neither of these units was ever operated. Both of these were designed essentially as desulfurization units with removal efficiencies of up to 98%, which at the time of design conformed to existing power station emission regulations. They did not address some of the other species, such as nitrogen compounds and halides, nor for that matter mercury. Furthermore, the sulfur removal efficiencies made them unsuitable for most chemical applications.

Nonetheless, the potential in terms of efficiency improvement remains and continues to provide an incentive for research and development to find appropriate systems.

8.2.5 Further Developments

One current program is addressing some of these issues. It includes the use of membrane technology for bulk desulfurization, zinc oxide or similar chemisorption process for fine sulfur removal, a sodium carbonate-based sorbent for HCl removal, and a high-temperature molecular sieve for ammonia removal (Gupta 2001). However, the use of polymer membrane technology is likely to limit its ability to operate at high temperature and thus any efficiency gains on this basis. Mercury capture is not specifically mentioned as being part of this program, although on the basis of currently published flowsheets it could possibly be incorporated as a separate stage.

8.2.6 Biomass Syngas Treating

Treating the syngas generated from biomass has special problems—particularly those associated with the presence of tars in the gas. Attempts have been made to reduce the tar content by cracking (Morris and Waldheim 2002; Bajohr etal. 2002). Other attempts have been made at using an oil wash (Boerrigter, den Uil, and Calis 2002; Hofbauer 2002). To date, success has been limited to achieving a quality suitable for power applications. Considerable work is still required to achieve a chemicals application syngas quality.

8.3 CATALYTIC GAS CONDITIONING

8.3.1 CO Shift

Besides having an important influence on the composition of the raw syngas from the gasifier itself, the CO shift reaction

$$CO + H_2O \leftrightarrows CO_2 + H_2 \qquad -41\,\text{MJ/kmol} \qquad (2\text{-}7)$$

can be and is operated as an additional and separate process from the gasifier at much lower temperatures in order to modify the H_2/CO ratio of the syngas or maximize the total hydrogen production from the unit. As can be seen from the reaction 2-7, one mole of hydrogen can be produced from every mole of CO. The reaction itself is equimolar and is therefore largely independent of pressure. The equilibrium for hydrogen production is favored by low temperature.

The CO shift reaction will operate with a variety of catalysts between 200°C to 500°C. The types of catalyst are distinguished by their temperature range of operation and the quality (sulfur content) of the syngas to be treated.

High Temperature (HT) Shift

Conventional (high temperature) shift uses an iron oxide–based catalyst promoted typically with chromium and more recently with copper. The operating range of these catalysts is between 300 and 500°C. Much above 500°C sintering of the catalyst sets in and it is deactivated. HT shift catalyst is tolerant of sulfur up to a practical limit of about 100 ppmv, but it is likely to loose mechanical strength, particularly if subjected to changing amounts of sulfur.

An important aspect in the design of CO shift in the gasification environment, where inlet CO contents of 45% (petroleum residue fed) to 65% (coal) are common, is the handling of the heat of the reaction, particularly under end-of-run conditions where an inlet temperature of 350°C or more may be necessary. On the one hand, the reaction must be performed in several stages to avoid excessive catalyst temperatures and to have an advantageous equilibrium. On the other hand, optimum use must be made of the heat.

One such arrangement is shown in Figure 8-12. Desulfurized syngas containing about 45 mol% CO, which leaves the AGR at about 54 bar and ambient temperature, is heated and water saturated at a temperature of about 215°C by water that has been preheated with hot reactor effluent gas. The saturated gas is further preheated to the catalyst inlet temperature of between 300°C and 360°C. The steam loading from the saturator is such that only the stoichiometric steam demand for the reaction is required to be added from external sources. In the first stage, the CO is reduced to a

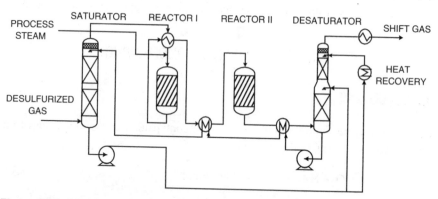

Figure 8-12. CO Shift with Saturator-Desaturator Circuit (*Source: Higman 1994*)

level of about 7–8 mol% at an outlet temperature of about 500°C. The outlet gas is cooled to a temperature of about 380°C in the gas and water preheaters before entering the second catalyst bed. Here the residual CO is reduced to about 3.2 mol%. The gas is then cooled in a direct-contact desaturator tower. There are a number of different designs, particularly for the first reactor, that incorporate the gas-gas heat exchanger as an internal. In such reactors, the exchanger is arranged centrally inside an annular catalyst bed with an axial (Lurgi) or axial-radial (Casale) gas-flow pattern. Alternative methods of controlling the catalyst outlet temperature include interbed condensate injection (e.g., Toyo). The use of an isothermal steam raising reactor has been proposed, and although such a solution has been employed in a steam reformer plant, none is recorded at the high CO inlet concentrations involved in a gasification plant.

Typical catalyst lifetime for the first bed in a gasification situation is two to three years, which is considerably shorter than for a steam-reforming situation. This is generally attributed to the high operating temperatures associated with high CO concentrations in the inlet gas. On a moles-converted basis over the lifetime of the catalyst, the performance in the gasification context is comparable with that of steam reforming.

Low Temperature (LT) Shift

Low temperature shift operates in the temperature range 200°C to 270°C and uses a copper-zinc-aluminum catalyst. It is used in most steam reforming-based ammonia plants to reduce residual CO to about 0.3 mol%, a requirement for a downstream methanator, but has generally not been applied in gasification-based units. On the one hand, it is highly sulfur-sensitive, and even with 0.1 ppmv H_2S in the inlet gas, will over time become poisoned. A second reason for its lack of use particularly in oil-gasification plants, is the effect of the higher pressure on the water dewpoint in the gas. Operation near the dewpoint will cause capillary condensation and consequent damage to the catalyst. With a dewpoint of about 215°C and a temperature rise of 25–30°C, there is not much margin for error below the upper temperature limit of 270°C when recrystallization of the copper catalyst begins. The first application of low temperature shift at high pressure was in Shell's Pernis gasification facility, which has now performed successfully for several years (de Graaf et al. 2000).

Medium Temperature (MT) Shift

An improved copper-zinc-aluminum catalyst able to operate at higher exit temperatures (300°C) than conventional LT shift has been developed, particularly for use in isothermal reactors. No application in gasification plants is known.

Raw Gas Shift

For applications where it is desired to perform CO shift on raw syngas, a cobalt-molybdenum catalyst, variously described as a "sour shift" or "dirty shift" catalyst, can be used. In some parts of the literature this catalyst is described as sulfur tolerant.

This is actually a misnomer, since the catalyst requires sulfur in the feed gas to maintain it in the active sulfided state. It is generally applied after a water quench of the raw syngas, which typically will provide a gas at about 250°C saturated with sufficient water to conduct the shift reaction without any further steam addition. For an ammonia application the raw gas shift is typically configured as two or three adiabatic beds with intermediate cooling resulting in a residual CO of about 1.6 or 0.8 mol%, respectively.

An important side-effect of the raw gas shift catalyst is its ability to handle a number of other impurities characteristic of gasification. COS and other organic sulfur compounds are largely converted to H_2S, which eases the task of the downstream AGR. HCN and any unsaturated hydrocarbons are hydrogenated.

Carbonyls are decomposed and deposited as sulfides, which increases the pressure drop over the bed. Selective removal of arsenic in the feed is also claimed (BASF undated).

8.3.2   COS Hydrolysis

In all synthesis gases produced by gasification sulfur is present not only as H_2S but also as COS. Typically a syngas from the gasification of a refinery residue with 4% sulfur may contain about 0.9 mol% H_2S and 0.05 mol% COS. While some washes such as Rectisol can remove the COS along with the H_2S, others, particularly amine washes, require the COS to be converted selectively to H_2S if the sulfur is to be substantially removed. This is best achieved by catalytic COS hydrolysis, according to the reaction

$$COS + H_2O \leftrightarrows H_2S + CO2 \qquad -30\,MJ/kmol \tag{8-3}$$

Commercially this reaction takes place over a catalyst at a temperature in the range of 160–300°C. Various catalysts are available including a promoted chromium oxide-alumina, pure activated alumina or titanium oxide [Higman, 2000 and Puertas Daza and Ray, 2004]. Lower temperatures favor the hydrolysis equilibrium. Typically the optimum operating temperature is in the range 150–200°C. Depending on process conditions the residual COS can be reduced to the range of 5–30 ml/Nm³. This catalyst also promotes the hydrolysis of HCN.

The catalyst operates in the sulfided state and is not poisoned by heavy metals or arsenic. Halogens in the gas will however reduce activity, selectivity and lifetime, a fact that needs to be addressed carefully in coal gasification applications. In applications downstream gasification of refinery residues nickel and iron carbonyls, which may have formed upstream, can decompose depositing nickel or iron sulfide on the catalyst bed, thus creating an increased pressure drop over the system. Typically a guard bed which can be taken off line during normal operation is installed upstream of the COS hydrolysis bed to catch these deposits.

For any application care must be taken to avoid catalyst degradation by liquid water. Catalyst manufacturers recommend that the bed temperature be maintained at least 20–30 °C above the water dew point under all circumstances.

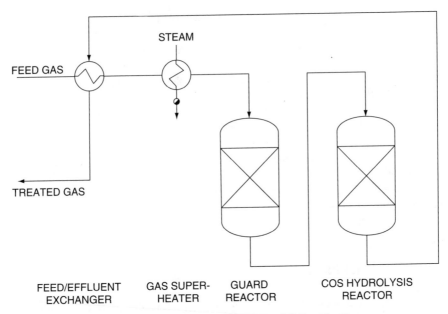

Figure 8-13. Typical COS Hydrolysis Flowsheet for Oil Gasification

A typical operation is shown in Table 8-3 [Higman, 2000].

Table 8-6 COS Hydrolysis References		Inlet	Outlet
CO_2	mol %	5.0	
CO	mol %	45.0	
H_2	mol %	45,0	
CH_4	mol %		
N_2	mol %		
H_2S	mol %	0,71	
COS	ml/m³	300	10
H_2O	kg/Nm³	0.086	
Pressure	bar	50	
Temperature	°C	170	
Dew point	°C	150	

8.4 SULFUR RECOVERY

The sulfur compounds from the feedstock of a gasification-based process are generally removed from the synthesis gas as a concentrated stream of hydrogen sulfide and carbon dioxide known as acid gas. Depending on the design of the upstream AGR unit, the acid gas may contain other sulfur species such as COS as well as ammonia and hydrogen cyanide. It is unacceptable to emit H_2S, a highly toxic, foul-smelling gas, to the atmosphere, so it is necessary to fix it in one form or other. There are essentially two alternative products in which one can fix the sulfur, either as liquid or solid elemental sulfur, or as sulfuric acid. The choice of product will depend on the local market. Where there is a strong local phosphate industry, then there will be a good local market for sulfuric acid. If this is not the case, then elemental sulfur will probably be the better choice, since bulk transport of this material is much easier than of the concentrated acid.

8.4.1 The Claus Process

The basic Claus process for substoichiometric combustion of H_2S to elemental sulfur was developed as a single-stage process on the basis of reaction 8-5 at the end of the nineteenth century. During the 1930s it was modified into a two-stage process in which initially one third of the H_2S was combusted to SO_2 and water, and in a second low-temperature catalytic stage, the SO_2 was reacted with the remaining H_2S to sulfur. Operating the second stage at a comparatively low temperature (200–300°C) used the more favorable equilibrium to achieve much higher sulfur yields than had been possible with the original process.

$$H_2S + 1\tfrac{1}{2}\,O_2 \longleftrightarrow SO_2 + H_2O \tag{8-4}$$

$$2\,H_2S + SO_2 \longleftrightarrow 2\,H_2O + 3/8\,S_8 \tag{8-5}$$

$$\overline{3\,H_2S + 1\tfrac{1}{2}\,O_2 \longleftrightarrow 3\,H_2O + 3/8\,S_8} \tag{8-6}$$

Today there are innumerable Claus processes available, all of them ultimately variants of the modified Claus process. A typical standard Claus process is shown in Figure 8-14. In the first combustion stage all the H_2S is combusted with an amount of air corresponding to the stoichiometry of reaction 8-6 at a temperature in the range 1000–1200°C. The thermodynamics of these three main reactions is such that about half the total sulfur is present in the outlet gas as elemental sulfur vapor, the rest as an equal mix of H_2S and SO_2. The hot gas is cooled by raising steam and the sulfur already formed is condensed out. The removal of sulfur at this point assists in driving reaction 8-5 further to the right in the subsequent catalytic stage. The gas is reheated and passed over an alumina catalyst at a temperature of about 200–300°C,

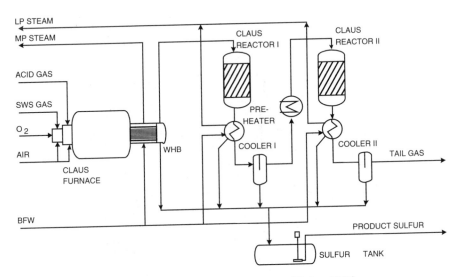

Figure 8-14. Typical Two-Stage Claus Unit (*Source: Weiss 1997*)

and cooled again to condense the sulfur formed. This may be performed a number of times to remove further amounts of sulfur. Typically, two (as shown in Figure 8-14) or three catalytic stages are used.

Oxygen Claus Processes

A standard air-blown Claus plant is limited in the dilution of H_2S possible in the acid gas. At concentrations less than about 25–30 mol% H_2S, the temperature in the furnace is insufficient to maintain the reaction. Although this is only rarely a limitation for plants gasifying refinery residues, gasification of low-sulfur coal can easily produce an acid gas with such a low H_2S concentration. Although it is possible to increase this concentration with some AGR systems, using oxygen instead of air as the oxidant in the Claus furnace often offers a more economic approach, especially since the oxygen demand for the Claus unit is only a fraction of that required for the gasifier itself. In addition to saving the cost of concentration within the AGR, the cost of the Claus unit itself is lower. The chief determinant for sizing the equipment in a Claus plant is the volume of gas throughput. Elimination of all or most of the nitrogen involved with air-blowing reduces the gas volume by between 30 and 60%, depending on the acid gas quality. The equipment is accordingly smaller.

There are a number of suppliers of oxygen-blown Claus technology, such as Lurgi, BOC, APCI/Goar. All use special burners to handle the oxygen. One example is the Lurgi OxyClaus burner, which is shown in Figure 8-15. The design of this multipurpose burner provides for operation on air, enriched air or pure oxygen. The burner has a series of acid gas burners arranged concentrically around a central burner muffle. Each acid gas burner consists of three coannular lances, with oxygen

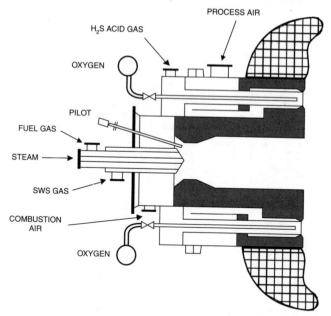

Figure 8-15. Lurgi OxyClaus Burner (*Source: Knab, Neumann, and Nehb 1994*)

being injected through the inner lance, acid gas through the intermediate one, and air
through the outer ring.

One aspect of processing acid gas in a Claus plant, which deserves special
mention, is the problems that can occur, when excessive amounts of ammonia or
HCN are present in the gas. These components must be oxidized fully to molecular
nitrogen as otherwise ammonium salts are formed downstream and can plug the liquid
sulfur lines. Ensuring complete oxidation is essentially a matter of ensuring that these
components are combusted stoichiometrically at a sufficiently high temperature,
eventhough the H_2S combustion is sub-stoichometric.

The main path of ammonia and HCN in a gasification unit is into the process
condensate, from which they are removed in a sour water stripper (SWS). A feature
of the Lurgi OxyClaus burner is the separate nozzle for the sour water stripper off-gas,
which allows this gas to be combusted with a different stoichiometry and therefore
a higher temperature than the bulk acid gas from the AGR unit.

8.4.2   Tail Gas Treatment

Environmental regulations require a high degree of sulfur removal in Claus units.
Typical are the German regulations (TA Luft), which require 97% sulfur recovery
for plants up to 20 t/d S, 98% for plants between 20 and 50 t/d, and 99.5% for larger
plants. The thermodynamics of the Claus reactions do not allow one to reach these

yields. It is therefore necessary to incorporate some form of tail gas treatment to achieve these yields.

Superclaus

One approach developed by Comprimo is the inclusion of a selective catalytic oxidation step after the second Claus stage, which goes by the name of Superclaus.

Hydrogenation and Absorption

There is an important class of tail gas treatment processes typified by Shell's SCOT (Shell Claus Off Gas Treating) process, although there are now a number of similar processes on the market. The principle arrangement of all these processes is similar and is portrayed in Figure 8-16. The tail gas is hydrogenated in a catalytic reactor, so that all sulfur species are converted to H_2S. The gas is then quenched with water and fed to an amine scrubbing unit to remove the H_2S. Since the proportion of H_2S in the gas at this point of the plant is low, it is important for the economics to choose a selective washing system, typically MDEA. The sulfur-rich gas from the regenerator is recycled to the Claus furnace. Alternatively, the regeneration can be integrated with that of a main H_2S amine wash.

Other systems using hydrogenation and subsequent H_2S removal are Beavon, which used a Stretford wash for H_2S removal (but now also uses MDEA), and others which use, for example, Flexsorb.

Subdewpoint Processes

An alternative to hydrogenation and absorption is to continue making use of the Claus reaction, but to do this at lower temperatures, at which the sulfur condenses out in the catalyst bed. Such processes include Amoco's cold bed adsorption (CBA) process and the SNEA-Lurgi Sulfreen process.

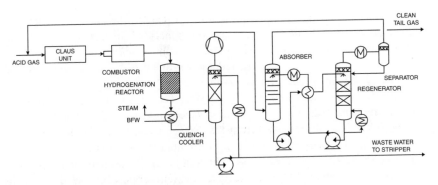

Figure 8-16. Typical Tail Gas Treatment Plant

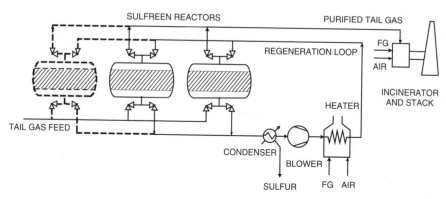

Figure 8-17. Sulfreen Tail Gas Process (*Source: Lell 1993*)

The Sulfreen process depicted in Figure 8-17 serves as a typical example of this class of tail gas treatment. The process is a cyclic one, which uses one reactor in the adsorption mode while the second is being regenerated. For large plants it can be advantageous to include a third reactor. During the adsorption step, tail gas is passed over the catalyst at a temperature of 120–140°C. The low temperature enables the Claus reaction to proceed further towards the production of sulfur, which is adsorbed on the catalyst. In the regeneration mode, hot tail gas is used to desorb the sulfur, which is then recovered in a steam raising sulfur condenser. The heat for the regeneration loop can be supplied by a separate fired heater (as shown) or by heat recovery from the thermal incinerator. Overall sulfur recoveries of 99.0 to 99.5% can be achieved with a standard Sulfreen unit, depending on the Claus feed gas quality. Additional enhancements can be incorporated to achieve higher recoveries, such as incorporating an upstream hydrolysis (Hydrosulfreen, 99.5–99.7%) or a downstream direct oxidation (Carbosulfreen > 99.8%) (Lell 1993).

8.4.3   Integration in an IGCC

In contrast to most chemical applications for gasification, the presence of CO_2 in IGCC syngas is not only no problem but even beneficial, in that it increases the mass flow through the gas turbine and therefore contributes to the energy output. This fact allows the integration of the tail gas treatment of the Claus unit into the main AGR system in a manner that eliminates the SRU tail gas as an emission completely. The principle is shown in Figure 8-18. The tail gas from the Claus plant, which contains both H_2S and SO_2, is hydrogenated and recycled back to the AGR, where the sulfur is again removed from the syngas. Clearly, for the economics of this system it is necessary to keep the recycle compression costs low. This puts a premium on the selectivity of the AGR, which in the IGCC scenario is in any case desirable, so as to keep the CO_2 in the recycle low. Furthermore, one needs to use an oxygen-blown Claus unit to avoid large quantities of nitrogen placing an unnecessary load on the recycle compressor. On the other hand, since the Claus tail gas is no

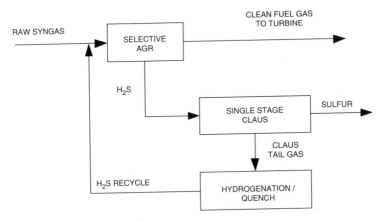

Figure 8-18. Integration of Claus Unit in an IGCC

longer an emission as such, there is no need to include more than one catalytic stage in the Claus unit itself. This type of integration has proved itself both in coal applications (Puertollano using an MDEA AGR) and for oil gasification (Sarlux using Selexol).

8.4.4 Sulfuric Acid

An alternative outlet for sulfur to the production of merchant sulfur is sulfuric acid. This is particularly interesting in locations where there is significant production of phosphate fertilizers, such as Florida, where the Polk IGCC unit processes its acid gas to sulfuric acid rather than sulfur. Other examples can be found in ammonia plants, which have formed the core of complex fertilizer production sites in Portugal and China, for example.

Wet Catalysis

Processing H_2S to sulfuric acid differs from conventional acid plants (whether using sulfur or metallurgical off gases as feed) in that water is already present in the converter. This water reacts immediately with the SO_3 formed there and therefore demands a totally different approach to the recovery of liquid acid compared with a standard double absorption unit.

The H_2S component of the acid gas is first combusted with excess oxygen to form SO_2 and water (see Figure 8-19). The excess is selected to produce a O_2/SO_2 ratio in the hot gas of between 1.1 and 1.3. The hot combustion gases are cooled in a waste-heat boiler to a converter inlet temperature of 440°C. The SO_2 and oxygen react over a vanadium catalyst according to the reaction

$$SO_2 + \tfrac{1}{2} O_2 \longleftrightarrow SO_3 \qquad (8\text{-}7)$$

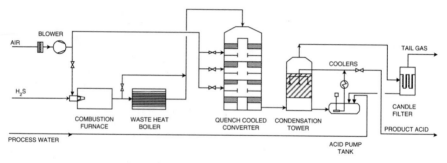

Figure 8-19. Wet Catalysis Sulfuric Acid Plant (*With permission: Lurgi*)

and the water reacts with SO_3 as follows

$$SO_3 + H_2O \longleftrightarrow H_2SO_4 \qquad (8\text{-}8)$$

The converter uses interbed quenching with cold air to control the temperature increase arising from these exothermic reactions. Depending on the feed gas quality, three or four catalyst beds are included.

Since the acid is already present in the gas phase leaving the converter, this is recovered by condensing at about 60°C. In a classic wet catalysis plant, the product is a 78% acid, below the 80–90% range that is critical for corrosion. The condensation is achieved by recycling this product acid to the condensation tower in a manner designed to minimize formation of acid aerosols. A candle filter is provided to remove any aerosols that are formed. If there is insufficient water in the gas leaving the converter, then make-up water is added to keep the acid strength at 78%—out of the corrosive range.

Advanced wet catalysis processes such as Concat (Lurgi) or WSA (Topsøe) are able to produce 93–98% strength acid, even with very weak feed gases and excess water. Concat achieves this, for example, by using a two-stage condensation section, the first of which operates at 180–3230°C so that a minimum of free water is present in the condensate, which is already about 93% H_2SO_4.

REFERENCES

Air Liquide. Available at: www.airliquide.com.

Air Products. Company brochure, 2001.

Allam, R. J., Russek, S. L., Smith, A. R., Stein, V. E. "Cryogenics and Ceramic Membranes: Current and Future Technologies for Oxygen Supply in Gasification Systems." Paper presented at IChemE Conference "Gasification for the Future", Noordwijk, April 2000.

Allam, R. J., Foster, P. F., and Stein, V. E. "Improving Gasification Economics through ITM Oxygen Integration." Paper presented at IChemE Conference "Gasification: The Clean Choice for Carbon Management", Noordwijk, April 2002.

Armstrong, P. A., Stein, V. E., Bennet, D. L., and Foster, E. P. "Ceramic Membrane Development for Oxygen Supply to Gasification Applications." Paper presented at Gasification Technologies Conference, San Francisco, October 2002.

Bajohr, S., Hoferer, J., Reimert, R., and Schaub, G. "Thermische und katalytische Umsetzung von Kohlenwasserstoffen in Rohgasen aus der Pyrolyse und aus der Vergasung von Biomassen," Paper presented at the DGMK Conference "Energetische Nutzung von Biomassen", Velen, April 2002.

BASF. BASF Catalyst K 8-11. Technical leaflet. (Undated.)

Boerrigter, H., den Uil, H., and Calis, H.-P. "Green Diesel from Biomass via Fischer-Tropsch Synthesis: New Insights in Gas Cleaning and Process Design." Paper presented at Pyrolysis and Gasification of Biomass and Waste Expert Meeting, Strasbourg, October 2002.

Butzert, H. E. "Ammonia/Methanol Plant Startup." *Chemical Engineering Progress* 72 (January 1976):78–81.

Collodi, G. "Commercial Operation of ISAB Energy and Sarlux IGCC." Paper presented at Gasification Technologies Conference, San Francisco, October 2001.

de Graaf, J. D., Zuideveld, P. L., van Dongen, F. G., and Hölscher, H. "Shell Pernis Netherlands Refinery Residue Gasification Project." Paper presented at IChemE Conference, "Gasification for the Future," Noordwijk, April 2000.

Grünewald, G. "Selective Physical Absorption Using the Purisol Process." Paper presented at 6th Continental Meeting of the European Chapter of the GPA, Bremen, 1989.

Gupta, R. "Emerging Technologies for the Production of Ultraclean IGCC Syngas," Gasification Technologies Conference, San Francisco, 2001.

Higman, C. "Gasification of Heavy Residues." Paper presented at IMP Gasification Seminar, Mexico City, 1994.

Higman, C. A. A. and Kalteier, P. "Carbonyl Sulfide Conversion" In *Ullmann's Encyclopedia of Industrial Chemistry*, 6th ed., Weinheim, Wiley VCH, 2000.

Hirtl, C. (Linde AG) Personal communication. September 2002.

Hofbauer, H. "Biomass CHP-Plant Güssing: A Success Story." Paper presented at Pyrolysis and Gasification of Biomass and Waste Expert Meeting, Strasbourg, October 2002.

Knab, H., Neumann, G., and Nehb, W. "Operating Experience with Lurgi OxyClaus Technology." Paper presented at Achema, Frankfurt, June 1994.

Kohl, A., and Nielson, R. *Gas Purification*. Boston: Gulf Publishing, 1997.

Korens, N., Simbeck, D. R., and Wilhelm, D. J. "Process Screening Analysis of Alternative Gas Treating and Sulfur Removal" SFA Pacific for U.S. Department of Energy, Mountain View, CA, December 2002.

Koss, U., Meyer, M., and Schlichting, H. "Zero Emissions IGCC with Rectisol Technology" Paper presented at Gasification Technologies Conference, San Francisco, October 2002.

Kriebel, M. "Gas Production" In *Ullmann's Encyclopedia of Industrial Chemistry*, 5th ed., vol. A12. Weinheim: VCH Verlagsgesellschaft, 1989, p. 265.

Kubek, D. J., Polla, E., and Wilcher, F. P. "Purification and Recovery Options for Gasification." Paper presented at IChemE Conference, "Gasification Technology in Practice," Milan, February 1997.

Kubek, D. J., Sharp, C. R., Kuper, D. E., DiDio, M., and Whysall, M. "Recent Selexol, PolySep, and PolyBed Operating Experience with Gasification of Power and Hydrogen. Paper presented at Gasification Technologies Conference, San Francisco, October 2002.

Lell, R. "New Technology for Sulfur Recovery from Tail Gases Based on the Sulfreen Process." Paper presented at "Sulfur 93" Conference, Hamburg, April 1993.

Miller, G. Q., and Stoecker, J. "Selection of a Hydrogen Process." Paper presented at NPRA Annual Meeting, San Francisco, 1989.

Morris, M., and Waldheim, L. "Update on Project ARBRE, UK." Paper presented at IChemE Conference, "Gasification: The Clean Choice for Carbon Management" Noordwijk, The Netherlands, April 2002.

Müller, H. (2002) "Sulfuric Acid and Sulfur Trioxide." In *Ullmann's Encyclopedia of Industrial Chemistry*. Weinheim: Wiley VCH, 2002.

Nagl, G. L. "Liquid Redox Enhances Claus Process." *Sulfur* 274 (May–June 2001).

Nehb, W. "Sulfur." In *Ullmann's Encyclopedia of Industrial Chemistry*, vol. A25. VCH, 1994.

Praxair. Available at: www.praxair.com.

Puertas Daza, V. and Ray, J. L., "Elcogas IGCC in Puertollano: Six Years Experience with COS Hydolysis Reactor Catalysts", IChemE Gasification Conference, Brighton, May 2004.

Sharp, C. R. "Recent Selexol and Membrane/PSA Operating Experiences with Gasification for Power and Hydrogen." Paper presented at Gasification Technologies Conference, San Francisco, October 2002.

Simbeck, D. "Industrial Perspective on Hot Gas Cleanup." Paper presented at 5th International Symposium, "Gas Cleaning at High Temperatures," Morgantown, 2001.

Smith, A. R., and Klosek, J. "A Review of Air Separation Technologies and Their Integration with Energy Conversion Processes." *Fuel* 80 (2001):115–134.

Stevens, D., Stern, L. H., and Nehb, W. (1996) "OxyClaus Technology for Sulfur Recovery." Paper presented at Laurence Reid Gas Conditioning Conference, Norman, Oklahoma, March 1996.

Supp, E. *How to Produce Methanol from Coal*. Berlin: Springer, 1990.

U.S. Department of Energy. "Piñon Pine IGCC Power Project: A DoE Assessment." DoE/NETL report 2003/1183, December 2002.

Weiss, M.-M. "Selection of the Acid Gas Removal Process for IGCC Applications." Paper presented at IChemE Conference, "Gasification Technology in Practice," Milan, February 1997.

Zwiefelhofer, U., and Holtmann, H.-D. "Gas Expanders in Process Gas Purification." Paper presented at VDI-GVC Conference, "Energy Recovery by Gas Expansion in Industrial Use," Magdeburg, Germany, 1996.

Chapter 9

Economics, Environmental, and Safety Issues

9.1 ECONOMICS

The economics of every major capital investment are individual to the project concerned. Given the broad range of applications and feedstocks, this is especially so for gasification. Nonetheless a number of trends can be identified.

Gasification is generally a capital-intensive technology, which has, however, the capability of working with cheaper or more difficult feedstocks than many alternatives.

Ammonia. The capital intensity of gasification is clearly visible in the data in Table 9-1 for a 1800 t/d ammonia plant located in Northwest Europe based on different technologies and feedstocks. The capital estimates and feed rates are those of Appl (1999).

This data can be presented in a manner that allows comparison of production costs with varying feedstock pricing for the different feedstocks, as in Figure 9-1. Based on these data ammonia production by gasification of heavy residue becomes competitive with U.S.$2.50/MMBTU natural gas, if the residue is valued at about U.S.$20/t. The natural gas price must rise to over U.S.$4/MMBTU before coal or petcoke can become a competitive feedstock.

Methanol. Higman made a similar study for methanol comparing natural gas and vacuum residue feed (1995). The results are in principle similar (Figure 9-2).

His conclusions are that under today's economic conditions, it would not be competitive to manufacture methanol by gasifying locally available resid compared with importing methanol produced from cheap natural gas in a remote location. "If, however, strategic or other considerations demand that production be located in one

329

Table 9-1
Cost of Ammonia Production with Different Feedstocks

Feedstock	Natural Gas	Vaccum Residue	Coal
Process	Steam reforming	Gasification	Gasification
Feedstock price, $/MMBTU	2.8	1.8	1.5
Total energy consumption, MMBTU/t NH$_3$	27.0	36.0	45.5
LSTK for plant, 10^6 $	180	270	400
Total capital*, 10^6 $	250	350	500
Feedstock and energy costs, $/t NH$_3$	75.60	64.80	68.25
Utilities, $/t NH$_3$	3.78	3.24	3.41
Maintenance, $/t NH$_3$	7.51	11.26	16.68
Personnel, $/t NH$_3$	6.67	8.34	12.51
Overheads, $/t NH$_3$	12.22	17.25	25.65
Financing costs**, $/t NH$_3$	71.74	100.43	143.48
Total costs, $/t NH$_3$	177.52	205.32	269.99

*Total capital (1998) includes lump-sum turn-key price for plant and storage, spare parts, catalysts, clients, in-house costs, offsites, working capital (3-months).
**Assumed debt/equity ratio 60:40; depreciation 6%, 8% interest on debts, 16% ROI on equity.

of the major industrial countries, then a fall in the residue price could make additional capacity of this sort more attractive."

Synfuels

The economics of gas-to-liquids (GTL) projects is dependant on both a cheap source of natural gas and the capital investment. The main incentive for such projects is to provide a means of bringing gas from remote or other locations where it has little value to the world energy market. For long-term contracts at such locations, natural gas prices of around U.S.$0.70/GJ are achievable—a very different situation from gas prices in Western Europe or North America.

The investment costs for GTL projects have dropped dramatically since the first-generation (Bintulu, Mossel Bay) projects with the introduction of "second-generation" technologies. Published data indicate specific capital expenditure of U.S.$20,000–30,000 per installed bpd of liquid product capacity. Clearly, for a remote location, the cost development of a local infrastructure will be a major uncertainty in such numbers and will be highly project-specific.

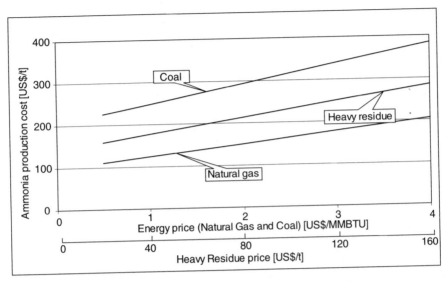

Figure 9-1. Production Cost of Ammonia from Different Feedstocks

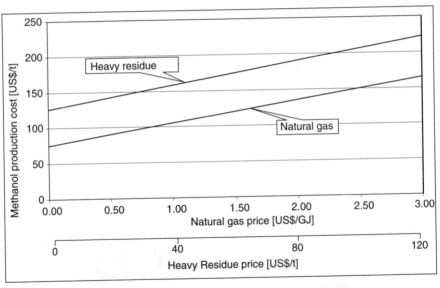

Figure 9-2. Production Cost of Methanol from Different Feedstocks

Taking this data for a 50,000 bpd project and a conversion efficiency of 8.5 GJ/bbl, production costs work out at about U.S.$23/bbl of refined product at the remote location. Given the premium quality of the product, this is a figure that can justify such projects at current oil prices, but the margins will not be spectacular. On the other hand, the existence of such plants and the experience gained with them would be a rewarding investment should oil prices rise significantly.

Power Production. As with chemical applications, the economics of power production using gasification technology is dependant on the clean utilization of cheap feedstocks. In many parts of the world the power industry is experiencing a period of change, brought about by privatization and deregulation, which does not make decision making easy. In a recent study conducted for the U.K. Department of Trade and Industry, it was found that even the cheapest technology, natural gas combined cycle (NGCC), is not viable at current U.K. electricity prices of around 0.02 £/kWh (~0.03 $/kWh) (Ricketts et al. 2002). However, such a situation cannot be expected to continue over a longer period of time, so there is considerable value to reviewing the factors, which can make gasification a competitive option. It must, however, be appreciated that there are two or even three markets that need to be considered separately. In the market for large utility plants, only coal and refinery residues can provide the feed volumes required. Biomass and waste are fuels that, for reasons connected with the logistics of the fuel supply, can only support small units (say <50 MW). At this scale, the cost of electricity production is higher than for utility-size plants; such projects can, however, attract financial support on the basis of environmental benefits or in the case of waste by charging a gate fee. The third market where gasification is showing promise is co-firing syngas from a biomass gasifier in a utility-scale plant, thus securing the benefit of scale without overloading the fuel supply logistics.

Figure 9-3 shows a typical investment cost breakdown for a coal-based IGCC. The most striking aspect of such a presentation is the approximately equal investment required for syngas production (ASU, AGR, and gasification) and for the conversion of the syngas into electricity. In a direct comparison with NGCC, this fact practically doubles the investment.

Typical investment costs for different types of new power plants have been summarized in Table 9-2.

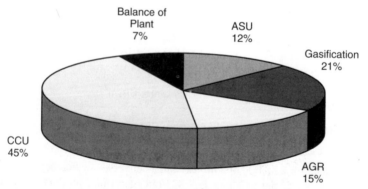

Figure 9-3. Capital Cost Breakdown for Coal-Fed IGCC (*Source: Adapted from O'Keefe, et al. 2001*)

Table 9-2
Comparative Costs for New Utility Scale Power Plant

Technology	MW	BTU/ kWh	%	$/kW
NGCC (F turbine)	239	7359	46.4	687
NGCC (G turbine)	323	6743	50.6	524
NGCC (H turbine)	335	6396	53.3	461
Subcritical PC boiler	398	9077	37.6	1129
Supercritical PC boiler	402	8568	39.8	1173
Ultra-supercritical PC Boiler	400	8251	41.4	1170
IGCC, oxygen blown	543	8522	40.0	1241
PFBC	425	8354	40.8	1190

Source: Lewandowski and Gray 2001

These figures do not take any account of environmental performance, and the typical efficiencies do not reflect the maximum achieved by current IGCCs such as Buggenum (41.4%, HHV basis). The data does show, however, that with the expectation of increasing legislative pressure to reduce mercury emissions or capture CO_2, IGCC is better placed to respond than combustion technologies. These possibilities have been discussed in Section 7.3.

Availability and Reliability. As discussed, gasification is a capital-intensive process that provides a means of utilizing cheap and sometimes unpleasant feedstocks. It is therefore vital to the successful economics of a plant to ensure that it is operating at a high rate of utilization. This quickly becomes obvious when studying the figures in the examples above. An analysis will show that 1% higher availability (i.e., three days less per year off stream) is worth almost 3% increase in efficiency. It is also worth about 30% more than a 1% lower investment cost. These are facts of life for any capital-intensive production facility, and a consciousness of this is important at every level of decision making.

Gasification has not always had good press when it comes to issues of reliability. This is partly because still today, almost all plants are one-of-a-kind units that do not incorporate the benefits of standardization, and partly because the performance of demonstration plants has been more widely publicized than that of commercial operations. The technology is, however, also demanding in terms of operation and maintenance (O & M) know-how and understanding, and a failure to recognize this at the inception of a project is a mistake that can be very costly in terms of downtime. On the other hand, there are many plants in different parts of the world that demonstrate that with the correct procedures in place, gasification can provide a reliability performance every bit as good as alternative technologies.

When discussing matters of reliability and availability, it is important to ensure that one is starting with comparable data. To assist in ensuring a common understanding, the Gasification Technologies Council has published a set of Guidelines for Reporting Operating Statistics for Gasification Facilities, based on the concepts of planned and unplanned outage and on-stream time. These guidelines are reproduced in Appendix D.

Data on commercial operation is limited, but Higman has reported on the operation of a number of liquid-feed plants (1994), including an example with a 98% on-stream factor. In another paper he reports on three Indian plants producing 110–123% of their annualized nameplate capacity (1998). Trapp (2001) reports an on-stream time of 97.8% for a coal-fed plant.

It should be recognized, however, that there is considerable difference in the performance of individual components depending on the feedstock. For resid burners, for instance, an inspection is required every 4000 hours with an anticipated repair interval of 8000–12000 hours. Compare this with the 2200 hour life reported for a coal-feed injector (Trapp 2001). Although this difference is largely related to the difference in the abrasion characteristics of the feed, there are some indications that flashing of the water from a slurry-feed exacerbates this situation (U.S. Department of Energy 2002, p. 25).

Similarly, the very different ash in resids and coals (both quantity and quality) has an important effect on refractory life. Refractories in resid service require minor repairs at 16,000-hour intervals with a full replacement on a 20,000–40,000 hour cycle (Higman 1994), whereas in coal service uncooled "refractory liners are reported to last on the order of 6–18 months" (U.S. Department of Energy 2002, p. 26).

Given this background, it is essential for the success of any gasification project to recognize all the relevant factors and build them into the design and O & M strategies from the beginning. It is not possible in a book of this nature to develop a universal algorithm to finding the appropriate strategy for any future project. It is important, however, to give an idea of successful strategies and the philosophies behind them.

Number of Trains. For very large plants, there may be a minimum number of trains dictated by the capacity of the largest available reactors. With the steady increase in unit capacities that has been visible over the last 20 years, this is only likely to be a major issue for IGCC applications.

A second consideration is the behavior of the overall plant if one reactor is out of operation. If, as is for example in an ammonia plant, there are a considerable number of centrifugal compressors, then there is an incentive to maintain the overall plant operation close to or above the surge limit of the compressors. The strength of this incentive is, however, also dependant on a number of factors. With resid feed only the downtime due to burner changes needs to be planned in. This is usually a short operation, say 8 to 12 hours if the reactor reheat is considered, so the production and energy loss is relatively small in a two-reactor configuration, especially if they are configured as two 60%–capacity units.

In a coal feed but otherwise similar situation, where the reactor is also refractory lined, one would possibly need to consider relining a reactor in between major turnarounds. This is an activity that can last as long as three weeks (again, including time for cooling down and reheating the reactor). This is intolerable, both because of loss of production as well as energy efficiency in a two reactor line-up. As a minimum, a spare offline reactor shell is required, which can be available and already lined prior to taking the operating reactor off stream. One can then simply swap the two reactors in the lined condition and restart, saving a considerable length of downtime. The lining repair can then be performed offline. We know of one plant that took this philosophy one step further and executed the drying out and part of the refractory preheating offline and swapped the (admittedly, relatively small) reactors hot, thus saving even more downtime. The implementation of such a strategy (whether the reactor swap takes place hot or cold) is, however, dependant on the layout and detail design of the facility, which must include access and other features to permit the quick removal of a reactor.

Spare Trains. In smaller plants, a strategy like the one outlined above may not be so appropriate. The saving in having two 50% or 60% versus two 100% capacity reactors decreases substantially with decreasing reactor size. Isolation of a reactor can only take place between relatively cold locations where valves can be used. This implies that, for example, a reactor including its syngas cooler has to be taken into account. Under such a situation, one can develop an operating strategy of keeping the spare reactor on hot standby and starting it prior to any planned refractory or feed injector repairs. This is the strategy employed by Eastman and is a key to their excellent reliability (Hrivnak 2001).

Operation and Maintenance. An additional key to achieving a high availability with a gasification plant is a high level of attention to detail in O & M activities (Hrivnak 2001).

9.2   ENVIRONMENTAL ISSUES

It is difficult to write much about the environment without risking becoming involved in political arguments. Although we have made no attempt anywhere in this book to disguise our views on energy policy matters, it is not our purpose here to propagate such views. This book is being written to convey the basic facts about gasification so that readers may develop their own ideas from the facts or use them in the shaping of a gasification project, should they already be involved in one.

Every country and many provinces or states have their own environmental regulations, so no general guidelines can be formulated that have universal application in detail. Important trends can, however, be observed from various countries, and reference will be made to these in the course of this section. Most environmental legislation can be grouped by the phase from which the pollutants have to be removed—gaseous, liquid, or solid. This section on the environmental impact of a gasification project will be grouped along the same lines.

9.2.1   Gaseous Effluents

The principal gaseous effluents arising from the use of fossil fuels are oxides of sulfur and nitrogen and particulate matter. Others that require attention are not fully combusted components, such as carbon monoxide.

Currently in gasification systems the first three components are all removed in the intermediate fuel gas, where sulfur and nitrogen are present as reduced species and in higher concentrations than would be the case in flue gas. Both features make removal of these species to low levels easier. In a comparison with ultra-supercritical PC technology using SCR and wet FGD, O'Keefe and Sturm 2002 shows that an IGCC power plant can achieve significantly lower emissions than the PC unit (Figure 9-4). In fact, his figures do not disclose the potential in the IGCC for further reductions should it be required, as is discussed under the specific pollutants below.

The U.S. air pollution regulations are at the time of this writing in a state of flux while the Clear Skies Initiative is being debated. Under the proposals made in the draft legislation, a coal-gasification–based IGCC would clearly fall under the same

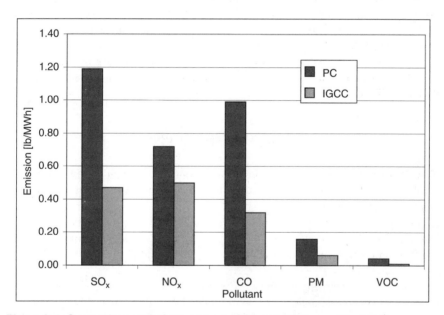

Figure 9-4. Comparison of Emissions from IGCC and PC Power Plants (*Source: O' Keefe and Sturm 2002*)

regulations as for any other coal-based technology, a situation that, at least as far as NO_x is concerned, is currently the subject of different legal interpretations. The national limits proposed are contained in Table 9-3.

German legislative limits for large liquid-feed power plants, which are largely drawn from European legislation, are shown in Table 9-4 with the comparison of values achievable from a residual-oil–based IGCC.

Sulfur emissions. Sulfur compounds have historically received the most attention when it comes to environmental impact because they are the major cause of acid rain. The fact that it is so much easier to remove sulfur as H_2S from high-pressure fuel gas than as SO_2 from the flue gas has been an important factor motivating the development of gasification-based power stations and the IGCC concept.

When serious interest in IGCC developed around 1980, it was already possible to remove 99% of the sulfur in an IGCC based power plant. In fact, the technology for sulfur removal at orders of magnitude lower than this had already been operating reliably in commercial situations for many years, whether downstream of coal gasifiers (e.g. Sasol) or of oil gasifiers (in e.g. ammonia plants). In comparison,

Table 9-3
Emissions Limits in U.S. Clear Skies Initiative

Pollutant	Emission limit
Sulfur dioxide	2.0 lb/MWh
Nitrogen oxides	1.0 lb/MWh
Particulate matter	0.20 lb/MWh
Mercury	0.015 lb/GWh

Table 9-4
Emission Limits and Oil-Based IGCC Performance

Pollutant	Emission Limits	Oil-IGCC	Regulation
SO_x, mg SO_2/m^3	400	20–40	13. BImSchV
Sulfur recovery, %	85	99.6–99.8	13. BImSchV
NO_x, mg NO_2/m^3	150	60–150	Umweltminister Konferenz 5.4.84
Particulates, mg/m^3	50	<0.5	13. BImSchV

state-of-the-art sulfur removal in a conventional power station was at this time only about 85–90%.

This situation has changed somewhat. Now 98–99% sulfur removal can also be achieved in conventional power stations with two-stage flue gas scrubbing. Dry processes can result in over 99.5% sulfur removal. Nonetheless it remains true that desulfurization of fuel gas can be achieved to any level desirable with proven technology.

For chemical applications such as methanol production, the sulfur levels in the syngas can be reduced to below 100 ppbv, which translates into some 10–12 ppbv in the flue gas or less than 0.001 lb SO_2/MWh. This is two orders of magnitude less than current expectations, as shown in Figure 9-4. Given today's regulations, this is more costly than necessary for power production, however it is an indication that gasification-based technology is ready to meet the tougher standards of the future.

Nitrogen. An important feature of coal as a fuel is its high organic nitrogen content of 1–2 wt%. Upon combustion a significant part of this nitrogen is oxidized to NO_x, which must be removed from the flue gas in an SCR unit where it is catalytically reduced with NH_3 to elemental nitrogen.

In gasification, virtually no NO_x is formed, but the organic fuel nitrogen is partially converted into HCN and NH_3. These components are removed in a water wash or— in the case of HCN—by catalytic conversion to NH_3 together with the COS. The gas combusted in the gas turbine is therefore essentially free of any nitrogen compounds except molecular nitrogen. NO_x from the gas turbine is therefore limited to thermal NO_x. Current burner designs are capable of values as low as 15 ppmv NO_x when firing syngas (Jones and Shilling 2002). Individual units are reported as having single-figure NO_x emissions (Hannemann et al. 2002). Should lower values be required, then it would be necessary to add an SCR, but at present such a demand would appear illogical, since the IGCC without the SCR can already achieve better results than any other coal-fired technology. Only firing with natural gas can achieve better values (9 ppmv). Flue gas recycle and or wet compression offer potential to reduce the thermal NO_x even further.

Mercury. Mercury is an element that is present in coals in very differing amounts and is difficult to remove. The risk to engineering materials and downstream plant equipment is described in Section 6.9.9.

Mercury emissions to the atmosphere, particularly from conventional coal-fired power plants, are causing increasing concern, and it will in the near future be subject to emissions regulations in the United States.

One of the difficulties with mercury capture from flue gas streams is the uncertainty about the distribution of the various species. In addition to elemental mercury vapor, it can exist in flue gas as an oxide, chloride, sulfide, or sulfate, the proportions depending on the levels of other contaminants in the coal. Thus not every mercury capture technology is suitable for every fuel.

The IGCC concept has a natural advantage over conventional combustion technologies in the removal of mercury. Removal to over 90% from synthesis gas has been demonstrated at Eastman Chemical Company's coal-to-chemicals facility in Kingsport, Tennessee, since 1983 using technology developed and used as standard for natural gas applications (Trapp 2001). In fact, only difficulties with measurement in the ppb range have prevented a determination whether even higher removal efficiencies are achieved or not. Certainly no product contamination has ever been detected. Against this is the continued recognition that "no single technology has been proven that can uniformly control mercury from power plant flue gas emissions in a cost-effective manner" (U.S. Department of Energy 2003). In a recent investigation into the costs of mercury removal from flue gas, EPRI estimated the costs of 90% sulfur removal at U.S.$2.80 to $3.30 per MWh (Chang 2001). Rutkowski, Klett, and Maxwell (2002) have compared this with the application of sulfur-impregnated activated carbon beds, as in the Eastman plant to a 250 MW$_e$ IGCC configuration. The resulting cost obtained was U.S.$0.254 per MWh.

An additional aspect to consider is the clear destination of the mercury on removal from syngas with activated carbon whereas research is still in progress to determine the fate of mercury removed from flue gas by various techniques. In conventional coal-fired power stations equipped with flue gas scrubbing, there is concern that mercury may end up in the gypsum wallboard, Portland cement, or manufactured aggregates that are produced. This issue is being addressed in leachability studies on a broad range of solid by-products and wastes (Schwalb, Withum, and Statnick 2002). Final clarity on this issue is unlikely, however, until commercial introduction of mercury capture.

Arsenic. Arsenic is currently not regulated as an emission, although there is increasing concern about it. Arsenic is only present in coals and mainly in those coals that have a high pyrite content. Under reducing conditions, compounds of these elements are volatile and are entrained in the gas of slagging gasifiers. When fuel gas treating is applied, compounds of these elements will eventually gather in the water treatment plant and end up in the settler/filter cake of the flocculation section that has to be considered as chemical waste. Alternatively, where raw gas shift is applied, it will deposit on the catalyst. In the case of flue gas treating, the nonvolatile As_2O_3 will deposit in any filter or end up in the gypsum.

9.2.2 Greenhouse Gases

The immediate and apparent effects that sulfur and particulate matter have on human health and SO_x and NO_x have on the world around us (forests dying from acid rain), have provided a strong motivation for the substantial progress that has been made in reducing these emissions from both the power and the transport sectors. In contrast, the potential damage of CO_2 emissions is a long-term issue for which the mechanisms are not fully understood, and this has led to the fact that no clear strategy has emerged to counter the problem despite the intensive debate over the last 10 or 15 years. Nonetheless, there is no disputing the correlations between global average temperatures and atmospheric CO_2 concentrations (as determined from Antarctic ice cores) over hundreds of thousands of years, and between anthropogenic emissions and atmospheric CO_2 concentrations over the last 250 years (Figures 9-5A and 9-5B). And whether we understand the mechanisms in detail or not, it is certainly wise to take measures to reduce man-made contributions to global warming.

Before discussing what place gasification could have in any such strategy for CO_2 emissions reduction, it may be useful to have a look at the overall greenhouse gas effect as related to the use of fossil fuels.

As can be seen from Figure 9-6, the two biggest contributors to CO_2 emissions are the electric power and transport sectors. Given the millions of small moving emission sources involved in transport, any significant reductions in CO_2 emissions is only likely to emerge through the change to a less carbon-intensive fuel such as natural gas or hydrogen. Although the use of natural gas (at least while it lasts) would still result in large, even if lower, emissions from millions of individual sources, the use of hydrogen offers the possibility of bundling CO_2 emissions at fixed locations in a manner that would allow fixation or sequestration. Further discussion of this aspect is discussed after first looking at the largest CO_2 emitter, the power sector.

A major contribution to lowering CO_2 emissions in the power industry could be realized almost immediately simply by increasing the efficiency of the power park. Replacement of a 30-year-old coal-fired unit operating at, for example, 32% efficiency by a modern plant (whether with IGCC or ultra-supercritical PC technology) using the same fuel with an efficiency of 43% would drop the CO_2 emissions per produced kilowatt-hour by 25%. The barriers here are not of a technological but of a financial nature given the limited capital and incentive for such replacement projects. Again, the reductions offered by cogeneration (or combined heat and power), although ultimately limited by the available heat sinks, face at present financial rather than technical hurdles.

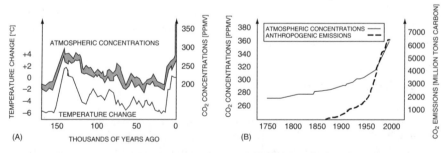

Figure 9-5. (A) Correlation Atmospheric CO_2 Concentration and Temperature (*Source: Simbeck 2002*); (B) Correlation between Anthropogenic CO_2 Emissions and Atmospheric CO_2 Concentrations (*Source: U.S. Oak Ridge National Laboratory*)

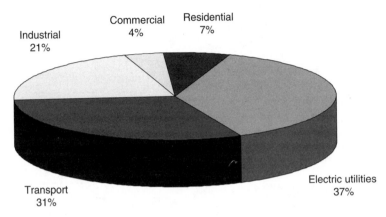

Industrial
21%

Commercial
4%

Residential
7%

Electric utilities
37%

Transport
31%

Figure 9-6. Share of U.S. CO_2 Emissions by Sector (*Source: U.S. EPA 2002*)

Beyond the above, serious consideration is now being given to recovery of the CO_2 from the energy conversion process and putting it to use or simply sequestering it. In such a scenario there are some natural advantages to gasification over combustion technologies. In fact, CO_2 capture and usage is already standard practice in almost all gasification-based ammonia plants, where the CO_2 is recovered from the synthesis gas and used to manufacture urea.

Another example of CO_2 recovery from a coal gasification plant is the Great Plains SNG plant, which sells CO_2 for enhanced oil recovery (EOR) in the Weyburn field in Canada. EOR is particularly attractive for CO_2 usage in that besides the avoidance of adding to atmospheric CO_2, it provides a means of extending the life of existing energy resources. EOR has been practiced in the Permian basin oil fields in Texas for over 30 years, although much of the CO_2 used comes from natural underground CO_2 reservoirs and therefore does not as such contribute to greenhouse gas abatement. Nonetheless, it points the way to one potential route for sequestration. Other alternatives have been or are being investigated as well, such as sequestration in underground saline aquifers (e.g., Statoil's Sleipner project in the North Sea), storage in coal seams, and others (White 2002; Beecy 2002).

As mentioned above, gasification-based processes have a natural advantage over combustion technologies when it comes to CO_2 capture. In one study based on using gasification with total water quench technology at 80 bar, the addition of CO_2 capture was shown to add only a small increment on capital costs (about 5%) and an efficiency penalty of only 2% (O'Keefe and Sturm 2002). The block flow scheme is shown in Figure 9-7.

This should be compared with the impact of CO_2 capture on a ultra supercritical PC combustion technology power block. This is extremely heavy even if one assumes the prior existence of FGD and SCR flue gas treatment, which are a prerequisite for CO_2 recovery from the flue gas, for which at present amine scrubbing is

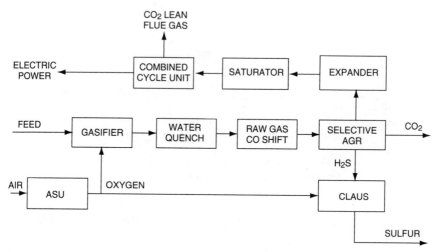

Figure 9-7. IGCC with CO₂ Capture

universally proposed. The key fact is that the volume of gas from which the CO_2 has to be extracted is in the case of combustion technology 150–200 times larger than is the case in an IGCC plant. The add-on capital cost has variously been estimated at 60–80%. The steam requirement for amine regeneration is lost to the final turbine stages, and this causes a drop in efficiency of some 9.5–14 percentage points (Simbeck 2002; Koss and Meyer 2002).

Clearly there is an incentive to find some use for the CO_2 and "recycle" it. Although there may be local markets where this is possible on a small scale, a review of the data in Table 9-5 will show that the "chemical" scope for such a solution on a global scale is very limited. Apart from oxygen and hydrogen, there are no potential reactants with relative mass flows approaching that of carbon. Even iron usage is an order of magnitude smaller. It is further observed that the only mass in the world that is of the same order of magnitude as fossil fuels is waste biomass flow (though this is not valid for the energy, which is an order of magnitude less) (Shell 2002).

The only major application of waste CO_2 is for the enhanced recovery of oil and gas, but most of the CO_2 will have to be sequestered underground in depleted gas fields, in aquifers, and in deep-sea basins (van der Burgt, Cantle, and Boutkan 1992).

An alternative option for removing CO_2 from flue gases from power stations that has been proposed is combustion with pure oxygen. The flue gas will then contain only water and CO_2, and in the case of coal and heavy oil fractions, also SO_2, NO_x, and other contaminants such as mercury. The advantage is that in this case all contaminants can be sequestered together, and apart from the ASU, no gas separations are required. However, with the present cryogenic separation of air, this solution is unlikely to become attractive for both economic and efficiency reasons. Moreover, in

Table 9-5
Important Mass Flows in the World

	Ton/y	Relative Mass Flow
Fossil fuels	10×10^9	100
Carbon	8×10^9	80
CO_2 emissions	30×10^9	300
Chemicals	3×10^8	3
Ceramic building materials	$10-15 \times 10^8$	10–15
Iron	5×10^8	5
Carbon for iron ore reduction	3×10^8	3
Carbon for SiO_2 reduction, 100,000 MW_e		
peak/year new installed (1 mm thick wafers)	1×10^6	0.01
Waste biomass	$5-10 \times 10^9$	50–100

Source: Shell 2002

the case of CC power stations, this route would require the development of gas turbines that are optimized to handle pure tri-atomic gases.

It should be noted that the IGCC variant provides a source of hydrogen that, with a combination of membrane and PSA technologies, can be extracted during periods of low electric power demand and stored for onward sale into the transport sector. The importance of this possibility is discussed further in Chapter 10.

Methane. On a mole for mole basis, methane contributes 20 to 25 times more to the greenhouse effect than CO_2. Anthropogenic methane from fossil sources enters the atmosphere in the form of vented or incompletely combusted associated gas, leaks in natural gas pipelines, and methane emissions related to coal mining.

The largest source of methane emissions resulting from fossil fuel production and refining is caused by venting associated gas. On average about 10% of the energy leaving an oil well is in the form of associated gas, and assuming that 5% of this gas is not combusted either because people do not want to see flares or because flares are not working properly, it is conservatively estimated that about 8% of the greenhouse gas effect related to the use of crude oil is due to methane emissions during production. Adding to this the effect of the CO_2 from the 95% of the associated gas that is properly combusted in operating flares, this figure increases to 15%. Schaub and Unruh (2002) have estimated that the quantity of associated gas flared in Nigeria alone would be sufficient to supply 30% of Germany's natural gas supply. Clearly, there is potential here for a reduction in greenhouse gas emissions, to which synfuels production via partial oxidation of the gas and subsequent Fischer-Tropsch synthesis could make a significant contribution. Global application of such a solution could reduce the 15% of the greenhouse gas effect related to the use of crude oil to a mere 2–3%.

9.2.3   Liquid Effluents

Whether a gasification complex is a net water consumer or producer will depend on the feedstock and the downstream operation. In general, when coal or petroleum residues are used as feedstock, there is always a net consumption of water, whereas with natural gas–based plants, there can be a net production of water. Similarly, the application, chemicals or power, will have an influence over the overall water balance of the plant.

Most gasification plants have a water wash at some part of the syngas treatment. For coal gasifiers, its main purpose is to remove ammonia and chlorides, though many other constituents of the gas are also captured. Whether this process water is acidic or basic will depend on the amounts of nitrogen and chlorine in the coal. Whatever the pH of this water, it will almost certainly require to be adjusted during the flocculation step of the overall water treatment. In oil gasification plants, the main objective is removal of soot, but the water also contains ammonia, hydrogen cyanide, and H_2S.

In practice, with appropriate design, an IGCC can be made with zero liquid discharge, whether using a dry feed technology as in Buggenum (Coste, Rovel, and George 1993) or a slurry feed as in Polk (U.S. Department of Energy 2000). In both these plants the final water treatment stage is a brine concentration and evaporation unit producing a solid waste salt.

Oil gasification units generally do not recycle the excess process water. In those applications, where the plant is located in a refinery, the water treatment is generally integrated into that of the overall refinery after ammonia, HCN, and H_2S have been removed in a sour water stripper. Typical emission limits and performance for a stand-alone plant are given in Table 9-6

Specific aspects of water treatment are addressed in the following sections.

Table 9-6
Emission Limits and Oil-Based IGCC Performance

Pollutant	Emission Limits	Oil-IGCC	Regulation
Vanadium (mg/l)	2	<2	City of Hamburg
Nickel (mg/l)	0.5	<0.5	Rahmen Abwasser VwV 1992
BOD_5 (mg/l)	25	<20	Rahmen Abwasser VwV 1992

Fluorine. Fluorine in the gas will dissolve in the wastewater from the water wash, from which it can be removed by adding calcium ions that will precipitate the fluorine as CaF_2. This salt will eventually end up in the same settler/filter cake as the heavy metal-containing precipitate from the flocculation unit.

Cyanide and Cyanometallates. A significant portion of the HCN produced in the gasifier is contained in process condensate. In any plant gasifying heavy oil, the excess water from the gasification section is usually stripped to remove free ammonia, H_2S, and HCN. Typically, residual HCN values of 10–20 mg/l can be achieved in a single-stage stripper. The residual HCN can be reduced to below 1 ppm in a biological treatment unit.

An alternative approach to cyanide and cyanometallates is to oxidize them. This is particularly appropriate when recycling all the process condensate for a zero-discharge system. Although aeration in a closed vessel is possible and has been employed, the required contact times are long and require large volumes. A more economic approach is the use of ozone as an oxidant, since this reacts rapidly with the cyanides (Coste, Rovel, and George 1993).

Heavy Metals Precipitation. Whether gasifying coal or heavy residues, there are heavy metals contained in the process condensate. Typically, these are subjected to treatment by flocculation and precipitation. The choice of flocking agent is determined by the metals to be removed, but is typically ferric chloride. The metals sludge from the precipitation stage is then thickened prior to filtration.

The excess water so treated would be "free of fluorides, cyanides and heavy metals and contain only dissolved soluble salts" (Coste, Rovel and George 1993). Sand filtration may be required to remove traces of metal hydroxides from the flocking step.

9.2.4 Solid Effluents

Solid effluents from gasification plants are essentially related to the ash in the feedstock, the quantities of which can vary from as much as 40 wt% in some coals to under 1 wt% for petroleum feeds. In many cases gasification plants are able to make the solid residue available in a form that can be used as a raw material in other industries.

In contrast to a PC boiler, gasification has no FGD sludge or gypsum disposal problem because the sulfur is all captured as elemental sulfur in the Claus plant. At present this is a saleable product, although one should be aware that, were a large part of the power industry to switch to gasification-based processes, this together with increasing sulfur production from oil refineries would oversaturate the market. Nonetheless, even in this case the amounts of material would be substantially smaller than with gypsum.

Dry Coal Ash. Dry ash from nonslagging gasifiers is essentially the same material as the product of a PC boiler processing the same feed. This ash can often be utilized in the cement or building industry. If this option is not available due to the nature of the ash or due to transport problems, the ash can become a major environmental liability that can be aggravated if the ash is leachable or caustic in nature. The ash is then often stored in large ponds, which have become more and more unacceptable. This problem can be largely avoided by selecting a slagging gasifier from which the quantity of dry ash waste is one to two orders of magnitude lower than from a dry ash gasifier.

Slag. Slagging gasifiers have in general an advantage over dry ash processes in that most of the ash components leave the gasifier as molten slag, which on being quenched turns into a fine inert gritty material that can be used as a replacement for sand and aggregate in concrete, for example. Van Liere, Bakker, and Bolt (1993) give the particle size of water-quenched slag as being about 78–80% in the range 0.5–4 mm in a paper in which they describe usage trials in Dutch road construction. Data from Geertsema, Groppo, and Price (2002) generally supports this size data for the actual slag component, although the material analyzed contained considerable quantities of unreacted carbon, which after separation are recycled to the gasifier or used as fuel in a PC boiler. They report the slag as being used for blasting grit and roofing granules. Amik and Dowd (2001) provide an analysis for slag produced at the Wabash River plant (Table 9-7), which essentially mirrors the ash content of the coal used. The slag contains trace metals such as lead, arsenic, selenium, chrome, antimony, zinc, vanadium, and nickel, which is captured in the nonleachable glassy matrix. The results of some leachability tests are included in the paper. They report marketing for asphalt, construction backfill, and landfill cover applications.

Table 9-7 Slag Analysis	
SiO_2	51.8 %
Al_2O_3	18.7%
TiO_2	0.9%
Fe_2O_3	20.3%
CaO	4.2%
MgO	0.8%
Na_2O	1.0%
K_2O	1.9%
Source: Amik and Dowd 2001	

Heavy Metals. The nature of any heavy metals leaving the plant as solid effluent will depend heavily on the feedstock. With a coal feed, most heavy metals will end up in the slag or fly ash of a gasifier. Heavy metals that are not removed together with the fly ash and slag will eventually end up in the wastewater from the water wash. As described on page 345 the heavy metals are removed from the water as a filter cake after treatment by flocculation and precipitation. With coal feeds this filter cake may contain such elements as arsenic, antimony or selenium. Unlike the slag, this concentrate is not inert and has to be considered as chemical waste.

In the case of oil feeds, the main heavy metals are vanadium, nickel, and iron. There are a number of processes available for recovering the ash for use in the metallurgical industry (see Section 5.4). The economics of such processes depends very much on the vanadium content in the ash.

Salt. All coal gasifiers will have dissolved salts in the treated wastewater. When this water cannot be disposed of, the water has to be evaporated and a salt mix is obtained that consists of over 98% NaCl. This product is often not suitable as road salt and may have to be considered as chemical waste.

9.2.5   Waste Gasification

For reasons that are mostly historical, emissions regulations for thermal treatment of waste are generally much more stringent than for, say, a coal-based power plant. Waste gasification plants, particularly slagging units, have the potential to provide a viable alternative to incineration. The current review in the United States on removing refinery wastes that are processed to syngas from the hazardous classification is, however, typical of the moving targets in this field. A number of plants, such as Schwarze Pumpe, provide a local or regional waste disposal center in which waste is gasified and the syngas used for production of methanol and electricity. Here the solid slag is nonleachable and meets the German municipal waste standards "Siedlungsabfall class 1" (Greil et al. 2002).

9.3   SAFETY

Issues of safety are associated with practically all industrial technologies, and understanding the appropriate measures for safety management is an important part of understanding the technology itself. In this respect, gasification is no different from many other technologies. The purpose of this section is to make those unfamiliar with gasification aware of those issues that may be considered specific to the technology. It is certainly no substitute for a project-specific safety manual and does not address issues common to any large-scale industrial plant. It does, however, point to other sources, which may help the reader develop his or her own ideas for a particular project in more detail.

Gasification plants are complex plants that produce a high-pressure toxic gas that is inflammable or even explosive in the presence of oxygen and an ignition source. Furthermore, the presence of pure oxygen, as is the case in most gasification processes, requires additional precautions. Generally, all these dangers are well taken care of in the process designs. Handled correctly during construction, operation, and maintenance, they pose no more problems to personnel or environment than many other industrial plants.

9.3.1 Start-Up

One of the potentially dangerous moments during start-up is the ignition of the coal or oil burners. The problem is much more complex than in regular atmospheric pressure furnaces, as ultimately the burners have to work at pressures ranging from 20–70 bar, and there are no burners that can operate properly over this whole pressure range. Where membrane walls are used, which cool very quickly in the absence of a flame, start-up burners often have to be used to cover part of this pressure range. At all times the situation should be avoided in which a mixture of a combustible gas and oxygen is present in the reactor.

9.3.2 Shutdown

After shutdown the gasifier is often nitrogen blanketed to avoid corrosion. When repairs have to be carried out inside the gasifier, one has to make sure that there are no other gases than air present. Drawing a good vacuum and breaking this with air is the best way to ensure that only air is present. This operation may have to be repeated several times to ensure that all noxious gases are removed from insulating materials, bricks, and dead ends in the plant. Even with all these precautions, air masks may be required under certain conditions.

9.3.3 Spontaneous Combustion

As in a conventional PC power plant, safety precautions in coal storage and handling are essential to avoid spontaneous combustion. This applies equally well to other feedstocks such as biomass and certain types of waste. The key in particular is to prevent fines from drying out, so that a first-in first-out inventory policy must be part of the safety procedures.

Besides the fuel itself, there are other potential sources of spontaneous combustion. Particular attention must be paid to FeS which may form as a product of corrosion. Another potential source is incorrect handling of catalysts, particularly unloading spent catalyst, which may not have been adequately oxidized in situ. Acting in strict conformity with the catalyst vendors' procedures is important.

9.3.4 Toxic and Asphyxiating Materials

Apart from the main syngas component, carbon monoxide, there are many other toxic gases present in a gasification complex, particularly if the end product is a chemical. Typical toxic gases present in synthesis gas can include compounds such as H_2S and COS as well as ammonia and HCN. The design of a plant must take account of this, and personnel must be trained in their safe handling. There are many public sources of safety information available on material safety data sheets. Many of these are available from Internet sources such as www.ilpi.com/msds, which has links to many international source sites. An up-to-date data sheet should always be available with the safety officer or other member of staff responsible for safety training.

Nitrogen. It may come as a surprise, but a large part of the accidents occurring in gasification plants are due to nitrogen that is produced as a (by-)product from oxygen plants and used for blanketing and transport of coal and further as a diluent for the fuel gas. In this latter function it reduces the stoichiometric adiabatic flame temperature and as a result the thermal NO_x.

The problem with nitrogen is that contrary to the fuel gas it has no smell, and even more problematic, it leads very fast to unconsciousness. Good ventilation of the plant is a very efficient precaution. Building a gasifier inside a closed structure because it may look better from a distance can be dangerous. Not only because of nitrogen, but also because we deal with pressurized gases containing CO, H_2S, COS, HCN, and NH_3. If the plant has to be visually enclosed, the best alternative is to have louver walls that guarantee good ventilation.

CO_2. Where concentrated CO_2 streams are present, one should be aware of its asphyxiating properties and the fact that it is heavier than air. The potential danger is not only because of leaks and open valves. Also, the gas in the stacks through which the CO_2 is vented should have sufficient buoyancy by ensuring elevated temperatures. For all large quantities, dispersion calculations should be made.

9.3.5 Oxygen

Oxygen makes up about 21% of our atmosphere and is essential to life. It is also an essential ingredient for combustion in which fuels are oxidized in an extremely exothermic reaction. If oxygen is present in concentrations significantly above 21%, then the combustion becomes much more vigorous, and materials such as metals that normally oxidize in a slow manner without fire risk (e.g., rusting of iron) can behave as fuels for fire. When handling oxygen it is therefore essential to take the necessary precautions to prevent oxygen fires.

The precautions necessary for safe operation of oxygen systems are well codified. Not only do all the leading industrial gas supply companies and gasification technology suppliers have their own strict safety regulations, but also trade associations such as Compressed Gas Association (CGA), British Compressed Gas Association (BCGA), and European Industrial Gas Association (EIGA), in which these companies and other major operators of oxygen systems are represented, publish codes of practice based on the joint experience of all their members.

Many substances such as oils and grease will combust spontaneously in the presence of pure oxygen. The energy of impact of small particles on many metals is sufficient to cause ignition. Fires initiated by both these causes are sufficient to ignite the primary material of construction. The principle means of combating these dangers lies in meticulous cleaning of the system prior to the introduction of oxygen. All safety guidelines for oxygen systems include recommendations for cleaning and inspection after cleaning, such as CGA publication G-4.1.

An important aspect of safety precautions for oxygen service is material selection and system geometry. Materials are selected to keep the ignitability of the material and its capability of sustaining a fire in an oxygen atmosphere to a minimum. Typically, copper-based materials (e.g., Monel) or stainless steels are used in high-pressure noncryogenic systems. Inside the coldbox, where low temperature suitability is also a criterion and pressures are limited, aluminum is also used. For pressures up to 40 bar, carbon steel may also be used. Cleaning in such applications is, however, extremely important.

System geometry is also important. Velocities in oxygen lines are generally kept low as a measure to limit the energy release on impact of any particles in the system. An additional measure to limit the ignition risk is to avoid sharp bends in piping where turbulence can increase local velocities much above these limits. For this reason also much attention is required to the design of valving and piping at pressure letdown stations.

An additional approach to safety in oxygen systems is to incorporate design features, which ensure that personnel is not put at risk and that any material damage in the event of a fire is kept to a minimum. This type of precaution is taken with oxygen compressors, which are enclosed by a fireproof wall. There is an extensive monitoring system that, on detection of a fire risk, will cause the machine to be stopped, depressurized, and flooded with nitrogen (e.g., EIGA code of practice). The cooling water circuit for intercooling is usually a closed-circuit system to avoid any potential corrosion and subsequent leaks on the intercoolers.

An important safety aspect to consider is the quality of the air entering the air separation unit. Modern molecular sieve PPUs will generally remove heavy hydrocarbons present in the air. Cases are known where the concentration of hydrocarbons in the atmosphere increased substantially over the life of the ASU and overloaded an internal hydrocarbon filter inside the coldbox, breaking through into the oxygen-rich environment of the LP column. In one case known to us, results of ethylene leakage from a nearby plant were detected in time and the filter was enlarged to cope with the new air quality. In another, the mechanism was more complicated and an explosion resulted (van Hardeveld et al. 2001). Such incidents do, however, illustrate the need to specify the feed air quality conservatively and with an eye to future developments.

REFERENCES

Agee, M. "The Natural Gas Refinery." *Petroleum Economist* (January 2002).

Amik, P., and Dowd, R. "Environmental Performance of IGCC Repowering for Conventional Coal Power Plants." Paper presented at Gasification Technologies Conference, San Francisco, October 2001.

Appl, M. *Ammonia: Principles and Industrial Practice*. Weinheim: Wiley VCH, 1999.

Beecy, D. "U.S. Carbon Sequestration RD&D Program." Paper presented at Gasification Technologies Conference, San Francisco, October 2002.

BP. *BP Statistical Review of World Energy*. London: BP PLC, June 2002.

Chang, R., and Offen, G. "Mercury Control Options." *Modern Power Systems* (November 2001).

Clayton, S. J., Siegel, G. J., and Wimer, J. G. "*Gasification Technologies: Gasification Markets and Technologies—Present and Future: An Industry Perspective*." U.S. DOE Report DOE/FE-0447, U.S. DOE:July 2002.

Compressed Gas Association. "Cleaning Equipment for Oxygen Service." Publication G-4.1, 4th ed. Arlington, VA, 1996.

Coste, C., Rovel, J. M., and George, J. M. "Effluent System in View of Both Zero Discharge and Hazardous Solid Waste Minimization." Paper presented at VGB Conference, Buggenum IGCC Demonstration Plant, Maastricht, November 1993.

European Industrial Gas Association. *Centrifugal Compressors for Oxygen Service. Code of Practice*. IGC Doc 27/01/E Brussels: EIGA, 2001.

Forte, R., Porter, F., Tavoulareas, E. S., and Ratafia-Brown, J. "Coal Gasification and Air Emissions." Paper presented at 19th Annual International Pittsburgh Coal Conference, September 2002.

Geertsema, A., Groppo, J., and Price, C. "Demonstration of a Beneficiation Technology for Texaco Gasifier Slag." Paper presented at Gasification Technologies Conference, San Francisco, October 2002.

Greil, C., Hirschfelder, J., Turna, O., and Obermeier, T. "Operating Results from Gasification of Waste Material and Biomass in Fixed Bed and Circulating Fluidized Bed Gasifiers." Paper presented at IChemE Conference, "Gasification: The Clean Choice for Carbon Management," Noordwijk, April 2002.

Hannemann, F., Schiffers, U., Karg, J., and Kanaar, M. "V94.2 Buggenum Experience and Improved Concepts for Syngas Application." Paper presented at Gasification Technologies Conference, San Francisco, October 2002.

Higman, C. A. A. "Gasification: An Indian Perspective." Paper presented at IChemE Conference "Gasification: The Gateway to a Cleaner Future," Dresden, September 1998.

Higman, C. A. A. "Methanol Production by Gasification of Heavy Residues." Paper presented at IChemE Conference, "Gasification: An Alternative to Natural Gas," London, November 1995.

Higman, C. A. A. "Perspectives and Experience with Partial Oxidation of Heavy Residues." Paper presented at AFTP Conference, "L'Hydrogène, Maillon Essentiel du Raffinage de Demain," Paris, 1994.

Hrivnak, S. "Fine-Tuning to Improve Availability and Reliability of Coal-Based Gasification." Paper presented at Gasification Technologies Conference, San Francisco, October 2001.

Jones, R. M., and Shilling, N. Z. "IGCC Gas Turbines for Refinery Applications." Paper presented at Gasification Technologies Conference, San Francisco, October 2002.

Koss, U., and Meyer, M. "Zero Emission IGCC with Rectisol Technology." Paper presented at Gasification Technologies Conference, San Francisco, October 2002.

Lewandowski, D., and Gray, D. "Potential Market Penetration of IGCC in the East Central North American Reliability Council Region of the U.S." Paper presented at Gasification Technologies Conference, San Francisco, October 2001.

O'Keefe, L. F., and Sturm, K. V. "Clean Coal Technology Options—A Comparison of IGCC vs. Pulverized Coal Boilers." Paper presented at Gasification Technologies Conference, San Francisco, October 2002.

O'Keefe, L. F., Weissman, R. C., DePuy, R. A., Griffiths, J., East, N., and Wainwright, J. M. "A Single Design for Variable CO_2 Capture." Paper presented at Gasification Technologies Conference, San Francisco, October 2001.

Ricketts, B., Hotchkiss, R., Livingston, B., and Hall, M. "Technology Status Review of Waste/Biomass Co-gasification with Coal." Paper presented at IChemE Conference, "Gasification: The Clean Choice for Carbon Management," Noordwijk, 2002.

Rutkowski, M. D., Klett, M. G., and Maxwell, R. C. "The Cost of Mercury Removal in an IGCC Plant." Paper presented at 19th Annual International Pittsburgh Coal Conference, September 2002.

Schaub, G., and Unruh, D. "Synthetische Kohlenwasserstoff-Kraftstoffe und Minderung fossiler CO2-Emissionen." VDI-Berichte Nr. 1704, 2002.

Schmidt, W. P., Winegardner, K. S., Dennehy, M., and Castle-Smith, H. "Safe Design and Operation of a Cryogenic Air Separation Unit." Paper presented at AIChE Annual Loss Prevention Symposium, Houston, April 2001.

Schwalb, A. M., Withum, J. A., and Statnick, R. M. "The Re-evolution of Mercury from Coal Combustion Materials and By-Products." Paper presented at 19th Annual International Pittsburgh Coal Conference, September 2002.

Shell. *Venster*. The Hague (November/December 2002), p. 9.

Simbeck, D. "Overview of Climate Change Policy and Technical Developments." Paper presented at Gasification Technologies Conference, San Francisco, October 2002.

Trapp, W. L. "Eastman and Gasification: The Next Step, Building on Past Success." Paper presented at Gasification Technologies Conference, San Francisco, October 2001.

U.S. Department of Energy. "Mercury Emissions Control." Available at: www.netl.doe.gov/coalpower/environment/mercury/description.html, January 2003.

U.S. Department of Energy. "Gasification Markets and Technologies: Present and Future." Report DOE/FE-0447, July 2002.

U.S. Environmental Protection Agency. "Inventory of Greenhouse as Emissions," April 2002.

U.S. Department of Energy. "The Tampa Electric Integrated Gasification Combined-Cycle Project: An Update." U.S. DOE Topical Report No. 19, July 2000.

van der Burgt, M. J., Cantle, J., and Boutkan V. K. "Carbon Dioxide Removal from Coal-Based IGCCs in Depleted Gas Fields." *Energy Conversion Management* 33 (5–8) (1992): 603–610.

van Hardeveld, R. M., Groeneveld, M. J., Kehman, J.-Y., and Bull, D. C. "Investigation of an Air Separation Unit Explosion." *Journal of Loss Prevention in the Process Industries* 14 (2001):167–180.

van Liere, J., Bakker, W. T., and Bolt, N. "Supporting Research on Construction Materials and Gasification Slag." Paper presented at VGB Conference, "Buggenum IGCC Demonstration Plant", Maastricht, November 1993.

Weigner, P., Martens, F., Uhlenberg, J., and Wolff, J. "Increased Flexibility of Shell Gasification Plant." Paper presented at IChemE Conference, "Gasification: The Clean Choice for Carbon Management," Noordwijk, 2002.

White, D. J. "CO_2 Capture and Sequestration: European Policy Drivers." Paper presented at Gasification Technologies Conference, San Francisco, October 2002.

White, D. J. "Gasification: The Key to a Cleaner Future." London: Financial Times Energy, 2000.

Chapter 10

Gasification and the Future

The future of gasification is intimately intertwined with the future of energy and energy policy. It is generally recognized that human development cannot continue to base its economy on fossil fuels in the present manner forever, even if viewpoints on the timescale do diverge, sometimes dramatically. This viewpoint is put most strongly by the advocates of what is called the "hydrogen economy." There is no doubt that the use of hydrogen in combination with fuel cells as a transport fuel will improve the microclimate of our conurbations significantly through the elimination of CO_2, NO_x, CO, and hydrocarbon emissions from motor vehicles. And this is a prospect that could become reality within the next 20 years. However, it is our opinion that those proponents, who present the hydrogen economy as a solution to the CO_2 emissions or "greenhouse gas" issue, overstate their case. The hydrogen that we will use in our fuel cells is not ready and waiting for us to collect. It is chemically locked into other substances, the principle of which are water and to a lesser extent natural gas.

The issue remains, therefore, how to unlock this hydrogen and make it available in a useable form. There are essentially three routes to hydrogen production: electrolysis of water, steam reforming of natural gas, and gasification—whereby the fuel for the gasification can be anything from coal to biomass. Thus, as can be seen from Table 10-1, unless the power for electrolysis is generated without CO_2 emissions, hydrogen production is inevitably associated with CO_2 production. Furthermore, it has to be recognized that with the possible exception of nuclear energy, no CO_2-free power-generation technology is in the medium-term going to produce hydrogen in the quantities required to supply our transport needs. We therefore need to look at the potential for a reduced CO_2 technology to help us on the way to a "no carbon" energy future.

It is our opinion that in the transition between fossil fuels and a fully "renewable world," gasification can play an important role. First, in the move toward a hydrogen economy, one can expect that the hydrogen will be produced directly from fossil fuels rather than by electrolysis. Furthermore, during this transition period the implementation of polygeneration units producing both power and hydrogen will allow a gradual change over from the former to the latter. This is the only way in which an investment in power generation can be utilized for hydrogen production at

Table 10-1
CO₂ Emissions for Hydrogen Production Technologies

Technology	kg CO_2 Emissions per Nm^3 H_2
Electrolysis with conventional coal combustion	2.6
Biomass gasification	1.7–2.0
Coal gasification	1.0–1.4
Steam reforming of natural gas	0.8
Electrolysis with nuclear power	0

moderate cost (Simbeck and Chang 2002). Second, gasification is a key technology for more efficient power generation from coal and heavy oils with the best environmental performance. And third, gasification provides the best option for producing concentrated carbon dioxide streams that may have to be sequestered during the transition in order to reduce the emission of greenhouse gases.

Of course, the above remarks are not restricted to coal but apply to the gasification of any fossil fuel. Furthermore, they apply also to what may in the very long-term become the most important feedstocks, biomass, and waste—that may, in a totally "sustainable future," have to take over the role of today's fossil fuels.

What is seldom mentioned is that even in a "sustainable world" not only energy is required but also carbon for organic chemicals including plastics. Although gasification of waste may supply part of this requirement, the make-up will have to come from biomass, which in this idealized "sustainable world" model is the only allowable source of concentrated carbon (van der Burgt 1997). The only way to produce organic chemicals from waste and biomass is first to gasify them in order to make synthesis gas. At present, there is no process available to do this efficiently, as all biomass and waste gasification processes to date have been developed for producing fuel gas and power. Hence, although in a more renewable world hydrogen (by electrolysis) and electricity may be available, gasification of waste and biomass—directly or indirectly via bio-oil—are at least required to make synthesis gas for organic chemicals. The first generation of downstream technologies to allow the use of syngas (via methanol) as an alternative source for ethylene and propylene instead of the conventional naphtha cracking is already in the demonstration stage (UOP 1997; Holtmann and Rothaemel 2001).

We therefore conclude that gasification can and will have an important role to play in the coming decades, both for power generation and for the production of bulk chemicals. In the more distant future it may also develop to become an important source of base materials for all organic chemicals. It is hoped that this book will contribute to the development of a better understanding of gasification processes and their future development.

REFERENCES

Holtmann, H.-D., and Rothaemel, M. "A Cost-Effective Methanol to Propylene Route." *Petroleum Technology Quarterly* (Autumn 2001).

Simbeck, D., and Chang, E. "Hydrogen Supply: Cost Estimate for Hydrogen Pathways— Scoping Analysis." National Renewable Energy Laboratory report. NREL/SR-540-32525 (July 2002).

UOP. "UOP/Hydro MTO Process: Methanol to Olefins Conversion." UOP Company leaflet, 1997.

van der Burgt, M. J. "The Role of Biomass as an Energy Carrier for the Future." *Energy World* 246 (February 1997).

Appendix A

Companion Website

For those interested in following up on some of the ideas expressed in this book, we have set up a website, www.gasification.higman.de, which includes both programs for typical gasification calculations and a literature data bank. The content of the website at the time this book went to press covers the following:

Computer Programs

Gasify.exe performs calculations for gasification for coal, oil, and natural gas. It is an equilibrium calculation based on the information contained in Chapter 2. There is a wide selection of calculation modes, such as fixed reactant quantities with variable gasification temperature, fixed gasification temperature with variable oxygen demand, and others.

Proxult.exe will convert proximate and ultimate coal analyses between "as received" (ar), "moisture free" (mf), "ash free" (af), and "moisture- and ash-free" (maf) basis, as described in Section 4.1. Further, LHV and HHV on any basis may be converted into each other. In the absence of heat of combustion data, the heating value will be generated using the proximate and ultimate analyses.

Heatloss.exe will calculate the loss of heat in a membrane wall under different internal conditions (slagging/nonslagging) using the criteria described in Section 6.4.

Quench.exe provides a means of back-calculating the temperature at which the CO shift reaction and the CO_2 methane reforming reaction have frozen during cooling, as described in Section 6.8.

Viscos.exe estimates the viscosity of residual oil fractions as described in Section 4.2.1.

All these programs are executable files and can be downloaded for use by owners of this book. Help files include a fuller description of the theoretical background than contained in the book.

Literature Data Bank

The literature databank gaslit.mdb contains over 500 entries on the topic of gasification, including all the literature cited in this book. It includes titles, authors, and keywords for all papers presented at the Gasification Technologies Council Conferences since 1998, all the IChemE Gasification Conferences, and a number of other important sources not normally accessible to nonparticipants.

The database includes a keyword-based search facility, with up to six specified keywords per article. A gasification technology or project can be specified independently of the keyword.

A complied version of the database gaslit.exe is downloadable for those who cannot run Microsoft Access.

Legal Note

The companion website will be maintained for at least one year after the initial publication of this book. No guarantee can be made for its continued availability thereafter. We have made every effort to ensure the accuracy and usability of these programs and any others that may be included on the website, but we do not assume any legal liability whatsoever for the consequences of their use.

Appendix B

Conversion Factors

Mass				
kg (kilogram)	lb (pound avdp)	Short ton	Long ton	Metric ton
1	2.205	$1.102*10^{-3}$	$9.842*10^{-4}$	$1*10^{-3}$
0.4536	1	$5*10^{-4}$	$4.464*10^{-4}$	$4.536*10^{-4}$
907.6	2000	1	0.8929	0.9072
1016	2240	1.120	1	1.016
1000	2205	1.102	0.9842	1

Volume		
m^3	ft^3	bbl
1	35.31	6.290
$2.832*10^{-2}$	1	0.1781
0.1590	5.615	1

Power		
kW	tcal/day	Btu/h
1	20.64	3412
$4.846*10^{-2}$	1	165.3
$2.931*10^{-4}$	$6.048*10^{-3}$	1

Energy or Work

J	kWh	Btu	cal	TCE LHV	TCE HHV	bbl OE, LHV	bbl OE, HHV
1	$2.778*10^{-7}$	$9.478*10^{-4}$	0.2388	$3.413*10^{-11}$	$3.281*10^{-11}$	$1.757*10^{-10}$	$1.642*10^{-10}$
$3.600*10^6$	1	3412	$8.598*10^5$	$1.228*10^{-4}$	$1.181*10^{-4}$	$6.325*10^{-4}$	$5.912*10^{-4}$
1055	$2.931*10^{-4}$	1	252.0	$3.600*10^{-8}$	$3.461*10^{-8}$	$1.854*10^{-7}$	$1.733*10^{-7}$
4.187	$1.163*10^{-6}$	$3.968*10^{-3}$	1	$1.429*10^{-10}$	$1.374*10^{-10}$	$7.356*10^{-10}$	$6.875*10^{-10}$
$2.931*10^{10}$	8141	$2.778*10^7$	$7.000*10^9$	1	$9.615*10^{-1}$	5.150	4.813
$3.048*10^{10}$	8467	$2.889*10^7$	$7.280*10^9$	1.040	1	5.356	5.005
$5.691*10^9$	1581	$5.394*10^6$	$1.359*10^9$	$1.942*10^{-1}$	$1.867*10^{-1}$	1	$9.346*10^{-1}$
$6.090*10^9$	1692	$5.772*10^6$	$1.455*10^9$	$2.078*10^{-1}$	$1.998*10^{-1}$	$1.070*10^{-1}$	1

Pressure

Pa	bar	atm	lbf/in²	mm Hg=Torr	in Hg	m water
1	$1.000*10^{-5}$	$9.869*10^{-6}$	$1.450*10^{-4}$	$7.501*10^{-3}$	$2.953*10^{-4}$	$1.020*10^{-4}$
$1.000*10^5$	1	0.9869	14.50	750.1	29.53	10.21
$1.013*10^{-5}$	1.013	1	14.70	760.0	29.92	10.34
6895	$6.895*10^{-2}$	$6.805*10^{-2}$	1	51.72	2.036	0.7037
133.3	$1.333*10^{-3}$	$1.316*10^{-3}$	$1.934*10^{-2}$	1	$3.937*10^{-2}$	$1.361*10^{-2}$
3386	$3.386*10^{-2}$	$3.342*10^{-2}$	0.4912	25.40	1	0.3456
9798	$9.798*10^{-2}$	$9.670*10^{-2}$	1.421	73.49	2.893	1

Flowrate		
m³/h	**ft³/min**	**bbl/day**
1	0.5886	151.0
1.699	1	256.5
$6.625*10^{-3}$	$3.899*10^{-3}$	1

Normal and Standard Volumes	
Nm³	**SCF**
1	37.22
0.0269	1

Note: Nm³ refer to 1.0132 bar and 0°C SCF refer to 30" Hg and 60°F

Density		
g/cc	**kg/m³**	**lb/ft³**
1	1000	62.43
$1*10^{-3}$	1	$6.243*10^{-2}$
$16.02 *^{-2}$	16.02	1

Energy per Unit Mass		
kJ/kg	**Btu/lb**	**kcal/kg**
1	0.4299	0.2388
2.326	1	0.5556
4.187	1.800	1

Energy per Unit Volume		
kJ/m³	**Btu/ft³**	**kcal/m³**
1	$2.684*10^{-2}$	0.2388
37.26	1	8.899
4.187	0.1124	1

Heat Capacity per Unit Volume

kJ/m^3.°C	Btu/ft^3.°F	kcal/m^3.°C
1	$1.491*10^{-2}$	0.2388
67.06	1	16.02
4.187	$6.243*10^{-2}$	1

Heat Capacity per Unit Mass

kJ/kg.°C	Btu/lb.°F	kcal/kg.°C
1	0.2388	0.2388
4.187	1	1
4.187	1	1

Conversion Efficiency versus Heat Rate

Efficiency	Heat rate	Efficiency	Heat rate	Efficiency	Heat rate	Efficiency	Heat rate
%	Btu/kWh	%	Btu/kWh	%	Btu/kWh	%	Btu/kWh
100	3412	75	4549	50	6824	25	13648
99	3447	74	4611	49	6963	24	14217
98	3482	73	4674	48	7109	23	14835
97	3518	72	4739	47	7260	22	15510
96	3554	71	4806	46	7418	21	16248
95	3592	70	4874	45	7582	20	17061
94	3630	69	4945	44	7755	19	17958
93	3669	68	5018	43	7935	18	18956
92	3709	67	5093	42	8124	17	20071
91	3750	66	5170	41	8322	16	21326
90	3791	65	5249	40	8530	15	22747
89	3834	64	5331	39	8749	14	24372
88	3877	63	5416	38	8979	13	26247
87	3922	62	5503	37	9222	12	28434
86	3968	61	5594	36	9478	11	31019
85	4014	60	5687	35	9749	10	34121
84	4062	59	5783	34	10036	9	37912
83	4111	58	5883	33	10340	8	42651
82	4161	57	5986	32	10663	7	48744
81	4212	56	6093	31	11007	6	56868
80	4265	55	6204	30	11374	5	68242
79	4319	54	6319	29	11766	4	85303
78	4374	53	6438	28	12186	3	113737
77	4431	52	6562	27	12637	2	170605
76	4490	51	6690	26	13123	1	341210

Conversion Heat Rate versus Efficiency

Heat rate	Efficiency	Heat rate	Efficiency	Heat rate	Efficiency	Heat rate	Efficiency
Btu/kWh	%	Btu/kWh	%	Btu/kWh	%	Btu/kWh	%
3412	100	15500	22.0	28000	12.2	40500	8.4
3500	97.5	16000	21.3	28500	12.0	41000	8.3
4000	85.3	16500	20.7	29000	11.8	41500	8.2
4500	75.8	17000	20.1	29500	11.6	42000	8.1
5000	68.2	17500	19.5	30000	11.4	42500	8.0
5500	62.0	18000	19.0	30500	11.2	43000	7.9
6000	56.9	18500	18.4	31000	11.0	43500	7.8
6500	52.5	19000	18.0	31500	10.8	44000	7.8
7000	48.7	19500	17.5	32000	10.7	44500	7.7
7500	45.5	20000	17.1	32500	10.5	45000	7.6
8000	42.7	20500	16.6	33000	10.3	45500	7.5
8500	40.1	21000	16.2	33500	10.2	46000	7.4
9000	37.9	21500	15.9	34000	10.0	46500	7.3
9500	35.9	22000	15.5	34500	9.9	47000	7.3
10000	34.1	22500	15.2	35000	9.7	47500	7.2
10500	32.5	23000	14.8	35500	9.6	48000	7.1
11000	31.0	23500	14.5	36000	9.5	48500	7.0
11500	29.7	24000	14.2	36500	9.3	49000	7.0
12000	28.4	24500	13.9	37000	9.2	49500	6.9
12500	27.3	25000	13.6	37500	9.1	50000	6.8
13000	26.2	25500	13.4	38000	9.0	50500	6.8
13500	25.3	26000	13.1	38500	8.9	51000	6.7
14000	24.4	26500	12.9	39000	8.7	51500	6.6
14500	23.5	27000	12.6	39500	8.6	52000	6.6
15000	22.7	27500	12.4	40000	8.5	52500	6.5

Care should be taken to make sure that the heat rate and efficiency are both based on the same type heating value (LHV or HHV).

Basic Data

Universal gas constant	R	8.31441	J/mol.K
Normal molar volume of an ideal gas		$22.414 * 10^{-3}$	m^3/mol
Normal temperature		273.15	K
Normal pressure		101325	$Pa\,(N/m^2 = kg/m.s^2)$
Standard acceleration due to gravity		9.80665	m/s^2

Appendix C

Emissions Conversion Factors

Emission Conversions					
SO$_2$		NO$_2$		Particulates /SO$_2$/NO$_2$	
ppmv	mg/Nm^3	ppmv	mg/Nm^3	g/GJ	lb/MMBtu
10	29	10	21	10	0.023
20	57	20	41	20	0.047
30	86	30	62	30	0.070
40	114	40	82	40	0.093
50	143	50	103	50	0.116
60	171	60	123	60	0.140
70	200	70	144	70	0.163
80	229	80	164	80	0.186
90	257	90	185	90	0.209
100	286	100	205	100	0.233
110	314	110	226	110	0.256
120	343	120	246	120	0.279
130	372	130	267	130	0.302
140	400	140	287	140	0.326
150	429	150	308	150	0.349
160	457	160	328	160	0.372
170	486	170	349	170	0.395
180	514	180	369	180	0.419
190	543	190	390	190	0.442
200	572	200	411	200	0.465

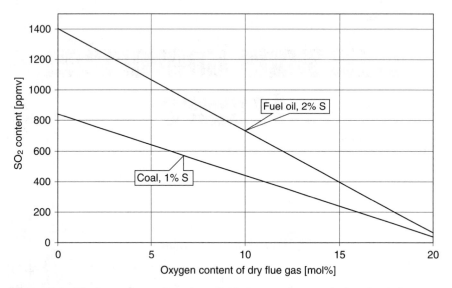

Figure C-1. SO$_2$ Content as a Function of the Oxygen Content in Dry Stack Gas

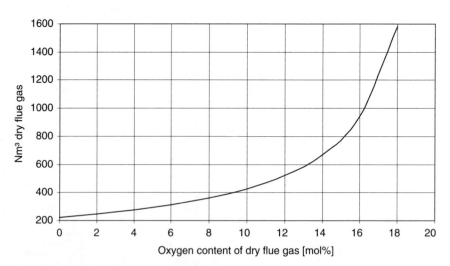

Figure C-2. Nm^3 Dry Flue Gas per GJ Fuel as a Function of the Oxygen Content

Appendix D

Guidelines for Reporting Operating Statistics for Gasification Facilities

(Courtesy of The Gasification Technologies Council, Rev. February 5, 2002)

The objective of these guidelines is to present a standardized way for reporting the operating statistics of gasification facilities. The statistics are primarily time-based, however, a single flow-based indicator is also included. An example is included.

Gasification Facility Units

The gasification facility is divided into two units so that the operating statistics can be reported for each of these critical areas of the facility. The units are defined as follows:

- Gasification (including ASU and Acid Gas Removal Unit)
- Product Units

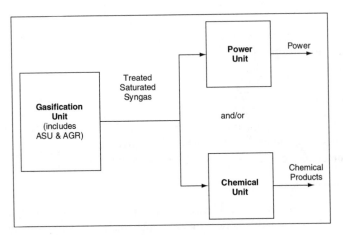

Figure D-1. Gasification Facility Units

369

○ Power production block, and/or

○ Chemical production block

Authors are also asked to indicate the specific configurations of the units with regard to back-up and multiple trains.

Unit Operating Statistics—Measured

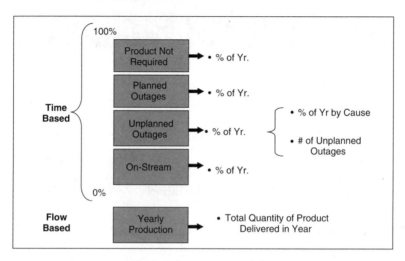

Figure D-2. Unit Operating Statistics—Measured

Definitions—Measured Statistics

• Product Not Required

  ○ % of year that the product from the unit was not required and, therefore, the unit was not operated. The unit was generally available to run and not in a planned outage or forced outage.

• Planned Outages

  ○ % of the year that the unit is not operated due to outages that were scheduled at least one month in advance. Includes yearly planned outages as well as maintenance outages with more than one month's notice.

• Unplanned Outages

  ○ % of the year that the unit was not operated due to forced outages that had less than one month's notice. Includes immediate outages as well as maintenance outages with less than one month's notice.

- On-Stream

  - % of the year the unit was operating and supplying product in a quantity useful to the downstream unit or customer.

- Yearly Production

  - Defined as the total quantity of product actually delivered from the unit in a calendar year. For the gasification unit the production is reported on the basis of total clean synthesis gas.

Unit Operating Statistics—Calculated

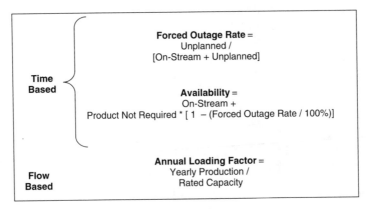

Figure D-3. Unit Operating Statistics—Calculated

Definitions—Calculated Statistics

- Forced Outage Rate

  - Defined as the time during which the down-stream unit or customer did not receive product due to unplanned problems divided by the time during which they expected product, expressed as a percentage.

- Availability

  - Defined as the sum of the time during which the unit was on stream plus an estimate of the time the unit could have run when product was not required, expressed as a percentage of the year. Assumption is that unit could have operated at the same Forced Outage Rate when product was not required.

- Annual Loading Factor

  - Defined as the yearly production of the unit divided by the rated capacity, expressed as a percentage.

- Rated Capacity

  - Defined as the design quantity that the unit would produce at the design rate over the calendar year when operated in an integrated manner. Calculated by multiplying 365 times the average annual daily design rate. Note that the Design Production can change over time as the plant is de-bottlenecked or re-rated.

Example

- Operating Unit is a gasification train that is designed to make 200 MMscfd of syngas.
- Measured Unit Operating Statistics for this example:

  - Product Not Required = 10% of year

  - Planned Outages = 8% of year

  - Unplanned Outages = 4% of year

    - Breakdown of the 4% by Cause
    - Report # of interruptions

  - Onstream = 78% of year

  - Yearly production = 55,000 MMscf of syngas

- Resulting Calculated Unit Operating Statistics:

  - Forced Outage Rate = 4%/[78% + 4%] = 4.9%

  - Availability = 78% + 10% * [1 − (4.9%/100%)] = 78% + 9.5% = 87.5%

  - Rated Capacity = 365 d * 200 MMscfd = 73,000 MMscf

  - Annual Loading Factor = 55,000 MMscf/73,000 MMscf = 75.3%

Appendix E
Basis for Calculations

All calculations were based on a typical internationally traded coal of which the properties are given in Table E-1. All flow schemes are based on 100 kg maf coal, and the relevant mass and energy values are all based on this quantity. Taking the mass values per second we get the energy flows in MW.

For drying the coal from 12.5 to 2% moisture 31 MJth are required. As this heat is supposed to be supplied by burning part of the clean fuel gas, this corresponds to an energy penalty of about 19 MJe.

For heating the water for the optimal coal-water slurry feed gasifier to 325°C 121 MJth/100 kg maf coal are required.

For a classical Texaco gasifier, the carbon conversion has been set at 95%. In all other cases this has been set at 99%.

The energy required for the production of oxygen was taken as 46 MJe/kmole. If not indicated otherwise, the purity of the oxygen is 95% mole.

For making 300°C process steam for dry-coal feed gasifiers, 3 MJth/kg are required.

The heat loss from the gasifiers has been taken as 0.5% of the coal LHV for the coal-water slurry feed gasifiers and as 2% of the coal LHV for the dry coal feed gasifiers. In the latter case it has been assumed that this heat becomes available for making steam.

For all compressors and turbines, an isentropic efficiency of 90% has been assumed. For the adiabatic compression of air to 32 bar, this corresponds to 15.6 MJe/kmole air. For wet air compression the energy data were 10.7, and 13.1 MJe/kmole wet air for compression to 32 and 64 bar, respectively. For nitrogen compression the same figures have been taken as for air compression.

For gas quenches a recycle gas compressor is required. The energy consumption is 11 MJe for 100 kg maf coal intake, except for the two-stage dry-feed gasifier where it is only 7 MJe for 100 kg maf coal intake. The reason for this lower figure is that the gas has to be quenched from 1100 to 900°C instead of from 1500 to 900°C.

In case of the Tophat cycle, corrections have been applied for the approach temperatures in the recuperator. As a standard 25°C was taken for the temperature difference between the turbine outlet and the humidified air leaving the recuperator. For higher temperature differences the efficiency bonus was one percentage point per 25°C.

For the own energy consumption of the power stations, 2 percentage points were assumed in case no CO_2 was removed. In case of CO_2 removal from the fuel gas, this figure was increased to 3 percentage points. For flue gas treating the latter penalty was also used, but it was further increased by an additional penalty of 109 MJe that are required to increase the pressure of CO_2 from the about 0.1 bar at which it is available in the flue gas to the about 6 bar at which it is present in the fuel gas. This figure corresponds to an additional 3 percentage points penalty in the station efficiency, bringing the total energy consumption for gas treating for this case to 6 percentage points.

It has been further assumed that heat above 250°C can be converted into power with an efficiency of 45% by means of a steam cycle.

Table E-1
Coal Properties

	Moisture and Ash Free	Ash Free	Moisture Free	As Received
Proximate analysis				
Fixed carbon	0.5908	0.5109	0.5400	0.4725
Volatile matter	0.4092	0.3539	0.3741	0.3273
Moisture	0	0.1352	0	0.1250
Ash	0	0	0.0859	0.0752
Total	1.0000	1.0000	1.0000	1.0000
Ultimate analysis				
Carbon	0.8166	0.7062	0.7464	0.6531
Hydrogen	0.0568	0.0642	0.0519	0.0594
Oxygen	0.0983	0.2050	0.0898	0.1896
Nitrogen	0.0171	0.0148	0.0157	0.0137
Sulfur	0.0113	0.0097	0.0103	0.0090
Ash	0	0	0.0859	0.0752
Total	1.0000	1.0000	1.0000	1.0000

Note: Lower Heating Value (LHV) maf coal 33.25 MJ/kg and of the a.r. coal 32.87 MJ/kg.

Nomenclature

A	pre-exponential factor
c_A	concentration of component A
$c_{A,0}$	initial concentration of component A
C	carbon content (maf) basis
E	activation energy for reaction
k_f	rate constant (f=forward, r=reverse)
k_m	mass-related rate constant
$K_{p,T}$	equilibrium constant of a reaction at temperature T
N	number of reactor stages
P	total pressure
P_A	partial pressure of component A
r_f	reaction rate (f=forward, r=reverse)
r_m	mass related reaction rate
R	universal gas constant
t	time
T	temperature
$^s v_A$	volume fraction of component A in stream s
v	Kinematic viscosity
τ	residence time

List of Names and Abbreviations

Names of Processes, Plants, and Companies

Over the course of time a number of technologies have changed owners or names, to the extent that this book could read like a Russian novel, where all the main characters go by at least three names. In order to simplify matters, we have used one name consistently throughout this book. In general, we have used the current names, but where this could cause confusion, we have used the name more generally associated with the particular technology.

Name Used in this book	Other Names or Owners
GSP	A gasification process originally known as GSP (Gaskombinat Schwarze Pumpe) and later purchased by Noell. Noell was bought by Babcock Borsig and traded under the name of BBP. This technology is now owned by Future Energy GmbH, which has revived the old GSP name.
E-Gas	This technology was originally developed by Dow Chemical, which later grouped its gasification technology assets into Destec Energy Inc. In 1997 Destec was sold to NCG Corporation, which changed its name to Dynergy in 1998. Since purchasing the technology from Dynergy in 1999, Global Energy marketed it under the name of E-Gas. It is now owned and marketed by ConocoPhillips.
Texaco	The technology developed by Texaco has been acquired by GE and is now marketed under the name of GE Energy.
BGL	The BGL technology (see page 94ff.) and the Lurgi circulating fluid bed gasification (page 105) have been acquired by and are marketed by Envirotherm GmbH, a unit of Allied Resource Corporation.

Borsig	A designer of syngas coolers and originally part of the Deutsche Babcock (later Babcock Borsig) Group. The company is now independent and has taken over the syngas cooler designs of Steinmüller, also part of the Babcock Borsig group.
Standard Fasel Lentjes	Syngas cooler and HRSG supplier in The Netherlands. Formerly Werkspoor, and then Bronswerk.
SilvaGas	A biomass gasification process originally developed by the Battelle Memorial Institute and now marketed by FERCO under the name of SilvaGas
Buggenum	Coal-based IGCC built by Demkolec and now owned by Nuon Power

Abbreviations

As in any other walk of life, the gasification industry has developed its own jargon and abbreviations. The following lists those abbreviations used in this book and the location where one can find a more detailed explanation or definition.

Abbreviation	Description
AGR	Acid Gas Removal
af	ash free
ar	as received
ASU	Air Separation Unit
BCGA	British Compressed Gas Association
BFW	Boiler Feed Water
BGL	British Gas Lurgi
BOD	Biological Oxygen Demand
BS	(1) Bituminous Solids
BS	(2) British Standard
CAPEX	Capital Expenditure
CC	Combined Cycle
CCP	Clean Coal Power R&D Company
CFB	Circulating Fluid Bed
CFD	Computational Fluid Dynamics
CGA	Compressed Gas Association
CGE	Cold Gas Efficiency
CHAT	Cascaded Humidified AirTurbine
CHP	Combined Heat and Power
COD	Chemical Oxygen Demand
CSTR	Continuously Stirred Tank Reactor
CW	Cooling Water

Abbreviation	Description
DEA	Di-Ethanol Amine
DGA	Di-Glycol Amine
DIN	Deutsches Institut für Normung
DIPA	Di-Isopropyl Amine
DMPEG	Dimethyl ether of Polyethylene Glycol
DoE	Department of Energy (U.S.)
EIGA	European Industrial Gas Association
EOR	Enhanced Oil Recovery
EP	Elevated Pressure
EPA	Environmental Protection Agency (U.S.)
EPRI	Electric Power Research Institute
FCC	Fluid(ized) Catalytic Cracking
FICFB	Fast Internal Circulating Fluid Bed
FT	Fischer-Tropsch (synthesis)
GSP	Gaskombinat Schwarze Pumpe
GTL	Gas to Liquids
HAT	Humid Air Turbine
HHV	Higher Heating Value
HP	High Pressure
HRSG	Heat Recovery Steam Generator
HTW	High Temperature Winkler
HWEP	Hot Water Extraction Process
IGCC	Integrated Gasification Combined Cycle
IGT	Institute of Gas Technology
ISO	International Standards Organization
KBR	Kellogg Brown and Root
KRW	Kellogg Rust Westinghouse
KT	Koppers Totzek
LHV	Lower Heating Value
LNG	Liquid Natural Gas
LNW	Liquid Nitrogen Wash
LOX	Liquid Oxygen
LP	Low Pressure
LPG	Liquid Petroleum Gas
LSTK	Lump Sum Turn Key
maf	moisture- and ash-free
MDEA	Methyl Di-Ethanol Amine
MEA	Mono-Ethanol Amine
mf	moisture-free
MHF	Multiple Hearth Furnace
MM	million
MP	Medium Pressure

Abbreviation	Description
MPG	Multi-Purpose Gasification
MSW	Municipal Solid Waste
NG	Natural Gas
NGCC	Natural Gas Combined Cycle
NMP	Normal Methyl Pyrrolidone
OPEX	Operational Expenditure
PC	Pulverized Coal
PFBC	Pressurized Fluid Bed Combustion
PFR	Plug Flow Reactor
POX	Partial Oxidation
PPU	Pre-Purification Unit
PSA	Pressure Swing Adsorption
PV	Photo Voltaic
RDF	Refuse Derived Fuel
RSC	Radiant Syngas Cooler
SAFT	Stoichiometric Adiabatic Flame Temperature
SAGD	Steam Assisted Gravity Drainage
SARU	Soot-Ash Removal Unit
SCGP	Shell Coal Gasification Process
SCOT	Shell Claus Off Gas Treatment
SCR	Selective Catalytic Reduction
SGP	Shell Gasification Process
SMDS	Shell Middle Distillate Synthesis
SNG	Synthetic Natural Gas or Substitute Natural Gas
SRU	Sulfur Recovery Unit
TEA	Tri-Ethanol Amine
UCG	Underground Coal Gasification
UVV	Unfallverhütungsvorschriften (German safety regulations)
VOC	Volatile Organic Compounds

Index